AF736014

Food Science and Technology Series

TRADITIONAL CHINESE FOODS: PRODUCTION AND RESEARCH PROGRESS

FOOD SCIENCE AND TECHNOLOGY SERIES

Food Science and Technology: New Research
Lorenzo V. Greco and Marco N. Bruno (Editors)
2008. ISBN: 978-1-60456-715-1

The Price of Food
Meredith N. Fisher (Editor)
2009. ISBN: 978-1-60692-440-2

Food Processing and Engineering Topics
Maria Elena Sosa-Morales and Jorge F. Velez-Ruiz (Editors)
2009. ISBN: 978-1-60741-788-0

Traditional Chinese Foods: Production and Research Progress
Li Zaigui and Tan Hongzhuo
2009. ISBN 978-1-60692-902-5

Food Science and Technology Series

Traditional Chinese Foods: Production and Research Progress

Li Zaigui
and
Tan Hongzhuo

Nova Science Publishers, Inc.
New York

Copyright © 2009 by Nova Science Publishers, Inc.

All rights reserved. No part of this book may be reproduced, stored in a retrieval system or transmitted in any form or by any means: electronic, electrostatic, magnetic, tape, mechanical photocopying, recording or otherwise without the written permission of the Publisher.

For permission to use material from this book please contact us:
Telephone 631-231-7269; Fax 631-231-8175
Web Site: http://www.novapublishers.com

NOTICE TO THE READER

The Publisher has taken reasonable care in the preparation of this book, but makes no expressed or implied warranty of any kind and assumes no responsibility for any errors or omissions. No liability is assumed for incidental or consequential damages in connection with or arising out of information contained in this book. The Publisher shall not be liable for any special, consequential, or exemplary damages resulting, in whole or in part, from the readers' use of, or reliance upon, this material. Any parts of this book based on government reports are so indicated and copyright is claimed for those parts to the extent applicable to compilations of such works.

Independent verification should be sought for any data, advice or recommendations contained in this book. In addition, no responsibility is assumed by the publisher for any injury and/or damage to persons or property arising from any methods, products, instructions, ideas or otherwise contained in this publication.

This publication is designed to provide accurate and authoritative information with regard to the subject matter covered herein. It is sold with the clear understanding that the Publisher is not engaged in rendering legal or any other professional services. If legal or any other expert assistance is required, the services of a competent person should be sought. FROM A DECLARATION OF PARTICIPANTS JOINTLY ADOPTED BY A COMMITTEE OF THE AMERICAN BAR ASSOCIATION AND A COMMITTEE OF PUBLISHERS.

LIBRARY OF CONGRESS CATALOGING-IN-PUBLICATION DATA

Zaigui, Li.
Traditional Chinese foods : production and research progress / Li Zaigui.
p. cm.
Includes bibliographical references and index.
ISBN 978-1-60692-902-5 (hardcover : alk. paper)
1. Food industry and trade--China. I. Title.
TP369.C6Z35 2009
664.00951--dc22

2008055323

Published by Nova Science Publishers, Inc. ✦ New York

CONTENTS

PREFACE

It is generally admitted that the expression 'traditional food' refers to a product with specific raw materials, and/or with a recipe known for a long time, and/or with a specific process. China has a wealth of traditional foods such as Chinese steamed bread, Chinese noodles, Chinese rice noodles, Starch noodles (Vermicelli), Tofu, Sofu (soybean cheese), douchi (fermentation soybean), Chinese vinegar and many other foods. These traditional foods are an important component of Chinese people's diet and the basis for their food habits and nutrition. They also constitute an essential aspect of their cultural heritage and related closely to Chinese people's historical background and to the environment in which they live. During the last few decades, the development of international food trade and the extensive urbanization process which have affected life-styles to a large extent in many parts of the world have resulted in a sizeable decrease in the consumption of some kinds of traditional foods and a relative neglect in the cultivation of traditional food crops. Some traditional foods had withered away or are withering away. The governing bodies of FAO have recommended that FAO give due consideration in its programme to the promotion of the production and consumption of traditional foods worldwide. Several studies and projects have been initiated by FAO and EU in different parts of the world to survey existing traditional foods and food crops, especially Chinese traditional foods. Accordingly, China government, academia and industry all begin to give more attentions to own traditional foods, study their nutritional values and identify ways and means of promoting their production and consumption. In recent years, as a result of food globalization, the consumption of traditional foods has increased considerably and many of these foods are concurrent with easy-to-prepare, processed, semi-processed and high-tech foods. For example, tofu is sold in almost all of supermarket even in west countries. It was decided therefore that a book should be carried out to document existing Chinese traditional foods in China and to assess their nutritional value and contribution to the diet.

Among many new works on food, however, few studies address the Chinese foodways, despite their enormous and continual influence on local food habits around the world. Even classic works on Chinese food provide us with only basic information about China itself, or interpret Chinese foodways in the restricted local food scene and within Chinese history. This book however provides an up-to-date reference for traditional Chinese foods and a detailed background of history, quality assurance, and the manufacture of general traditional food products. It contains topics not covered in similar books. It is divided into 8 chapters. We shall highlight the main point in each of the chapters, with emphasis on additional background information that connects the individual chapters to others and to the overall theoretical concerns as well.

Chapter 1 by Li Zaigui and Bi Ying (China Agricultural University), "Chinese steamed bread", looked into the development of staple traditional food mantou (Chinese Steamed Bread, CSB) in: (1) The Definition, Categories and Consumption of CSB; (2) Materials for the production of CSB; (3) Situation and its development of processing technology for CSB making; (4) Researches on the requirements of flour quality for different kinds of CSB; (5) Methods which can improve CSB production including addition of different kinds of flour or additives; (6) Quality and properties of CSB.

Chen Jie (Henan University of Technology) in Chapter 2, "Chinese noodles", detailed (1) history and development of noodles; (2) Raw materials for noodles making; (3) Processing technology and equipments for different kinds of noodles such as fine dried noodles, instant noodles and long life noodles; (4) Researches on noodles processing.

Liang Jianfen (China Agricultural University) in Chapter 3, "Chinese rice noodles", brought follows information on rice noodles: (1) Origin, history and classifications of rice noodles; (2) Materials for rice noodle; (3) Processing procedures and (4) Quality evaluation.

In Chapter 4, "Starch noodles (Vermicelli)", Tan Hongzhuo (Academy of State Administration of Grain) summarized the current knowledge on: (1) Definition, naming, history and categories of starch noodles; (2) The morphological, physico-chemical, thermal, rheological, characteristics and molecular structure of materials for starch noodles including mung bean starch, pea starch, common bean starch, sweet potato starch, potato starch, corn starch; (3) The traditional and modern processing technology for starch noodles; (4) structure and nutrition of starch noodle; (5) quality evaluation for starch noodles, and (6) quality improvement for starch noodles.

In Chapter 5, Li Jun (The Chinese Academy of Agricultural Sciences) and Qian Keying (China Agricultural University) analyzed the recent developments of "Tofu". This chapter including: (1) Definition, Origin, history and Categories, production and consumption of tofu; (2) Material for tofu producing; (3) Processing technology of tofu; (4) Researches and progress on processing, quality and nutrition of tofu.

Fan Junfeng (Beijing Forestry University) in Chapter 6, "Sofu (soybean cheese)" provided analysis of (1) introduction; (2) The classification of sufu; (3) Processing Development in sufu manufacture; (4) Enzymes Produced during Fermentation; (5) The characteristics of sufu and (6) Microbiological aspects of sufu.

In Chapter 7, Li Zaigui and Li Dongwen (China Agricultural University) presented a fermentation soybean- "douchi". It consisted from: (1) Introduction; (2) Materials for the production; (3) Processing technology of douchi; (4) Researches on douchi.

Finally, in Chapter 8, Lin Qin (Shanghai Institute of Technology), Chou Ju and Jiang Da (China Agricultural University) gave a detailed account on Chinese vinegar including to: (1) Introduction; (2) Raw Materials for vinegar processing; (3) Nutrition and taste of vinegar; (4) Manufacture of Chinese Vinegar; (5) Research and technological advances in vinegar; (6) Quality standards of vinegar in China.

Together the chapters presented here provide a wide-ranging conspectus of the variety of traditional Chinese foods.

Li Zaigui, with the help of Tan Hongzhuo, edited all of the parts. The work of Ms. Wang Aili, Ms. Li Lu and Ms. Yang Hong are also helpful.

We believe that "*The Production and Research Progress on traditional Chinese food*" make a particularly strong subject of study to increase our understanding of the globalization trend in Chinese foods distribution and consumption. While it is recognized that the information contained in this document is far from being exhaustive, as there are many traditional Chinese foods that are not cited in the literature, it is hoped that its publication will encourage nutritionists, food scientists, and food technologists in the region to give this subject more attention and to develop appropriate technologies for the induction and commercial distribution of traditional Chinese foods. It also is our sincere hope and expectation that it will serve as an essential reference on the manufacturing of traditional food products, for professionals in government, industry, and academia.

In the last part of this introduction, we thank all the contributors for sharing their experience in their fields of expertise. They are the people who made this book possible and many references are detailed after each chapter.

Dr. Li Zaigui and Dr. Tan Hongzhuo

Chapter 1

MANTOU (CHINESE STEAMED BREAD, CSB)

There are two kinds of staple foods in China: wheat and rice. The annual production of wheat and rice has been about 100 million tons and 200 million tons in recent years. Wheat originated in the Central region, and was introduced to China in the Neolithic Age. The inscriptions on the bones and tortoise shells dating from the Shang Dynasty (1751–1122 B.C.) indicated that wheat was already widely grown throughout the Henan province in central China. People used stone mortars to grind wheat into flour and made wheaten food by hand.

Bing was the common name of cooked wheaten foods in ancient times. There was further development of wheaten foods during the Han Dynasty (206 B.C. to 220 A.D.). The writer Liu Shi reported on *shou mian* (a kind of fermentation dough) in his work *Shi Ming*. This indicated that, at that time, dough fermentation technology was already in use. The Chinese had mastered flour fermentation techniques by using the easily fermented rice soup as a catalyst. Later, bases were used to neutralize the fermentation process when making dough. It was said that, during the "Three Kingdoms" (221–263 A.D.), steamed bread was first made and similar products were then introduced to Japan, Korea, and Southeast Asian countries. Steamed bread has evolved continuously throughout Chinese history so that today there are many styles of steamed bread.

The most common food made from flour would be Mantou, Chinese steamed bread (CSB). Chinese steamed bread, a kind of Chinese traditional fermented food based on wheat flour, has been consumed for at least 2,000 years in China. It is a staple food for the Chinese people, especially in northern China where it is eaten at almost every meal and also has been gaining popularity in southern China in recent years. Today, the industrialization of CSB production in China has the same trend of development as western-style bread production in western countries. Although there are similarities between CSB and western-style bread, the processing of CSB is quite different from that of western-style bread. The processing of CSB uses a method that produces a product with a dense crumb and a thin smooth white skin rather than the brown crust of traditional western bread.

1. INSTRUCTION

1.1. The Definition of CSB

Chinese steamed bread is a leavened wheat flour product, which is cooked by steaming in a steamer. The most common type of steamed breads, weighing about 100 g, is either round or roughly cylindrical in shape, white in color, and has a smooth, shiny, surface devoid of a crust. The crumb texture varies from dense to open, and the flavor varies to suit local tastes. One piece of dough can be used to make different forms of steamed products such as steamed bread, steamed bun, and steamed twisted roll. Steamed products can be made with or without fillings. The products without filling are called steamed bread, or *mantou* (Figure 1-1), and with fillings are called a steamed bun (*baozi*). Other forms of steamed products include twisted rolls in various shapes (*huajuan*).

In the national standard of "Chinese steamed bread made of wheat flour" which was issued at the beginning of 2008, CSB was defined as "wheat flour and water as raw material, microzyme as leavening and steamed food" (Sun, 2008). From the definition, baozi with fillings is not CSB and manju in Japan (the character of manju in Japanese is the same as that of mantou in Chinese) is also not CSB.

1.2. Categories of CSB

There are three main styles of steamed bread in China and East, Southeast Asian countries as northern, southern and Guangdong styles. The northern style, preferred in northern China, has a very cohesive and elastic eating quality, a higher arch domed shape and dense structure. The southern style has a soft, elastic, and medium cohesive eating quality, a lower arch domed shape and open structure. The Guangdong style, which is popular in the very southern part of China, and East, Southeast Asian countries, has an open structure, a sweet taste, and a very soft and elastic, but not a cohesive eating quality. People usually consume this style of steamed bread as a snack.

Figure 1-1. A view of steamed bread making in a small countryside shop.

Steamed bread is a staple food in the wheat-growing area of northern China, representing approximately 45% of flour produced in this region. In contrast, a lesser proportion is used in the south, where rice and noodles are more popular. People in the south often consume CSB for breakfast. The dough of CSB for northern- and southern-style steamed bread is made of flour, water, and yeast while for the Guangdong-style steamed bread, up to 25% sugar, 10% fat and 1.2% salt are added.

1.3. Consumption of CSB

Wheaten foods have had a very important role in the diet and culture of Asian countries since very early times. Today, steamed bread is a common food in China and the East, Southeast Asian regions. Millions of people consume it regularly. The commercial production of frozen steamed bread, creating more convenience for consumers, has raised their popularity even further.

Over the past two decades, the rapidly growing economies in China and East, Southeast Asian countries have led to an improvement of living standards. The demand for convenience and quality of steamed bread is increasing. Many innovative products have been developed, particularly among those distributed to supermarket chains. Sold both fresh and frozen, an enormous variety of types is available. For example, layered steamed breads with chocolate or taro colorings have been widely marketed. Whole meal steamed bread has also recently appeared in markets. In addition, there are some new types of steamed breads made from mixtures of wheat flour with other flours such as buckwheat, millet, sorghum, black rice, or maize flour. These new products are marketed as health foods and are sold in northern China. There is increasing production of steamed bread, buns, and rolls in factories equipped with modern machines.

Steamed bread is the most important food in the main growing areas of wheat. In Henan Province, for example, steamed bread is the main staple for over 90% of the residents and nearly 100% for county-side residents.

About 20 years ago, almost all steamed breads were prepared by hand and in the home, but now 90% of the steamed bread that is sold is prepared by machine in the city while the conditions in the countryside have not been improved much (Sun, 2008). The first automatic production line for steamed bread was established in China at the beginning of 1980, but now, the manufacturers of steamed-bread-making machines could be found all around China.

2. Materials for the Production of CSB

The materials for steamed bread making are simply wheat flour, yeast and water while in some cases adding sugar, especially for Guangdong style steamed bread. Yeast includes enzyme and traditional starter culture ('Jiaotou' in Chinese). Although someone reported the steamed bread was better using Jiaotou than that using enzymes, but Jiaotou is inconvenient and makes it difficult to control the quality of steamed bread so it is just used in homes or in a few small shops.

2.1. Wheat Flour

Wheat flour is the most important material for CSB making and accounts for about 60% of product in weight. The effects of wheat flour on the quality of steamed bread are very complex and still not clear even though there were many studies done on the subject. It is acceptable that protein, lipid, starch and water are all related with the crystalline network forming during steaming. Protein was considered to be the most important factor affecting the quality of steamed bread, but the role of starch in flour has been reconsidered and reaffirmed recently.

2.1.1. Carbohydrate Composition

The main composition of flour is carbohydrate. It includes starch and non-starch polysaccharides. In a modern milling factory, the crude fiber content of flours can be kept so low that only traces of it (under 0.5%) remain in the final product.

Starch is present in dough in the native state where it appears as distinct semi-crystalline granules. During dough preparation, starch absorbs up to about 46% water. It was suggested to act as inert filler in the continuous protein matrix of the dough, while some researchers described dough as a bicontinuous network of starch and protein. Other studies reported that the rheological behavior of wheat dough is influenced by the specific properties of the starch granule surface and by the presence of amylolytic enzymes.

Due to the combined effects of heat and moisture during the steaming process, the starch granules gelatinized and swelled. However, their granular identity is retained. A small amount of starch (mainly amylose) is leached into the intergranular phase. Furthermore, due to phase separation, amylose and amylopectin are not homogeneously distributed in the granules: the centre of the large granules is enriched in amylose, while the outergranule layers are enriched in amylopectin. Part of the solubilised amylose forms inclusion complexes with both added (if any) and endogenous wheat polar lipids, as evidenced by the V crystal type of fresh crumbs.

In cereal science, non-starch polysaccharides (NSP) is a generic term for arabinoxylans (AX), β-glucan, cellulose and arabinogalactan-peptides, i.e. polysaccharides that differ from amylose and amylopectin either by the nature of their composing monosaccharides and/or by the nature of their linkages. Water-extractable arabinoxylans (WE-AX) added to dough increase dough consistency and stiffness and decrease mixing time. On the same dough consistency basis, WE-AX addition increases baking absorption but does not affect mixing time, lowers the energy input to achieve optimal mixing and enhances resistance to extension and decreases extensibility. WE-AX of high average molecular weight (Mr 201,000–555,000) exerts greater effects on baking absorption and development time than that of lower molecular weight counterparts (Mr 50,000–134,000). Addition of water-unextractable arabinoxylans (WU-AX) has similar effects as that of WE-AX, but does not alter dough extensibility properties. A positive correlation between flour WU-AX level and baking absorption was equally shown for endogenous WU-AX through fractionation-reconstitution bread-making experiments. Using this approach, extensibility decreased and resistance to extension increased with the increasing of WU-AX content of flour. This would feed the hypothesis that the WU-AX rich cell-wall fragments interfere with optimal gluten formation during dough mixing. WE-AX functioned somewhat as gluten during fermentation as it slows down the diffusion rate of carbon dioxide out of the dough, thus contributing to gas retention.

However, they lack elastic properties. Presumably, WE-AX increases dough foam stability because it increases the viscosity of the dough aqueous phase which in its turn stabilizes the gas cells liquid films. Others attributed the positive impact of WE-AX to the formation of a secondary, weaker network enforcing the gluten network. Upon addition of WU-AX, gas retention and evolution of dough were similar to those of the control dough. This observation is in contrast to the postulated negative impact of WU-AX which suggests that they: (i) destabilize gas cells by forming physical barriers for gluten during dough development, (ii) absorb a large amount of water which consequently is not available for gluten development and film formation, (iii) perforate the gas cells which causes them to coalesce.

It is assumed that, during the initial phase of baking, AX affect bread making by mechanisms equal to those observed for fermentation. Stabilization of gas cells by WE-AX will prolong the oven rise and improve bread characteristics (crumb firmness, structure and texture, loaf volume), while WU-AX enhance gas cell coalescence and decrease gas retention, resulting in poorer bread quality. Indeed, fractionation-reconstitution experiments demonstrated that loaf volume was increased both when decreasing the WU-AX content and increasing the level of WE-AX of medium and high molecular weight in dough.

2.1.2 Protein Composition

It is said that the protein of wheat flour decides the suitability of steamed bread making, and the medium protein content is most suitable. But a wide scope of wheat flour with low, medium or high protein content are used in steamed bread making in different areas. For example, the soft wheat with wet gluten content 21~24% is the main kind of wheat flour in Anhui province. While the wet gluten content of flour for CSB making may be over 30% in Shandong province. So not only the content but also the character relate to the properties of CSB making.

Table 1-1. Overview of the different groups of wheat proteins (Dong et al., 2005)

Osborne fraction	Solubility behavior	Composition	Biological role	Functional role
Albumin	Extractable in water	Non-gluten protein (mainly monomeric)	Metabolic and structural proteins	Variable
Globulin	Extractable in dilute salt	Non-gluten protein (mainly monomeric)	Metabolic and structural proteins	Variable
Gliadin	Extractable in aqueous alcohols	Gluten proteins (mainly monomeric gliadins and low molecular weight glutenin polymers)	Prolamin-type seed storage proteins	Dough viscosity/ plasticity
Glutenin	Extractable in dilute acetic acid	Gluten proteins (mainly high molecular weight glutenin polymers)	Prolamin type seed storage proteins	Dough elasticity/ strength
Residue	Unextractable	Gluten proteins (high molecular weight polymers) and polymeric non-gluten proteins (triticins)	Prolamin-type (gluten) and lobulin-type (triticin) seed storage proteins	Variable

Suitable protein content of flour is significantly related to the color, structure and smoothness of surface, taste and volume of CSB (Dong et al., 2005). If the dried protein content of flour was over 13%, the surface of CSB would crinkle and the color became gray. But if that is lower than 10%, the surface and color of CSB would be smooth and white, but the construction, texture and taste will be affected negatively. It is also said that the suitable protein content of flour for southern-style CSB is a little lower than that for northern-style CSB.

Osborne introduced a solubility-based classification of plant proteins using sequential extraction in the following series of solvents: (1) water, (2) dilute salt solution, (3) aqueous alcohol and (4) dilute acid or alkali. Using this Osborne classification scheme, wheat proteins were classified in albumins, globulins, gliadins and glutenins, respectively (Table 1-1). From a functional point of view, two groups of wheat proteins should be distinguished: the non-gluten proteins, with either no role or just a minor role in CSB making, and the gluten proteins, with a major role in CSB making.

The producing quality of wheat flour is largely determined by its proteins. Both quantity and composition (quality) of proteins are important for wheat quality. The observation and producing performance of wheat flour is linearly related with its protein content though different linear relationships exist for different wheat varieties. Notwithstanding some roles of different non-gluten proteins (e.g., certain enzymes, enzyme inhibitors, lipid-binding proteins and possibly also triticins) in the producing process are observed, the main quality determinant of the producing process is the gluten proteins. Indeed, the unusual properties of the gluten proteins allow wheat flour to transform into the dough with suitable properties for production. Gluten proteins undergo various changes during the different steps of CSB making, although the nature of these changes, like the native gluten protein structure itself, is poorly understood.

The gliadin/glutenin ratio of gluten proteins is very important. This is a direct consequence that, within the viscoelastic gluten protein network of dough, gliadin and glutenin showed different roles. Due to their large size, glutenin polymers form a continuous network that provides strength (resistance to deformation) and elasticity to the dough. On the other hand, the monomeric gliadins are believed to act as plasticizers of the glutenin polymeric system. In this way, they provide plasticity/viscosity to wheat flour dough. For bread making, an appropriate balance between dough viscosity and elasticity/strength is required. Up to a certain limit, higher dough strength increases loaf volume of CSB just as that of western style bread. The second factor in gluten protein quality is the quality of its glutenin fraction (extractable as well as unextractable). Though differences in gliadin properties might also have some effects, it is now generally believed that differences in glutenin properties are more important in explaining gluten protein quality during production. Although a lot of questions still remain because of the lack of detailed knowledge about the molecular structure of glutenin and its contribution to elasticity, it can be assumed that differences in glutenin functionality during production result from differences in (i) composition, (ii) structure and/or (iii) size distribution of the glutenin polymers (Veraverbeke and Delcou, 2002) (Figure 1-2). Firstly, differences in glutenin composition may result in differences in the non-covalent interactions that determine the elasticity of glutenin. Each wheat variety contains 3~5 different high molecular weight glutenin subunits (HMW-GS) and about 7~16 different low molecular weight glutenin subunits (LMW-GS). Knowing that more than 20 different HMW-GS and more than 40 different LMW-GS have been detected so far in different wheat varieties, explains an enormous variation in glutenin composition between different wheat varieties. Secondly, although it is hard to hypothesize on this matter because of the poor knowledge of the structure of glutenin, it can be assumed that (even subtle) differences in the structure of glutenin largely affect glutenin functionality in bread making. To a certain extent, differences in the structure of glutenin may also result from differences in glutenin composition. For example, if the glutenin structure is indeed branched, as suggested from its rheological behavior, GS composition may determine the degree of branching since some GS would allow for branching while others would not. Thirdly, based on polymer theories, only the polymers above a certain size would contribute to the elasticity of the glutenin polymer network. This corresponds well with several reports in the literature on positive correlations between dough strength/bread making performance and levels of the unextractable/least extractable glutenin fractions and/or the largest glutenin polymers. As with the glutenin structure, differences in the glutenin size distribution may also (at least partly) be attributed to differences in GS composition. Size differences of GS, resulting in variations in, e.g., HMW-GS/LMW-GS ratio, and/or different numbers of cysteine residues available in GS for cross-linking, influencing, e.g., the ratio of 'chain terminator' GS (only one cysteine residue available for cross-linking) to 'chain extender' GS (two or more cysteine residues available for cross-linking), may significantly affect glutenin size distribution.

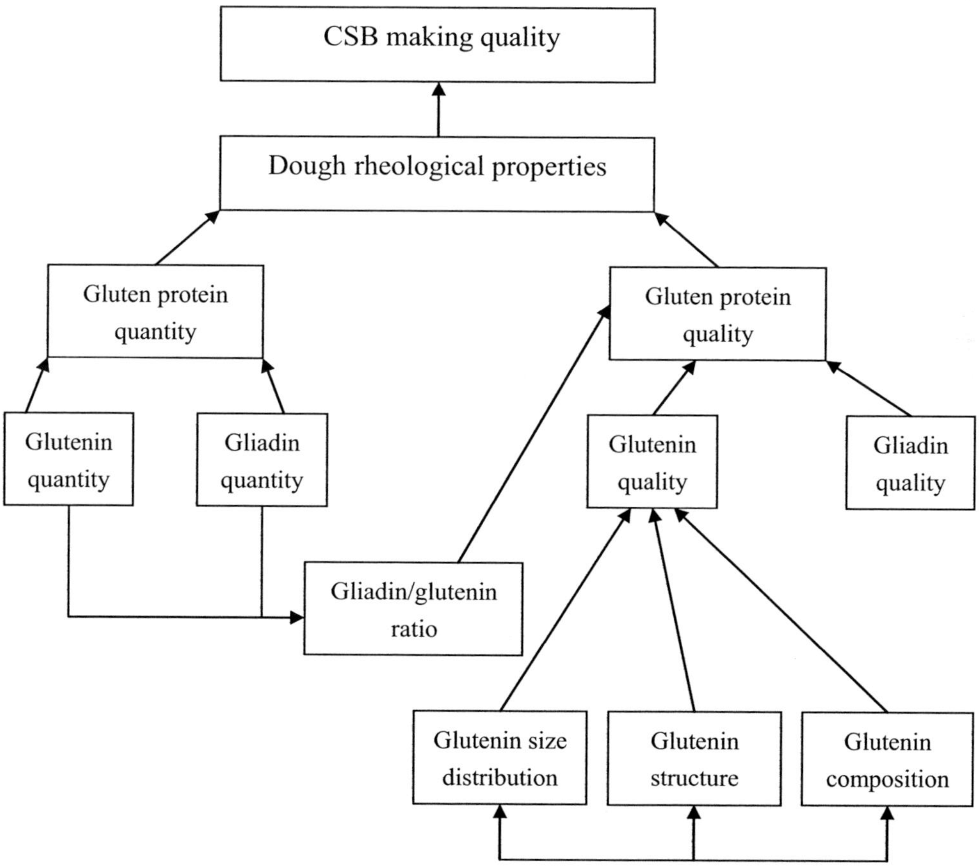

Figure 1-2. Factors governing CSB making quality and wheat dough rheological properties.

During the production process, dramatic changes occur in the gluten proteins that are probably a combination of changes in protein surface hydrophobicity, sulphydryl/disulphide interchanges and formation of new disulphide cross-links. As a result of these heat-induced changes as well as those of the starch, the typical foam structure is formed.

2.1.3. Lipids

It is well known that flour lipids, in particular the non-starch lipids (NSL) fraction, significantly affect the production quality of CSB. Starch lipids are too strongly bound in the starch granules and are essentially unavailable to affect dough processing before starch gelatinization occurs. When non-polar wheat lipids are added back to defatted flour, bread loaf volume is reduced. This observation has been ascribed to free fatty acids. Polar lipids can have a similar detrimental effect, but at higher concentrations, they increase loaf volume. In addition, the ratio of non-polar to polar lipids and the galactolipid content of the free NSL are strongly correlated with loaf volume. Presumably, lipid functionality is related to their effect on the stability of the gas cells. In this respect, the positive influence of the polar lipids is attributed to their ability to form lipid monolayers at the gas/liquid interphase of the gas cells, thus increasing the gas retention of the dough. Furthermore, polar flour lipids positively contribute to dough handling properties as well. In addition, during dough mixing, two processes occur which affect the lipids and hence the bread making performance of the flour.

First, most of the free NSL 'bind' to gluten or the starch granule surface and, as a consequence, their extractability is reduced. Secondly, polyunsaturated fatty acids are oxidised by wheat lipoxygenase, yielding hydroxyperoxides and free radicals. These compounds can oxidise other constituents, such as proteins and carotenoids, thus affecting dough rheological properties and crumb colour.

2.1.4. Milling Methods of Flour

Components of flour affect the CSB making properties, while milling methods also have an obvious influence on the quality of CSB and CSB making properties.

We milled 3 kinds of wheat (strong wheat 8901, medium wheat Nanyang White Wheat (NYWW), and weak wheat Australia White Wheat (AWW)) with debranning or conventional milling and investigated the variation in components and properties of CSB making (Sun et al., 2007). As shown in table 1-2, the ash and pericarp contents of most of the samples from debranned flour were higher than that of flour. The mean pericarp particle size in the conventional flour was larger than that from debranned flour except for some of the second flour in the extent of debranning about 4.5%. Thus a high pericarp and ash content affects the flour quality, and a smaller pericarp size has a negative impact on steamed bread height. Fortunately, Debranned flour mixed with water was whiter and brighter compared to conventional flour.

The starch damage of the conventional flours was higher than that of debranned flours, moreover the mean particle size of conventional samples were smaller than that of debranned flour.

Damaged starch hydrates easily and is more susceptible to enzymatic hydrolysis. A certain level of damaged starch is beneficial because of the increase of baking absorption and gassing power of the dough. However, excessive starch damage can over-hydrate the dough, accelerate enzymatic action, and lead to inferior baking performance. Flour has the best baking performance when the starch damage is between 4.5–8.0%.The results demonstrated that starch damage decreased markedly in debranned flour. The 7% starch damage in all AWW flours is acceptable. The starch damage of NYWWD□ and NYWWD□ was the lowest, about 4.5%. The particle sizes of NYWWD were a little larger than that of NYWWC, but the starch damage of NYWWD was clearly lower. The starch damage of 8901C was higher than 8.3%, while for 8901D it was below 7 %.

As shown in Figure 1-3, the effects of milling methods on the quality of CSB. The quality scores of AWWCII and 8901CII were higher than that of AWWDII and 8901DII. However, steamed breads made from debranned second flour, had clearly improved quality scores, volume, volume/weight and structure (height, skin color, skin structure and interior) of NYWWII (Figure 1-4) and 8901□. The shape and structures of steamed breads from AWCII and 8901CII were better than that of debranned flour. The method of milling did not show a significant effect on the texture of steamed bread, except for the second flour from NYWW.

Research found that debranning had only slight effects on the quality of top flour in terms of gluten index, maximum resistance, starch damage, particle size, falling number, flour color and pasting properties. The low gluten index ($r=-0.66$, $p<0.05$) and large pericarp size of NYWWDII and 8901DII improved the volume of steamed bread and resulted in higher steamed bread quality ($r=0.89$, $p<0.001$) and whiter skin color ($r=0.624$, $p<0.05$). Hence it can be concluded that debranning improved the quality of second flour from NYWW and 8901, and in addition improved the performance of the flour in steamed bread making.

Table 1-2. Flour quality for debranning and conventional flour (Sun et al., 2007)

Name	Protein content	Ash	Pericarp content (%)	Mean Pericarp size	Falling number (s)	Particle size (μm)	Starch damage (%)
AWWDI	8.34i	0.61fg	1.54a	0.048d	510b	75.90b	6.83e
AWWCI	8.05j	0.54h	0.96d	0.069ab	467c	58.10e	7.95b
AWWDII	9.34h	0.83d	1.54a	0.029f	518b	76.28b	7.25cd
AWWCII	9.22h	0.75e	0.97d	0.039e	401d	65.13c	7.44c
NYWWDI	11.39g	0.60g	0.88d	0.059c	429d	63.84cd	4.56g
NYWWCI	11.26g	0.59g	0.60e	0.038ef	400d	62.38cd	5.86f
NYWWDII	12.41f	0.96b	1.19c	0.029f	476c	58.88de	4.68g
NYWWCII	13.27c	1.03a	1.21c	0.020g	426d	56.13e	6.63e
8901DI	12.75e	0.63f	1.31b	0.062bc	605a	88.30a	6.94de
8901CI	13.06d	0.60g	0.92d	0.077a	512b	75.09b	8.31b
8901DII	14.29b	1.04a	1.25bc	0.036ef	524b	89.80a	7.01de
8901CII	14.56a	0.92c	0.89d	0.030f	516b	75.44b	8.80a

D: debranning; C: conventional; I: top flour; II: second flour.

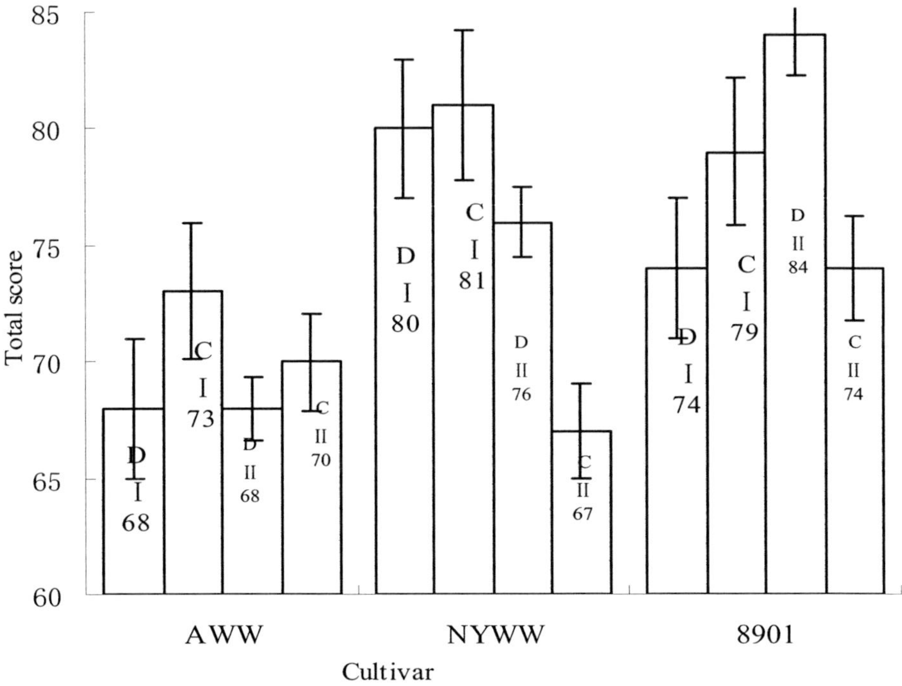

Figure 1-3. The total score of CSB made from flours milled with different methods (Sun et al., 2007).

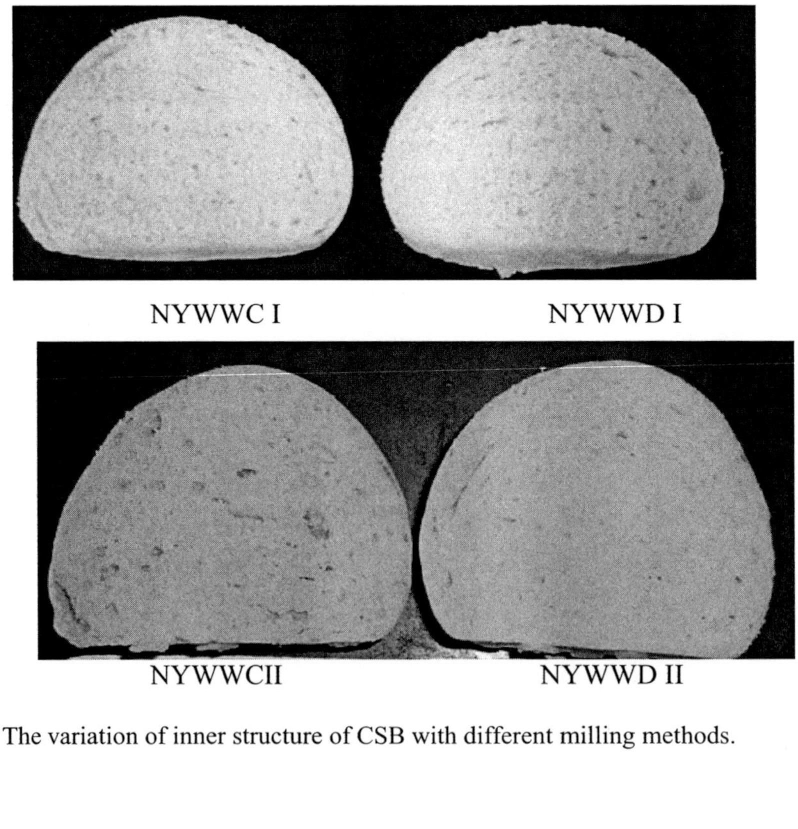

Figure 1-4. The variation of inner structure of CSB with different milling methods.

2.2. Yeast

Yeast cells metabolize fermentable sugars (glucose, fructose, sucrose and maltose) under anaerobic conditions producing carbon dioxide as a waste product, which acts as a leavening agent and enhances dough volume. Yeast also supports both gluten network and aromatic compounds production. Active cells of yeast are available as a compressed cake or in dried form. The compressed cake contains approximately 70% moisture so it is highly perishable unless refrigerated. Active dry yeast is produced by extruding cake yeast in fine strands, which are dried to low moisture content. Instant yeast is made from more active strains of yeast and dried faster and to a lower level of moisture. Although active dry yeast has a long shelf-life at room temperature, it must be hydrated before being incorporated with other ingredients. In contrast, instant yeast can be incorporated with flour and other ingredients without prior hydration. The actions of yeast may be shown in a simplified form as follows:

$$C_6H_{12}O_6 \rightarrow 2C_2H_5OH + 2CO_2$$

That is to say, in the actions, the simple sugar of yeast would form ethyl alcohol and carbon dioxide.

The yeast preparation is available in several forms. The yeast cream can be used directly, although that is highly perishable. Most commercial bakers use compressed yeast cultures. These are produced by pumping the yeast cream through a filtration press or vacuum filter to remove most of the water. The yeast is collected in the form of moist cakes, separated by wax paper. Compressed yeast cake still has high moisture content, and requires refrigeration for a few weeks. However, because the cells are metabolically active, once they are introduced into the dough, fermentation can occur very quickly. Compressed yeast can be further dried to about 90% solids to provide dry active yeast. This is the form that is familiar to consumers who make homemade CSB, but small manufacturing operators or consumers use a dry yeast preparation, too, when compressed yeast is not available. Preparation of dry active yeast takes six months or more at room temperature. They require a hydration step, and in general, are not as active as compressed yeast, although the improved drying technologies have greatly enhanced the activity of dried yeast. In addition, dry active yeasts can be "instantized" so that the rehydration is easy.

2.3. Water

Water is necessary for the formation of dough and is responsible for its fluidity. It assists the dispersion of yeast cells and is the medium for food transportation to the yeast through cell membranes. Water is also needed for starch and sucrose hydrolysis. The water is necessary to activate enzymes that bring on the formation of new bonds between the macromolecules in the flour, and alter the rheological properties of dough. The amount of added water is related to the moisture content and the physicochemical properties of the flour. The properties of the dough will vary according to the level of added water. The dough will be firm, difficult to mould if the addition of water is not enough and it will result in small volume and poor external appearance of CSB. While the dough would be soft, it also would be difficult to mould if the addition of water exceeded the needed quantity, and resulted in low quality of CSB. The 'optimum' level of water is really the maximum quantity we can get into the dough and still be able to mould the pieces and give bread of acceptable quality.

3. Processing Technology

Several ingredients can be used for the production of steamed bread, the most important of which are flour, yeast and water. As soon as dough is properly prepared and steamed, a product with superior quality and sensory features could be expected. However, fresh production is the one with a short shelf-life and a number of chemical and physical alterations occur during storage, known as staling. As a result of these changes, steamed bread quality deteriorates gradually as it loses its freshness quickly compared with western-style bread. The pleasant aroma vanishes and the flavor brings out a stale feeling. Those preservation problems in combination with the increasing market demands and the complexity of the traditional procedure, which requires night or early morning labor, led to the evolvement of several technologies in order to improve the preservation of CSB. Meanwhile, several additives were introduced in order to increase shelf-life and enhance its quality, conservation, sensory perception or even nutritional value.

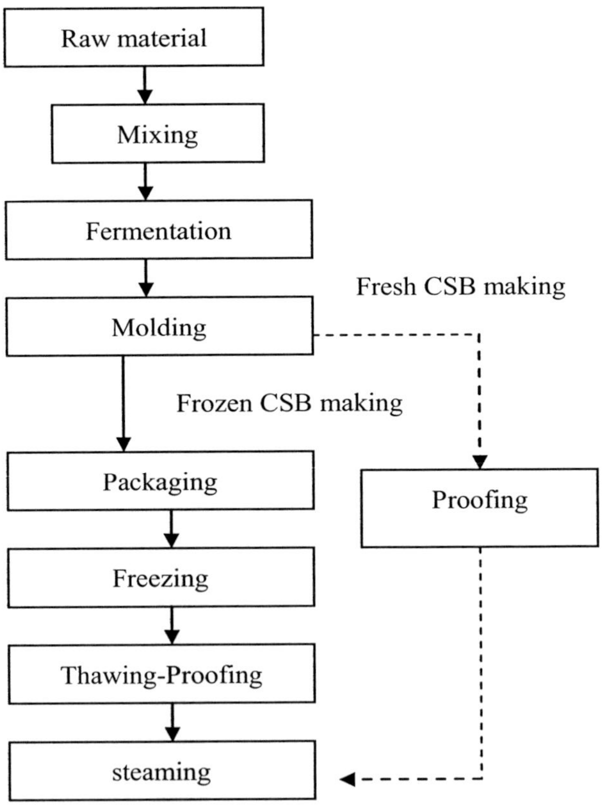

Figure 1-5. Process flow diagram of frozen dough CSB and fresh CSB making.

The processing of CSB is almost the same as that of bread. The process includes mixing, fermentation, remixing, molding, proofing, steaming, cooling and packing. Of course, processing is related significantly with the quality of CSB and there are many researches on the effects.

The processes of tradition and frozen CSB are just a little different from each other as shown in Figure 1-5.

Over the past few years, the CSB industry has exploited the advantages and applications of the freezing technology and developed a special interest in it in order to cover its customers (consumers, food service) needs for products with increased shelf-life. Frozen dough CSB is expected to be characterized by quick preparation time and affordable price, and look and taste as if they were freshly homemade. But the application of frozen dough for Chinese steamed bread making still needs to be studied.

3.1. Optimization of Laboratory Processing Procedure of CSB

To evaluate the quality of CSB or improve the processing technology, a laboratory processing procedure of CSB is necessary. Some of the steps may be completely mechanized but most of them are still manual work.

At first, Huang et al. (1993) studied the optimized processing procedure by response surface methodology for northern-style CSB. Compressed yeast (4.5 g) was dispersed in a volume of water 20 g less than that of 70% Farinograph water absorption. Flour (200 g) was then mixed with the yeast/water slurry (30°C) in a 300 g Farinograph bowl and fermented (32°C, 85% RH). The fermented dough was placed in the mixing bowl, additional flour (100 g) and water (20 g) were added and the dough was remixed and sheeted (20 times) by passing through a pair of rolls (diameter 11 cm; gap, 7.2 mm; and 11 rev/min). After each pass, the dough was folded end-to-end and re-sheeted in a unidirectional manner. The dough piece was then divided (100g dough pieces) and gently shaped by hand to form a rounded dough piece with a smooth upper surface. The dough piece was placed into an Extensograph rounder (smooth surface facing up), rounded (20 times), placed into a tray, proofed for 20 min and steamed for 20 min. In the process, the mixing time is 3/4 of Farinograph dough development time and the remixing time is 1/4 of Farinograph dough development time.

Huang et al. (1998) also researched the optimization of a laboratory processing procedure for southern-style CSB in 1998. For southern-style CSB, the mixing time and remixing time is 50% and 180% of Farinograph dough development time, and the fermentation time and the proofing time is 150 min and 35 min, respectively. And the amount of flour used for mixing is 240 g and the additional flour for remixing is 60g.

The manual method of CSB is also shown in a National Standard of the People's Republic of China named as wheat varieties for specific end-uses (China State Bureau of Technical Supervision, R.P. China, 1998, GB/T 17320-1998B). The formula consisted of 100 g flour, 1 g instant active dry yeast. The instant active dry yeast was dissolved in different volumes of water. The volume of water was determined through experiments (about 80% of water absorption capacity of Farinograph). The ingredients were put into an aluminum basin orderly, kneaded by hand until optimum dough consistency appeared. After resting for 15 min at room temperature, the dough was molded into a near hemisphere-like shape with a height of 60 mm on a smooth surface. Then the dough was kneaded by hand again for 3 min and proofed for about 60 min in an incubator at 33 °C and 85% RH. After that, the dough was shaped and put into a steamer with boiling water, and steamed for 20 min.

But manual methods can only be used in some parts of the countryside or in research. Most of CSB are processed by a mechanized method especially in the city so the effects of the mechanized method and the relationship between the quality of material and product focusing on the northern-style CSB will be introduced.

3.2. Mixing

Dough is produced when all the ingredients of the formula, introduced in a desirable sequence, are mixed together for a certain period of time. The major purposes of mixing are to blend the ingredients into a quasi-homogeneous mixture, to develop the gluten matrix in wheat dough, and to incorporate air. In the first step of mixing, the proteins are hydrated, and then hydrated proteins interact with each other. In addition to protein interaction, other flour components including lipid, non-starch polysaccharides and starch also participate in the formation of the gluten matrix. The viscoelastic properties of dough are primarily the result of a continuous protein phase that, in fully developed dough, surrounds the starch granules. The chemical bonds that stabilize gluten proteins in dough are covalent and secondary bonds. The covalent bonds are disulfide bonds, which form inter- and intramolecular crossbonds in the proteins during dough formation by the sulfide-disulfide interchange. The secondary bonds are hydrogen, hydrophilic, hydrophobic, and ionic bonds and polar interactions.

The mixers are divided by the design of the beater arms in the chamber. The two main variations are based on roller bars and an elliptical-shaped beater. In both cases the mixing action is strongly influenced by the relatively small size of the gap between the outer edge of the beaters and the sides of the bowl. The main action tends to be one of stretching and folding of the dough. The dough is picked up by the mixer blades and thrown against the outer side of the bowl.

In order to get suitable CSB dough, there are some things to which producers should pay more attention. Firstly, the amount of flour added into the mixer should be definite.

Secondly, it is generally considered more desirable to start with drier dough and adding water as you work to complete the kneading process. This is because flour does not absorb water (hydrate) quickly. But professional producers are experienced at knowing how much water to add. By adding water in two phases, the dough may be mixed inadequately. In order to obtain a desirable structure of dough, flour should be mixed with the required amount of water. When the amount of water added is relatively less, the transformation of starch into gelatin may be incomplete. As a result, CSB could easily dry and scrapes. At the same time, the CSB will become stale quicker. On the contrary, when excess water is used, it is not entirely constrained during starch gelatinization and a certain amount of water is still in free statement which makes crumbs moist and sticky. The water-holding capacity of flour depends on its type, origin and other properties. Zhang (2005) investigated the effects of addition of water on the quality of CSB. As shown in table 1-3, addition of water had negative effects on the height, appearance and whiteness of CSB but is good to skin, inner structure and volume in a suitable range.

Chen et al. (2005) also researched the effect of water addition on the CSB processing. They adjusted the addition of water according to FWA values of six cultivars of wheat flour as table 1-5 and had the relationship between water addition and CSB quality as table 1-5 and 1-6. In table 1-5 and table 1-6, strong wheat was Gaocheng 8901 and Zhengmai 9023, medium wheat included Zhongyou 9507 and Ningchun 4, while weak wheat included Jing 411 and Jingdong 8.

They indicated that the effects of water addition were different from the strong, medium and weak wheat. For example, the effects for medium flours were not so obvious, but increasing of water addition for strong flours almost improved the total quality, and even decreased the total quality for weak flours (table 1-6).

To same flours, the addition of water for southern-style CSB must be larger than that of northern-style CSB. The addition of water in dough for CSB is lower than that of western bread making by 20–35%, hence dough for CSB making is stiffer and firmer than the dough for western bread making.

Thirdly, the mixing time is decided by flour characteristics and mixing speed. Mixing should be quick, homogenous and temperature controlled. Lower mixing speed required a longer mixing time in order to develop the gluten structure for strong flours. Dough is kneaded after the initial mixing to realign the strands of gluten and form the network structure of the dough. Dough kneaded properly is shiny and elastic. If a little piece is pulled off, considerable extensibility will be observed. If dough was still undermixed, starch and proteins would be distributed unevenly, and compact protein masses are stretched out into sheets during mixing. When dough is overmixed, gluten proteins become stressed, few disulfide bonds would be broken to form thiyl radicals and gluten proteins are partially depolimerized, resulting in a greater solubility and less extractability of lipids. Overmixing usually results in sticky dough, partly because mechanical forces applied to the dough decreased the molecular weight of the protein. Prolonged mixing can enhance the effects of oxidants on disaggregation of large protein aggregates, probably because of oxidation of more -SH groups.

Lastly, the mixing environment including RH and temperature should be determinate.

Table 1-3.The effects of addition of water on the quality of CSB (Zhang, 2005)

Ratio of added water(water/flour,w/w)	Skin	Inner structure	Height	Appearance	Volume	Whiteness
44%	8.2	10.7	4.3	8.5	10.1	50.7
46%	8.5	11.6	4.1	8.4	10.8	50.3
48	8.7	11.9	3.9	8.3	12.2	50.2
50	8.5	12.2	3.5	8.0	13.2	49.1

Table 1-4. Additional water for 6 samples at five WA levels (Chen et al., 2005)

Sample	70 % FWA	75 % FWA	80 % FWA	85 % FWA	90 % FWA
Gaocheng 8901	45.16	52.11	58.16	65.12	71.17
Zhengmai 9023	47.15	54.12	60.19	67.16	74.12
Zhongyou 9507	39.10	45.11	51.11	57.12	63.13
Ningchun 4	38.16	44.16	50.16	56.17	62.17
Jing 411	34.15	40.13	46.10	51.18	57.15
Jingdong 8	33.15	39.12	44.19	50.16	56.12

FWA — Farinograph water absorption1.

Table 1-5. Correlation coefficients between water addition and CSB quality (Chen et al., 2005)

Parameter	Water addition		
	Strong wheat	Medium wheat	Weak wheat
Loaf volume	0.775**	0.885**	0.958**
Weight	0.957**	0.904**	0.983**
Width	0.881**	0.929**	0.905**
Height	0.928**	0.744**	0.927**
Specific volume	0.602	0.750*	0.830**
Spread ratio	0.936**	0.807**	0.915**
Skin color	0.048	0.337	0.676*
Crumb color	0.370	0.001	0.724*
Shininess of surface	0.238	0.331	0.663**
Smoothness	0.877**	0.536	0.810**
Structure	0.768**	0.202	0.814**
Stress Relaxation	0.523	-0.291	0.904**
Total score	0.518	-0.313	0.962**

*and** indicate significance at 5 % and 1 % probability levels respectively.

3.3. Fermenting

For CSB production, three fermentation methods could be used. In rural China, the steamed bread was produced from a starter "mother dough" (Jiaotou). Traditional starter culture (Jiaotou) could be made from wheat flour, wheat bran, corn flour, or rice flour by solid fermentation or by submerged fermentation. Jiaotou is still widely used in the northern Chinese countryside today. Jiaotou is usually produced in the natural environment, where the microorganism compositions are complicated. Only in the past thirty years, however, have the lactic bacteria and yeasts that participate in the sourdough fermentation been identified. It is remarkable that the interaction between these organisms is not stable. It was considered that the main mixed microflora inherent in it were yeasts and lactic acid bacteria (LAB). The co-cultured organisms might compete for growth nutrients or produce metabolic products that would inhibit each other's growth. LAB might produce acid that inhibited the growth of yeasts. Furthermore, mutual influence of the microorganisms on each other's metabolism might lead to different profiles of important flavor compounds in CSB. On the basis of carbohydrate fermentation and assimilation, 4 bacteria isolated and 3 yeasts isolated from 12 pieces of traditional Jiaotou were identified to species level (Ding et al., 2007). Three bacteria were identified as *Bacillus cereus, Brevibacillus brevis, Acinetobacter twoffii*. The source of *Bacillus cereus, Brevibacillus brevis, Acinetobacter twoffii* in Jiaotou was assumed to be casual contamination from the environment. The species of isolated yeasts were identified as Saccharomyces cerevisiae, which was widely used in brewing and genetics research. Carbohydrates were fermented by yeasts to produce carbon dioxide and a little ethanol. The carbon dioxide formed gas bubbles which contributed to the interior structure of CSB, similar to the interior structure of bread. Ding et al. (2007) reported that the pH value changed during dough fermentation by Jiaotou. During fermentation, the pH value remained in the first 3 hours. From 3 to 6 hours, the pH value decreased gradually from 5.7 to 5.5. In the last 2 hours, the value decreased sharply to 4.9. The LAB and yeast counts changed in the last 2 hours were the reason for the observed pH value decrease. Despite the relatively low level of LAB in the fermented dough by Jiaotou, these organisms were likely to be significant in flavor development in CSB. After 5 hr fermentation, the pH gradually declined, especially for the last 2 hr, while the yeast counts remained un-fluctuated, other than the increase in the preceding 5 hr. This reduction in pH value as a result of the organic acids production by LAB and yeasts was almost likely to have reflected the cause of suppression of yeast population in the dough.

The starter produces the product that is high in acidity, giving it a distinct tang. This pre-ferment sits at room temperature, to ferment and develop flavor until it is used to make the rest of the dough. Long fermentation develops extra gluten strength for the dough, adds depth and complexity of flavor, and increases the shelf-life of products. There are several different types of pre-ferments, which vary mainly by the amount of water they contain. For CSB, the way to prepare Jiaotou is to save a small piece of dough from one batch to add at the end of mixing the dough (during the last couple of minutes, since the structure is already developed). Jiaotou used to keep for more than 24 hr under a comfortable temperature and RH. But the disadvantage is the producing condition and values of pH of Jiaotou are difficult to control so experience is heavily needed.

In recent years, commercial dry yeast products were used instead of Jiaotou for the industrial production of CSB. Fermentation is what happens when yeast comes in contact with the flour and water. Then the carbon dioxide is held in by a network of gluten strands, or protein, formed by kneading together the flour and water, and it leavens or causes the products to rise. Temperature control is a very important factor in fermentation. Yeast is active at temperatures around 30°C. At warmer temperatures, the yeast is more active and grows and multiplies more quickly. The fermentation process itself produces heat. When fermentation takes place at too high a temperature, unpleasant flavors are produced. Length of fermentation is another important factor that determines both the flavor and the texture of products. If the dough ferments for too long, the yeast and bacteria will consume all of the sugar in the flour. Compared with Jiaotou, the advantages of commercial yeast products would include more rapid dough fermentation, less acid produced and purer flavor. The overall procedure involves mixing all of the ingredients and then allowing the dough to ferment for several hours.

Compared to Jiaotou, commercial yeast had a shorter dough fermentation time (usually no more than 2 hr), while the pH value of dough could be kept almost unchangeable during the fermentation. But CSB processed with yeast is short in fermentation flavor and has poor re-steaming ability compared with that of Jiaotou. Although the quality of steamed bread is still acceptable, the process lacks flexibility and is sensitive to time. In other words, the dough must be steamed soon after fermentation. Otherwise, prolonged fermentation will result in the forming of excess air cells and weaken the structure of the CSB.

Table 1-6. The quality of CSB made from different flours at five water addition levels (Chen et al., 2005)

Sample	WA	Weight	Volume	Width	Height	SV	SR	SC	CC	SH	SM	ST	SR	TS
Strong	70%	98.4c	232.5b	8.8b	6.0a	9.5b	7.3a	4.4a	3.4a	4.3a	5.8c	5.1d	23.0b	62.6c
	75%	99.6c	260.0ab	8.8b	5.9ab	12.7ab	5.0ab	4.0ab	3.0ab	4.1a	6.5bc	6.9c	32.0a	74.2ab
	80%	102.1b	283.8a	8.8b	5.7bc	14.8a	3.0bc	3.4b	2.5b	4.3a	7.5ab	7.8b	33.5a	76.8ab
	85%	103.8b	291.3a	9.1ab	5.6cd	15.2a	1.5bc	3.9ab	2.8ab	4.3a	8.1a	8.7a	34.0a	78.4a
	90%	105.8a	291.3a	9.4a	5.5d	14.5a	0.0c	4.4a	2.8ab	4.3a	8.0a	7.6bc	30.0a	71.5b
Medium	70%	95.9c	237.5c	8.5d	5.7a	11.0b	5.0a	3.6a	1.6a	4.1a	6.4b	6.8a	23.0b	61.5b
	75%	97.3c	242.5c	8.6cd	5.6ab	11.2b	3.5ab	3.5a	2.1a	4.0a	7.0b	7.2a	26.0b	64.5b
	80%	99.4b	252.5c	8.9bc	5.5ab	11.8b	1.5bc	3.1a	1.9a	4.3a	8.5a	7.9a	30.3a	69.1a
	85%	101.8a	272.5b	9.0b	5.4ab	13.5ab	1.0bc	3.4a	1.6a	4.1a	7.8ab	8.2a	25.3b	64.8b
	90%	105.3a	291.3a	9.5a	5.4a	14.8a	0.0c	2.8a	1.9a	3.9a	7.5ab	6.8a	19.5c	57.0c
Weak	70%	94.6d	230.0b	8.1c	5.8a	10.4a	9.0a	2.5a	0.9a	5.0a	8.8a	8.8a	31.0a	76.3a
	75%	96.1c	233.8b	8.5bc	5.7a	10.4a	5.5b	2.3a	0.8a	5.0a	8.5a	8.8a	31.5a	72.8a
	80%	97.6c	248.8ab	8.6b	5.5ab	11.9a	2.0c	1.6a	0.6a	5.0a	7.8a	6.0b	29.5a	64.4b
	85%	100.0b	260.0a	8.8b	5.4b	12.5a	1.0c	2.0a	0.4a	4.4a	7.8a	6.1b	26.8b	60.9bc
	90%	103.2a	267.5a	9.3a	5.1b	12.5a	0.5c	1.6a	0.3a	3.9b	6.5b	6.2b	24.5c	55.9c

Notes: WA=water addition (% farinograph water absorption,), SV=specific volume, SR=spread ratio, SC=skin color, CC=crumb color, SH=shininess, SM=Smoothness, ST=Structure, SR=Stress Relaxation, TS=Total Score. The values followed by a different letter are significantly different at 5% probability level.

Now, a new method called “remixed fermentation” is widely used in the processing for industrial production. There is the sponge and dough method where mixing of ingredients is performed in two steps. The leavening agent is prepared during the first step. Yeast, flour and water are mixed together. The mixture is left to develop for a few hours and, afterwards, it is mixed with the rest of the ingredients. In this case, only part of the ingredients are fermented, the sponge. Sponge fermentation times may vary considerably, as their composition does. The key features of sponge and dough processes are: First, the two-stage process in which part of the total quantity of flour, water and yeast are mixed to form homogeneous soft dough that is called sponge. Secondly, the resting of the sponge is formed, in bulk for a prescribed time, mainly depending on flavour requirements. Thirdly, the sponge is mixed with the remainder of the ingredients to form homogeneous dough. Lastly, the final dough is processed immediately, although a short period of bulk fermentation may be given. The sponge contributes to flavor modification and the development of the final dough. The process of flavour development in the sponge, though complex, is observed as an increase in the acidic flavour notes arising from the fermentation by added yeast and other microorganisms naturally present in the flour. To maintain the right flavour profile in the finished product, the sponge fermentation conditions are closely controlled to avoid unwanted flavours. During the sponge fermentation period, there will be a decrease in pH value with increasing fermentation. Under these conditions, the rheological character of the gluten formed during initial sponge mixing changes and the sponge becomes soft and loses much of its elasticity. The low pH of the sponge and its unique rheological characters are carried through to the dough where they have the effect of producing a softer and more extensible gluten network after the second mixing. In many cases, the addition of sponge changes the rheological character of the final dough sufficiently so that dividing and moulding can proceed without further delay.

Ding et al. (2007) also compared the quality between Jiaotou fermented CSB and commercial yeast fermented CSB by sensory analysis and a texture analyzer. In the sensory evaluation, the interior structure of Jiaotou fermented CSB was superior to that of the commercial ones. As to the cohesiveness, the Jiaotou fermented CSB was also superior to the commercial ones. There were no significant differences in stickiness between the two kinds of CSB. The special flavour was formed during dough fermentation by the microorganisms from the starter culture Jiaotou. The results of the texture analyzer test illustrated that the hardness, gumminess and chewiness of Jiaotou fermented CSB were significantly higher than the commercial ones ($p<0.05$). The springiness of Jiaotou fermented CSB was lower than that of commercial CSB, but not significantly different. The springiness of the texture analyzer was similar to elasticity by sensory analysis.

Even though Jiaotou may contribute to the quality of CSB, the difficulty in controlling the process and the much longer time needed for the processing are the biggest problems of mechanized processing of CSB. So the Jiaotou is seldom seen in the CSB industry.

Fermentation time is significantly related with the activity of yeast and is variable. The suitable temperature and relative humidity for fermentation of CSB dough are 30~35 °C and 70~85% respectively.

3.4. Rounding and Molding

Dough is divided into pieces with certain weight and is molded to obtain a desirable shape. Dividing and molding modify the structure of gas cells as they induce the coalescence of small cells into larger ones and contribute as well to the final development of the gluten network. After dividing, the dough pieces are commonly worked in some way to change their shape before proof. The most common shaping is by rounding. The action of mechanical first molding places the dough under stress and strain which may lead to damage of existing gas bubble structure in the dough. A portion of the gas is also lost during the dividing and rounding steps. Thus, not only does the dough have a chance to rest, but the fermentation also continues, adding a bit more gas into the dough. The dough pieces are delivered into a molding system that first sheets the dough between rollers then rolls the dough into the desired final form.

The most common equipment for the CSB molding is called lying CSB molder. It has parallel rollers. The sets of parallel rolls round dough pieces passed through at high speed. Sheeting reduces the thickness of the dough pieces. The gap between successive pairs of rolls decreases and on leaving the last gap, the dough piece has an ellipsoid shape. The other equipment is the plate CSB molder. The dough is squeezed and chipped, then rounded on a plate. The whiteness of CSB rounded by the plate molder is worse than that of a lying molder. So the application of lying ones is more common. Turntable or screw model molders have also been developed recently. Figure 1-6 is a scheme of screw model molder of CSB which is widely used for its high efficiency.

In regard to frozen CSB, the dough shape is influential on its stability and final quality. Round pieces are considered less satisfactory than slabs or cylinders. Furthermore, excessive molding can cause heat generation and enhance fermentation prior to freezing.

Figure 1-6. Scheme of a screw model molder of CSB.

3.5. Proofing

Proofing is mainly attributed to the action of yeast, which contributes many changes in the dough maturing or ripening. Properly matured dough exhibits optimum rheological properties (optimum balance of extensibility and elasticity) as well as good machinability and is necessary to process CSB with a desirable volume. During dough maturing, several reactions occur. As the yeast is fermented, alcohol and carbon dioxide are produced. Because alcohol is water-miscible, appreciable amounts of alcohol influence the colloidal nature of the flour proteins and alter the interfacial tension within the dough. Additionally, carbon dioxide dissolves partly in the aqueous phase of the dough and forms weakly ionizable carbonic acid, which lowers the pH of the system. Carbon dioxide production also contributes to dough distension which depends on the situation of gas cells. Growth of gas cells depends in part on cell size. Greater pressure is needed to expand a small gas cell than a larger one, and it is possible that the smallest bubbles will not expand at all. Gas cell stabilization and gas retention are of considerable interest as they fairly determine structure and volume of CSB.

For frozen dough CSB, thawed dough pieces should be left to proof before being steamed, just like the fresh ones, either for a certain period of time or until they obtain the desirable volume. In the case of frozen fermented dough, gas cell structure significantly affects frozen storage stability. A dough that contains a large number of small bubbles with a narrow size distribution and thick walls will be more stable than a dough that contains bubbles with less uniform size distribution and thin walls surrounding the larger bubbles. Proof time for frozen-thawed dough is necessarily longer than that for conventional dough. This is due to the lower dough temperature at which thawed pieces reach the proof box, and to a certain loss of dough gas retention power and yeast activity caused by the freezing process.

Figure 1-7. Mechanized steaming line of CSB.

3.6. Steaming

Steaming is the last and more impressive part of the CSB making procedure. It results in a series of physical, chemical and biochemical changes, which include volume expansion, evaporation of water, formation of a porous structure, denaturation of protein, gelatinization of starch and so on. The microstructure of flour is continuously modified during all these processes until the structure of the final product is fixed. These changes are dependent on the temperature, humidity and duration. The role of steaming is to alter the sensory properties of foods, to improve palatability and to extend the range of tastes, aromas and textures in foods produced from raw materials. Steaming also destroys enzymes and microorganisms though recontamination of cereal products may occur after. The flavor is mainly formed during fermentation and steaming. During fermentation, a number of alcohols are formed, including ethanol, isoamyl, amyl alcohols and isobutyl alcohol. However, much of these alcohols are lost to environmental space during steaming. A large number of organic acids are also formed and several carbonyl compounds have been identified in CSB, which are believed to be important flavor components.

CSB had been steamed in bamboo container just as shown in Figure 1-1 and now stainless steel mechanized line (Figure 1-7) is used in large CSB maker.

3.7. Cooling and Packaging

Once out of the steamer, the steamed bread is susceptible to microbial spoilage. Therefore, cooling must be performed under conditions in which exposure to airborne microorganisms, particularly mold spores, is minimized. The CSB also must be cooled enough so that condensate will not form inside the package, a situation that could also lead to microbial problems. Various cooling systems are used, including tunnel-type conveyers in which slightly cool air passes counter-current to the direction of the bread, as well as forced air, rack-type coolers.

Packaging materials and shapes vary according to products specifications. Materials usually applied to frozen dough products are plastic (films, membranes, etc.) and aluminum. In any case, packaging must form an effective, functional barrier to contamination and have sufficient impact and compressive strength to withstand the stresses which it is likely to meet. It must perform satisfactorily in storage and transport. This requires good crush resistance, resistance to variable moisture conditions, no embrittlement at the low temperatures experienced by frozen foods, and sufficient scuff resistance to avoid deterioration of the surface appearance of printed matter. When designed for frozen dough, it must perform a number of functions, such as contain, protect, identify, and merchandize the food. A good packaging material must keep loss of moisture to a minimum. Films to be used for frozen dough should posses the following characteristics: good moisture protection, good oxygen-barrier characteristics, physical strength against brittleness and breakage at low temperature, stiffness to work on automatic machinery, and good heat seal-ability.

3.8. Freezing

Frozen dough CSB needs more processing procedures such as freezing and thawing in different steps. Meanwhile, freezing may have negative effects on the activity of yeast and the quality of CSB. These will be explained concisely.

Dough pieces, immediately or after a short fermentation period, are frozen and then stored at appropriate temperature. Freezing technology can be categorized as mechanical (blast, plate, spiral, impingent, immersion, belt or fluidized bed freezers) and cryogenic. Selecting implemented technology based commonly on product requirements and availability or cost. Freezing, as a method of preservation and extension of a food product shelf-life, involves mainly two intimate processes. First, temperature reduction and second, phase transition from liquid to solid. Freezing normally starts at -1 to -3°C and as the temperature drops most of the water in CSB becomes frozen.

3.9. Thawing

Frozen dough must be thawed before proofing. This process can be conducted under various time–temperature conditions. Thawing is necessary for best performance of the dough as it involves the rehydration of the system, mainly of the gluten matrix and yeast cells. The process can be completed either at a certain temperature or by stepwise temperature increase, which is more favorable for two reasons. Firstly, during thawing, condensation occurs on the dough surface, as dough is colder than the surrounding air. This results in spotting and blistering of the crust especially when there is a large difference in temperature between the dough surface and the surrounding air. A stepwise increase in temperature minimizes this effect. Secondly, excessively rapid thawing raises the temperature only to the outer regions of the dough, which becomes ready for proofing, while the centre of the dough still remains frozen.

3.10. Minimizing Freezing Damage

The quality of frozen dough products is influenced by dough formulation, as well as processing parameters such as dough mixing time, freezing rate, storage duration and thawing rate. It appears that these factors may act either independently or synergistically to reduce yeast activity which results in reduced CO_2 production or damage to the gluten network, gradual loss of dough strength, declined retention of CO_2 and poor CSB making performance. The resulting loss of dough strength can be attributed, firstly, to the release of disulphide reducing substances from dead yeast cells, and secondly, to the disruption of the gluten network by ice crystals. Yeast cryoresistance is strongly influenced by fermentation time prior to freezing, dough freezing and thawing rates, storage time and temperature fluctuations of frozen dough.

Yeast characteristics also play an important role in determining yeast viability and product quality. Yeast content should be higher than in conventional CSB making in order to overcome the prospective loss of activity during freezing, storage and proofing. Several ways to minimize the effect of freezing on dough are reported in literature such as the formation and use of yeast strains more resistant to freezing, the modification of the CSB making process or the introduction of suitable additives and ingredients for frozen dough.

Use of oxidizing agents, whether from natural or chemical origin, exerts an improving effect on dough rheology and on the overall quality of the finished product. An oxidant exhibits improving effect by increasing the loaf volume during the first few minutes of steaming. Surface-active and dough strengthening agents such as sodium stearoyl lactylate (SSL) or diacetyl tartaric acid esters (DATEM) of mono and diglycerides are also used to maintain loaf volume and crumb softness for frozen dough. Medium to strong flour is required for frozen dough production. Shortening addition can be within the range of 4–6% of the flour weight. Accumulation of less water in the formula has been recommended. Reduced water content limits free water in dough during freezing. This may constrain ice crystal formation and lessen its negative effects on the quality of frozen dough. Ice crystallization particularly affects proteins, lowering the gas retention properties of dough.

Mixing duration and dough temperature are important factors in frozen dough stability. Gluten network should be fully developed during mixing. In frozen dough production, dough temperature after mixing is normally lower, in the range between 19 and 22 °C. This modification should be made in order to minimize yeast activity before freezing. Dough resting is often avoided completely in frozen dough production to minimize fermentation before freezing though it is a needed procedure for convenience CSB making. However, some researchers consider short rest times (8-15 min) to be beneficial. Another factor that may affect frozen dough product quality is the influence of storage time and conditions on gluten structure. The structure of the gluten protein matrix appears to be disrupted during extended storage resulting in a weakening of dough strength properties, loss of gas retention properties and deterioration of product quality. Finally, of great importance on dough rheology and yeast activity are also temperature fluctuations during storage. Accurate control of storage temperature, rapid movement of products between stores and correct stock rotation and control can minimize these fluctuations.

4. Requirements of Flour Quality for Different Kinds of CSB

Wheat flour is the main component of CSB and the quality of flour greatly affects the properties of CSB making. It is generally said that medium flours are suitable to CSB making considering the content and properties of flour, but things may be more complex.

In fact, the minor wheat flour constituents such as ash, lipid, enzymes and non-starch polysaccharides also play a comparatively important role in the CSB making process. There are many researches focused on the relationship between the quality of flour and CSB and they have continued to increase in recent years as shown in Table 1-7.

Table 1-7. Researches on the quality of flours and CSB since 2000

Author	Title	Source
Sang Wei et al.	Evaluation on Quality Traits and Processing Quality Properties of Steamed Bread and White Salted Noodle of Commercial Wheat Flour in Xinjiang	Acta Agriculturae Boreali-Occidentalis Sinica, 2008(3):91-96
Su Dongmin et al.	Effects of Waxy Wheat Flour Blending on Rheological Properties of Dough and Quality of Mantou	Journal of Henan University of Technology(Natural (Science Edition), 2008(02):2-7
Liang Ling et al.	Processing Technology of Steamed Bread Using Wheat Flour of Xiaoyan 22	Journal of Triticeae Crops, 2008(01):61-65
Zhang Jingmei, Li Feng	Discuss Importance which the North-east Agriculture and Reclamation Wheat Reasonably Match	Grain Processing. 2008(01):27-29
Chen Zhichen	Effects and its mechanism of quality and size of flour on the CSB quality	Grain Processing. 2007(05):19-22
Shen Jiong; Wang Xuedong; Li Qinglong	Relationship between the Components of Wheat Flour and Quality of Chinese Steamed Bread	Journal of Wuhan Polytechnic University. 2007(01):19-22
Lin Jiangtao et al.	A Study on Physicochemical Properties of French Wheat and its Application in Chinese Flour Foods	Grain Processing. 2006(09):9-11
Yuan Jian et al.	Relationship of Wheat Flour Quality with Chinese Steamed Bread Quality	Food Science. 2005(12): 57-61
Zhang Zhao; Chen Zhengxing	Study on the making of low-protein steamed bread of wheat starch	Food Science and Technology. 2005(11):26-29
Dong Bin, Zheng Xueling, Wang Fengcheng	The Relationship of Wheat Flour Composition and Steamed Bread Quality	Journal of Cereals & Oils. 2005(02):12-14
Wei Yimin, Zhang Guoquan, Wolfgang Sietz	Relationship between Wheat Kernel Property and Steam Bread Quality	Journal of the Chinese Cereals and Oils Association. 2003, 18(6):40-43
Huang Jian et al.	Quality of fortified flour and its processing and cooking properties	Journal of Hygiene Research. 2003(01):75-77
Lu Jing et al.	Investigation on Correlation Between Quality Characters of Wheat and Processing Quality of Flour Food	Xinjiang Agricultural Sciences. 2002(05):290-292.
Zhang Xinzhong; Lu Jing; Wu Xinyuan	Study on Variety Quality of Wheat Special on Streamed Bread,Pulling Flour,Dumpling of Xinjiang	Xinjiang Agricultural Sciences. 2002(04):220-221
Qi Bingjian	Study on the new Qualityevaluation of Flour for Making Chinese Steamed Bread	Journal of Zhengzhou Grain College. 2002,23(04):38-42
Guo Boli et al.	Study on the relationship between wheat quality and their food quality	Journal of Northwest Sci-Tech University of Agriculture and Forestry. 2001(05):61-63
He Zhonghu	Wheat Production and Quality Requirements in China	Review of China Agricultural Science and Technology. 2000(03):62-68
Liu Aihua et al.	Investigation of Wheat Flour Quality for Northern Style Chinese Steamed Bread	JOURNAL OF THE CHINESE CEREALS AND OILS ASSOCIATION. 2000,15(02):10-15

4.1. Requirement of Flour Quality for Northern-Style Steamed Bread

The textures and taste of northern-style CSB are much different from that of southern-style CSB, so the requirements on wheat cultivars and flour quality are different, too.

Huang et al. (1996) studied the relationships between flour quality and steamed bread quality for 49 kinds of Australian and Chinese wheat. The result is summarized in Table 1-8. Protein content was significant correlated with specific volume score and spread ratio score. Both skin and crumb color scores were negatively correlated with flour color grade and ash content, but were positively correlated with dough strength. It was also observed that the longer the dough had to be mixed, the whiter the skin and the crumb became. Smoothness of CSB did not correlate well with flour quality and appeared to be more dependent on processing procedures. There was no significant linear correlation between protein content and total score of CSB. However, when protein content was below 10%, it was linearly and positively correlated with the total score of steamed bread. As protein content increased further (i.e. over 10%), the total score was not significantly influenced by protein content and was more dependent on other flour quality attributes. Overall, it was apparent that dough strength was more important than protein content in determining quality of steamed bread. Most parameters for dough strength (Farinograph dough development time, stability, mixing tolerance, Extensograph maximum resistance and extensibility) were significantly correlated with individual components of steamed bread quality. It was therefore clear that dough strength was a significant determinant of northern style steam bread quality. It contributed positively to steamed bread quality in terms of specific volume, skin and crumb color, structure and eating quality. Extensograph maximum resistance exhibited the highest correlation with total score of steamed bead. All RVA parameters (except peak time) of flour were positively correlated with specific volume. Flour peak viscosity showed the best correlation with specific volume, crumb structure and total score. A positive correlation was also observed between eating quality and flour peak viscosity. Other starch-related parameters, including starch damage, starch granule size distribution, maltose figure and falling number, showed minor correlations with steamed bread quality attributes.

Zhu et al. (2001) investigated the effects of flour protein content and composition of HMW glutenin subunits, using several kinds of Chinese and Australian wheat as samples, on northern style CSB quality. The results indicated that at high protein level, the Australian wheat cultivar, with the biotype processing HMW glutenin subunits 5+10, had a significantly higher total score than the biotype with subunits 2+12. However, no significant difference existed between the two biotypes at the low protein lever. The improvement of CSB quality with 5+10 biotype could be due to its higher protein content. However, a significant difference still existed even though the protein contents of samples with 5+10 and 2+12 biotypes were all at same high level, suggesting that there is other factor that could affect CSB quality. One explanation might be the difference of size distribution of glutenin polymers between the flours with these glutenin subunits.

Table 1-8. Correlation between flour quality and steamed bread quality attributes from 49 kinds of wheat samples (Huang et al., 1996)

	Specific volume	Spread ratio score	Skin color score	Crumb color score	Shininess score	Smoothness score	Structure	Eating quality score	Total score
Protein	0.49***	-0.73***	N.S.	N.S.	-0.36*	N.S.	N.S.	N.S.	N.S.
Ash	N.S.	-0.39**	-0.54***	-0.54***	N.S.	-0.32*	-0.30*	N.S.	-0.45**
Color grade	N.S.	N.S.	-0.54***	-0.51***	N.S.	-0.37**	-0.32**	N.S.	-0.38**
Development time	0.58***	N.S.	0.47***	0.56***	-0.35*	N.S.	0.68***	0.61***	0.65***
Stability	0.40**	N.S.	0.48***	0.60***	N.S.	N.S.	0.65***	0.52***	0.57***
Tolerance	-0.39**	N.S.	-0.33*	-0.50***	N.S.	0.33*	-0.81***	-0.60***	-0.60***
Maximum resistance	0.53***	N.S.	0.62***	0.71***	-0.31*	0.32*	0.74***	0.59***	0.70***
Extensibility	0.63***	-0.46***	0.42***	N.S.	-0.47***	N.S.	0.42**	0.43**	0.44**
Peak viscosity	0.55***	N.S.	0.33*	N.S.	N.S.	0.30*	0.52***	0.39**	0.47**
Final viscosity	0.40**	N.S.	0.32*	N.S.	-0.40**	0.33*	N.S.	N.S.	N.S.
Falling number	N.S.	0.42**	N.S.	N.S.	N.S.	N.S.	N.S.	N.S.	N.S.

*** P<0.001, ** P<0.01, * P<0.05, N.S. = Not Significant.

4.2. Requirements of Flour Quality for Southern-style CSB

Southern-style CSB is somewhat softer and whiter than northern-style CSB. The flour made from medium or soft wheat may be suitable to the preparing of southern-style CSB.

Fifty Australian and seven Chinese wheat flours were used to investigate the requirements of flour quality for southern-style CSB (Huang et al. 1996). Both protein quantity and quality were identified as major determinants for specific volume of steamed bread. Protein content had a more significant effect on the specific volume of this kind of steamed bread than that for northern-style ones. Dough strength showed an important role in determining the overall quality of steamed bread, but was not so significant effect as to that for northern style. External smoothness and protein content of southern-style CSB showed significant negative correlation, of the same for smoothness and flour dough strength.

Flour with medium protein content and dough strength, medium or high falling number, and low ash content was recommended for the producing of southern-style CSB. No significant correlation was found between flour RVA viscosity parameters and steamed bread quality.

4.3. Requirements Flour Quality for Guangdong-style CSB

The majority of Guangdong–style CSB is prepared with flour and sugar. Preferences in sweetness, cohesiveness, and structure range throughout different regions and countries. There are few reports about flour quality requirements of this kind of CSB. Flour quality specifications from several modern mills in Hong Kong, Shenzhen, and Guangzhou where the CSB of this kind was consumed mainly indicated that flours with a protein content of 7.5-9.0%, wet gluten of 19-22%, and ash of 0.45-0.55% is used for this product. So low protein content and ash may be requested for flour of Guangdong-style CSB so that the CSB becomes whiter and softer

4.4. Requirements of Flour Quality for Manual Process or Mechanized Process

Almost all of CSB were processed manually in about 20 years ago, but now, most of it is processed mechanically in city although the condition is different at all. Usually, stronger flour is desired for mechanized process while softer flour is suitable to for manual process. The different requirements of flour are mainly resulted from the different mixing strength. If the flour is too soft, the gluten of dough could not endure the mechanized mixing during dough preparing. If the flour is too strong, mixing will be difficult and dough quality may be uneven for manual process.

Research showed the qualities of CSB prepared by manual or mechanized methods with same kinds of wheat flour were different. He et al. (2003) selected 56 Chinese cultivars from major wheat growing areas, including North China Plain, Yellow and Huai Valley, Yangtze Region. Besides, there were 10 Australian cultivars grown in Anyang located in Yellow and Huai Valleys and Chengdu located in Yangtze region. The mechanized method was from Huang et al. (1993), also. Manual method is according to SB/T 10139-93. The cultivars from North China Plain possess slightly poorer CSB quality using manual method, while Australian cultivars perform slightly better than Chinese cultivars from all ecological zones with both manual and mechanized methods in Anyang. It is observed that Australian cultivars showed much better CSB quality (total score) than Chinese cultivars in Chengdu with both manual and mechanized methods. This may be due partly to the strong gluten and good sprouting tolerance (as indicated by falling number and RVA peak viscosity) of Australian cultivars. Chinese wheat is well known for its rather weak gluten type. In addition, not enough selection pressure for sprouting tolerance is applied in wheat breeding programs since quality became an important trait only recently. These results indicate that selection for strong gluten and good sprouting tolerance could improve the CSB quality for wheat from Chengdu and other southern China locations, despite the unfavorable environmental effects on CSB quality. The correlation coefficient between flour quality and CSB quality differed with bread making method (manual or mechanized). For example, protein content, dough strength parameters such as SDS sedimentation value, water absorption, development time, stability, extensibility, resistance, and extension showed positive correlations with loaf volume and elasticity under both manual and mechanized conditions, however, they showed significantly negative association with appearance and stickiness under manual conditions, and very small negative or even positive association with appearance and stickiness under mechanized conditions. The correlation coefficients between appearance and protein content, development time, stability, and extension area are –0.29, –0.62, –0.44, and –0.50 under manual conditions, and are –0.18, –0.21, –0.03, and –0.07 under mechanized conditions, respectively. This indicates that high protein content and good gluten quality will contribute positively to the improvement of loaf volume and elasticity regardless of processing methods, but they could produce CSB with poor appearance and stickiness under manual conditions. As good CSB quality requires high loaf volume, good elasticity, and smooth appearance, and good stickiness, the above results suggest that quality requirements for CSB will be highly dependent on processing methods, i.e., medium protein content and weak-to-medium gluten strength will be desirable for the manual method, while medium protein content and medium-to-strong gluten type will be favorable for mechanized conditions. A good example is that the samples with medium protein content and medium gluten strength showed outstanding CSB quality in Anyang under both processing conditions. However, the one with strong gluten strength performed outstanding for CSB quality under mechanized condition, but had poor CSB quality under manual condition due to the poor appearance. Over strong gluten cultivars which performed poor CSB quality largely due to the poor appearance and dark color under both processing conditions.

5. Other Methods to Improve the Production

Beside wheat flour, water, processing technology and yeast, other efforts and researches are being done to improve the quality of CSB because of the effect of quality on the market is increasing. The efforts include the use of new additives, yeast and other kinds of flours.

5.1. Additives

Yeast was the only additive used in making of CSB about 20 years ago, but now, additives such as emulsifier, whitener, and strengthener are widely used in CSB making.

5.1.1. Emulsifiers

Emulsifiers are substances possessing both hydrophobic (lipophilic) and hydrophilic properties. Emulsifiers are routinely used in breadmaking as dough strengtheners and/or crumb softeners. The latter enhance crumb softness and retard bread staling.

For Chinese steamed bread, there are similar results in the usage of emulsifiers as those of bread making. Zhang et al. (2007) reported the effects of DATEM on the qualities of CSB. Diacetyl tartaric acid ester of monoglycerides (DATEM) is a kind of anionic oil in water (O/W) emulsifier. The hydrophile–lipophile balance (HLB) value of DATEM is 8-10. Acceptable daily intake (ADI) is 0-50 mg/kg. Lethal dose 50% (LD_{50}) for DATEM is 10 g/kg of body weight. DATEM is often used to increase the volume of bread, and was reported to be an effective bread emulsifier in many countries. DATEM can enhance the resistance of dough to collapse and improve gas retention of dough, so it is also a kind of dough strengthener. The commercial DATEM is a mixture of several components, including DATEM, monoacetyl tartaric acid ester of monoglycerides, acetic acid, esterification products of tartaric acid and acetic acid. The influence of DATEM on dough quality varies with the components of DATEM. If there are amounts of hydrophilic radicals such as diacetyl radical and hydroxyl in DATEM, they will interact with large amounts of water, which is favorable for the water retention of dough. As a result, the aging speed of bread is reduced. It was reported that the DATEM components with two carboxyl groups had the lowest baking activity, but they were most active in dough and gluten rheology. In addition, DATEM can interact with proteins intensively, especially glutenin, so gas retention of dough is improved and the formation of gluten-starch-fat network structure is accelerated. People use DATEM in the producing process of CSB just for a few years. Zhang' research indicated except for specific volume, spread ratio, smoothness, DATEM almost affected all the CSB characteristics significantly. The most important aspects including skin color, skin structure, inner structure and total score varied obviously with the variation of DATEM addition. The results above are probably due to the reactions of DATEM and fat, protein and carbohydrate molecules in dough which will stabilize and strengthen the gluten structure. With the increase of DATEM addition, skin color, skin structure, inner structure and total score of CSB were improved obviously at first, but if the quantity was over 0.10%, the scores of all these items decreased a little. Therefore, the optimal quantity of DATEM added in CSB was 0.10%. That's probably because, with the increase of DATEM, the emulsification effect is enhanced gradually, but when the addition is over 0.10%, the emulsification effect of DATEM can't be enhanced any longer. Compared to the control, the CSB with 0.10% DATEM had smaller specific volume, but the inner structure was improved obviously with even air holes and scores of shininess increased.

The sensory analysis results showed that the optimal quantity of DATEM was at the 0.10% level. The effects of DATEM on Farinograph property of CSB dough are complex. As the DATEM increased, development time was prolonged; stability time was almost changeless when the additions of DATEM were less than 0.08%; soften degree decreased and evaluation value increased with the increase of DATEM. The effects of DATEM on Extensograph property are also complex. Compared to the control, extensiveness and powdered strength decreased but extended resistance decreased at first, and then increased while extended ratio changed little. It was concluded that gas retention and structure of CSB dough were improved effectively while gluten strength increased after DATEM was added. The effects of DATEM on the rheological properties of dough were complex, so attention should be paid to the quantity of DATEM during CSB making. As a surfactant, DATEM can attach to the surface of starch particles and form an indissolvable film, so the water absorption ratio of dough will decrease. However, the effect of DATEM on water absorption ratio is considerably weak which is in accordance with the trial results above. DATEM can react with fat, protein and carbohydrate molecules in dough, therefore the gluten structure is stabled and strengthened. DATEM can also react with amino acids and form hydrogen bonds, which leads to the strengthening of gluten network structure, consequently improves gas retention of gluten. In fact, owing to the existence of diacetyl residues, anion residues in DATEM can effectively neutralize the cation residues in gluten, therefore the charge quantity of gluten is reduced, which is in favor of the gluten conglomeration, consequently improves the strength and gas retention of gluten. In addition, the results agreed with the conclusion that DATEM addition is especially suitable for the European and Asian wheat flour which are of low protein content and week gluten strength.

5.1.2. Enzymes

Oxidising agents are used to improve the gas retention abilities of the dough. The functions of the oxidant are complex and at the protein molecule level are related to the formation thought to be mostly related to "cross-link" of proteins. This would be the equivalent of typing knots in the ends of short pieces of string to gradually form a net. The contribution of oxidants to bread quality is significant. By improving dough development we will get larger product volume and improved softness. Enzyme active materials have become important following the limitations placed on the use of oxidants.

Su et al. (2005) studied the effect of endoxylanases on making performance of CSB. The endoxylanase includes two kinds, solid-phase enzyme (S-XYL) and liquid-phase enzyme (L-XYL). AX is cell wall, non-starch polysaccharides of cereal and an important source of dietary fiber. Wheat flour contains 1.5–2.5% total AX, of which one-third to half is WE-AX and the other is WU-AX. WE-AX leads to a highly viscous solution, whereas WU-AX has a strong water-holding capacity. AX can be hydrolyzed by endoxylanases (XYL). Endoxylanases (EC 3.2.1.8) have a strong impact on AX structure and functionality. They attack the AX backbone in a random manner, causing a decrease in the degree of polymerization of the substrate and liberating oligomers, xylobiose and xylose with retention of their configuration. The final product depends mainly on enzyme level, action time, substrate type, pH and other factors. Endoxylanases can be obtained from cereal, bacterial and fungal sources. In general, the mode of action of endoxylanases and the hydrolysis products vary not only with the source of the enzymes, but also between structurally different enzymes within one organism. In bread making, the endoxylanases are almost routinely used in bread improver mixtures, to improve dough handling properties.

The changes in water absorption of the flour samples caused by addition of S-XYL and L-XYL at 300 mg and 700 mg per 100 g of flour were assessed. Both enzymes significantly decreased the values of absorption, mixing time and stability time. There were significant ($p<0.01$) differences between S-XYL and L-XYL in the effects that they exerted on the water absorption; but all the values of the L-XYL added flour were lower than that of the S-XYL added flour. This indicated that L-XYL can rapidly and easily interact with AX of the flour, especially with WU-AX of the flour, yielding fast degradation or solubilization of AX. It is seen that L-XYL and S-XYL slightly decreased the time to peak and the peak height of the sample. Adding yeast 0.5% hardly showed any change as compared to the control sample. For flour samples, there were no significant differences in peak height, but when L-XYL was added the developing slope, it tended to be lower than the other three samples and when S-XYL was added the weakening slope, it tended to be further lower than the other three samples. In order to investigate the effects of adding yeast, S-XYL, L-XYL and fermentation time on the properties of dough for CSB, TPA of wheat dough was performed respectively at 0th, 5th, 10th, 15th, and 20th min on all the four samples for dough hardness, adhesiveness, springiness, cohesiveness, gumminess, chewiness, and resilience. Mechanical primary and secondary parameters measured before and after different time of fermentation were significantly correlated between hardness and adhesiveness, hardness and gumminess, hardness and chewiness, adhesiveness and gumminess, adhesiveness and chewiness, gumminess and chewiness except for springiness, cohesiveness and resilience versus other variables. Some significant correlations found between primary texture parameters were similar to the former reports, especially for hardness and adhesiveness, correlation coefficient (cf) –0.81**, –0.91**, –0.86**, –0.89** and –0.87**, respectively at 0th, 5th, 10th,15th and 20th min of the total four samples for dough textural values. In general, harder dough almost related to poorer cohesive dough with lower adhesiveness. There was no correlation between hardness and cohesiveness, and cohesiveness and adhesiveness.

The effects of adding both the endoxylanases on dough properties were clear. It was apparent that the decrease in the hardness of dough by adding S-XYL and L-XYL, respectively was more than that of the control sample and the sample with added yeast. The fermentation of samples led to softer dough. The longer the time of fermentation, the lower the hardness of dough is. The trend of the hardness of dough of four samples is the same. The values of the L-XYL added dough is always lower than that of the S-XYL added dough. Adhesiveness, an important parameter of dough properties, is usually evaluated as the maximum tensile force developed during adhesion or as the cohesive rupture between two flat plates and the sample. It is also measured by the negative curve of the force versus distance obtained experimentally as the plate separation increases. As a consequence of dough softening, adhesiveness of the two samples with added S-XYL and L-XYL decreased. Longer the time of fermentation, lower the adhesiveness of dough. The values of adhesiveness of L-XYL added dough is the lowest among four samples. It implied that adding S-XYL or L-XYL into flour may improve the machiability of CSB dough. Knowledge about the mechanism of action of endoxylanases in dough fermentation is still limited, probably because gas production, gas retention, dough stability during fermentation and other typical characteristics are not easily measured as farinograph and mixograph parameters. Both treatments of S-XYL and L-XYL strongly affected the dough making process of Chinese steamed bread. They caused substantial softening of dough consistency during mixing, especially after fermentation and proofing. In the preliminary experiments, with the increase of enzyme dosage, the dough became difficult to handle. The most marked effect of the L-XYL addition was the reduction in mixing time by about 27.5%. On the other hand, the main effect of adding S-XYL and L-XYL was shortening of fermentation time. It indicated that the addition of S-XYL and L-XYL reduce the water holding capacity of dough by hydrolyzing WE-AX and ES-AX to small AX fragments and increasing solubilization of WU-AX, which results in a significant drop in dough viscosity and leads to progressive slackening and softening of the dough. At the same time, this also results in redistribution of previously bound water and releases free water into the dough, taking into account the lower water content of dough of CSB, which is beneficial to yeast activity. Specific volume is the most important quality parameter for CSB. As a result of the addition of S-XYL and L-XYL the specific volume of CSB increased from 8% (for S-XYL addition) to 18.7% (for L-XYL addition). Spread ratio is an important quality parameter too. In this case, spread ratio ranged from 1.79 (for S-XYL addition) to 1.92 (for L-XYL addition) compared with control samples. The higher spread ratio indicates more spread of dough during steaming operation. It can be partly due to the action of enzymes in reducing the viscosity of dough, in which the action of L-XYL can be stronger than that of S-XYL. Addition of S-XYL and L-XYL from 16.8% (for S-XYL addition) to 26.4% (for L-XYL addition) influenced crumb softness by decreasing trend, but there were no significant differences in elasticity. For CSB, skin and crumb whiteness is given close attention by customers. When SXYL or L-XYL was added into dough, it caused a decreasing trend in L* value, a slightly increasing tendency in a* value and decreasing trend in b* value. Other sensory parameters such as structure and appearance of CSB showed no significant differences. However, a noticeable improvement in the overall acceptance of S-XYL or L-XYL supplemented CSB was observed in a range from 3.9% (for S-XYL addition) to 9.2% (for L-XYL addition).

5.1.3. Lipids

Pomeranz et al. (1991) studied the role of lipids in steamed break making. The results indicated that defatting significantly reduced volume and softness of steamed bread. The overall quality of the defatted flour could be reconstituted, albeit incompletely, by adding the extracted lipids. Doubling the amount of free lipids from 0.86, that is to say, added about 1.72% free lipid to the original or to the defatted flour, had little or no consistent effect on the volume of the steamed bread. Doubling the amount of added free lipids made the bread crumb softer.

The lipids are usually added to the flour with the company of emulsifiers. In above research, representatives of two types of emulsifiers were added to regular and defatted flours. With the control flour, 0.1% ethoxylated monoglycerides (EMG) or 0.2% lecithin had no significant effect on CSB volume or softness. In defatted flour, however, 0.2% lecithin or 0.1% EMG alone increased volume and made the crumb softer. The results indicated that lecithin or EMG could replace, in part at least, the polar lipids that were removed by defatting with petroleum ether and that are required in production of CSB. Addition of corn oil to the control flour from 1 to 2%, in the presence of 0.2% hydroxylated lecithin or 0.1% EMG, lowered the volume of steamed bread. An increase in corn oil added to defatted flour increased the volume of steamed bread in the presence of 0.2% lecithin but not in the presence of 0.1% EMG. Adding 0.1% EMG and especially 0.2% lecithin to defatted flour baked with 1 or 2% corn oil increased volume and softness of steamed bread.

5.2. Other Kinds of Flour

Specific volumes of Chinese steamed bread did not significantly decrease with the proportions of waxy flour increasing (Qin et al 2007). There was no significant difference on scent score among all waxy flour blends, which indicated that scent was not affected by adding waxy flour. Addition of waxy flour decreased the appearance, color, texture, elasticity, stickiness, and total score. However, all values were not significantly different when the proportions of waxy flour were below 10%. The results showed that the waxy flour was not appropriate to make Chinese steamed bread when the proportions were 15%. In other words, waxy flour proportions below 10% could be used to make Chinese steamed bread without decreasing their qualities significantly.

Farinograph properties were not as important as extensograph ones for Chinese steamed bread made from waxy flour blends. The peak viscosity, starch content and amylose content (of starch) were positively correlated to the scores of Chinese steamed bread. Waxy wheat possessed lower peak viscosity and no amylose, and therefore, waxy wheat flour was not suitable for fresh Chinese steamed bread making.

The firmness of Chinese steamed bread measured in 15 min after being removed from the steamed car showed that addition of waxy flour at lower levels (proportions lower than 20%) resulted in softening of Chinese steamed bread, and the firmness increased when the proportions of waxy flour were more than 25%. After storing at -18 °C for three days, the firmness of Chinese steamed bread increased slightly compared with fresh ones with same waxy flour proportions; however, the trend of firming influenced by waxy flour did not change. These results indicated that the low proportions of waxy flour had the function of decreasing the firmness of Chinese steamed bread, but excessive waxy flour resulted in compacting of the internal structure. Moreover, frozen storing conditions increased the firmness of re-steamed Chinese steamed bread, and thus, it is worth using lower proportion of waxy flour blends (maybe 20%) in frozen steamed bread making.

The peak viscosity, breakdown, setback, gluten index, starch, amylose, and damaged starch contents were negatively correlated to fresh Chinese steamed bread firmness significantly. The effects of peak viscosity, setback, starch, amylose, and damaged starch contents on the firmness of frozen Chinese steamed bread were negative, and the other physicochemical properties such as farinograph and extensograph had no correlations to it. Waxy wheat flour has no amylase in its endosperm, and possesses lower peak viscosity and setback; thus, adding waxy wheat flour into normal ones can decrease the firmness of frozen Chinese steamed bread.

The shelf life of Chinese steamed bread was 1-3 d at room temperature, and it is shorter with the increase of temperature and relative humidity. Chinese steamed bread became firmer when stored at room temperature or at -18°C with consequent decrease in the qualities. Adding an appropriate proportion (< 20%) of waxy flour into normal ones decreased the firmness of re-steamed Chinese steamed bread stored at – 18 °C. Gelatification was performed mainly by amylose, and amylose was easier to form regular structure than amylopectin. Chinese steamed bread made from blends with excessive amylopectin could not hold sufficient air during steaming, and its internal structure was compacted; therefore, the firmness of fresh and frozen Chinese steamed bread will be increased when the proportions of amylopectin increased by more than 25%.

Waxy flour was not suitable for steamed bread making, however, was capable of improving the qualities of frozen steamed bread with lower proportions. Flour blends with 10-15% waxy wheat flour were the best proportions to decrease the firmness of frozen Chinese steamed bread without decreasing its eating qualities. These results showed that the use of waxy flour on frozen steamed bread was feasible.

6. Quality and Properties of Products

6.1. Nutrition of the Product

The major constituents of wheat are the carbohydrates and proteins. Other components such as lipids and vitamins may be of great significance in human nutrition because of the large contribution of wheat to the diet. The carbohydrates of wheat include simple sugars, more complex oligosaccharides such as fructans, starch and the cell wall polysaccharides, all of which are of nutritional value. Benefits that may result include reduced cariogenic bacteria (dental health) and lower energy value. Fructans may be considered to be important to human nutrition because of their possible role as soluble fibre. Starch, as the major component by weight, may have a great impact on nutritional quality. The cell wall polysaccharides may also be important as either soluble or insoluble fibre, depending on the composition of the polysaccharides in the product. Soluble fibre may reduce the risk of heart disease while insoluble fibre contributes to reduced risk of colonic cancers. The lipids are generally of limited importance in human nutrition but may be important in animal diet.

About thirty years ago, rice and white flour were considered “fine foods”, which most common folks are not able to have at every meal. Its counterpart, the “rough foods”, were the real main dietary components of the Chinese, including corn, sorghum, buckwheat, oats, yams, beans and so on. So many people, especially for urban citizens, preferred to add these “rough foods” into “fine foods”. For example, something has a great amount of fibre like bran, and the other food has antioxidant properties as oat. The addition makes CSB be more suitable for modern people diet. We use the application of soybeans in the flour as the example.

The health benefits of soybeans have been recognized for millennia. Soy foods and their isoflavones appear to have clear protective effects related to coronary heart disease and probable protective and therapeutic effects related to osteoporosis. The effects on the kidney are clear, and these protective effects are under study. While the greatest interest may lie in their chemopreventive effects related to cancer, much more research is required. The effects of soy foods on cognitive function are unclear and also require further research. The use for menopausal symptoms appears promising, and postmenopausal women who can not or choose not to take hormone replacement therapy may be ideal candidates for daily soy food use. It should also be mentioned here that some concern has been raised with regard to allergies to soy foods and soy-based infant formulas, as well as potential soybean antinutrients; however, soy protein is ranked 11th in allergenicity, with 0.5% of young children having an allergy to soy. The incidence of soy protein allergy among older children and adults is extremely rare. As mentioned earlier, the FDA's approval of a health claim based on the association between consumption of soy protein and a reduced risk of coronary heart disease has significantly increased the demand for soy ingredients by the food industry. The government-approved health claim adds legitimacy to soy protein products, has helped to increase the awareness of soy foods, and has created an incentive for food processor to add soy protein to foods; consequently, the number of soy food products available has increased significantly.

In recognition of the many advantage of soy flour as a protein supplement, considerable effect has gone into blending soy flour with wheat flour. These mixtures, when suitably fortified with vitamins and minerals, have great potential for feeding people of all ages in developing areas of the world. In addition to their significant levels of high-quality protein, the calcium and phosphorus levels of soy ingredients are also considerably higher than those of any other cereal grain, and soy flour is an excellent source of available iron. Furthermore, soy flour is considerably richer in vitamins than unenriched patent wheat flour and somewhat richer than enriched wheat flour. During the 1950s and 1960s, nutritionists sought to increase the adequacy and amount of dietary protein in developing countries. Soy-fortified grain products were seen as one way to accomplish this, and much work was done on incorporating soy into bread. Soy protein is relatively rich in lysine but poor in methionine. Wheat protein, on the other hand, is poor in lysine but rich in cysteine (which the body can convert to methionine). Thus, the combination of the two protein sources creates a better balance of these two essential amino acids, which the human body can not synthesize and must be obtained from our diet.

A 1979 study found the protein efficiency ratio (PER) of gluten to be 0.7 and soy protein 1.3, but the PER for an 88:12 blend of wheat and soy flours was 2.0. Because of its low PER value, soy protein was long considered to be a second-class citizen in terms of protein value. Since that study, however, nutritionists have turned to a different way of measuring protein nutritional adequacy, the protein digestibility corrected amino acid score (PDCAAS). In this test, casein is given the top score of 1.0; wheat gluten comes in at 0.25, and soy protein varies from 0.90 to 0.95, depending on its form. The protein in the 88:12 blend scores 1.0; thus, the improved nutrition achieved by combining the protein sources is apparent.

A bakery-related potential use for soy flour in combination with cereals is in the production of the so-called "composition flours". These are mixtures of flours, starches, and other ingredients that are intended to replace wheat flour, totally or partially, in bakery products. Extensive research projects aimed at the development of such flours have been sponsored by international and national development agencies or programs such as the Food and Agriculture Organization (FAO). The main reason for developing composite flours is to relieve the burden of importing this commodity by countries where wheat is not grown.

An important application of combining defatted soy flour and grits with cereals is the production of nutritionally balanced food, as all-purpose food, which can be distributed to undernourished populations or in food-shortage emergencies. Specialty breads including steamed bread can be made with 13-14% of protein by incorporating soy proteins into a formula along with vital wheat gluten and, if necessary, a lipid emulsifier. Without an emulsifier, incorporating high levels of soy protein depresses volume. Defatted, enzyme-inactive soy flour is darker and has a roasted flavor, thus it is the preferred material for breads in which a distinct soy flavor is required. Soy protein isolates have been used for protein fortification of specialty breads because of their high protein content and blandness. Xu et al. (2007) reported that when the addition of soy flour was 2% (based on the wheat flour weight), the quality of CSB was improved. When the addition was over 6%, the inner structure would be destroyed through having a lot of loosened hole.

6.2. Microbiological Safety of the Product

CSB has a higher water activity than bread. So the microorganisms, particularly fungi, will grow and produce visible and highly objectionable appearance defects. Steaming ordinarily kills potential spoilage bacteria, the exception being spore-forming bacilli. Thus, when moulds are present in products, it is invariably a result of post-processing contamination. Fungal spores are particularly widespread due to their presence in flour and their ability to spread throughout the production environment via air movement. When the steamed bread leaving the steamer, their transit through the cooling and packaging operations leaving plenty of opportunity for infection, either indirectly by airborne spores or directly by contact with contaminated equipment. The bacteria can spoil the steamed bread via production of an extracellular capsule material that gives the infected steamed bread a mucoid or ropy texture. There are also wild yeasts capable of causing flavor defects.

Microbiological spoilage is most often associated with fungi, and occurs when fungal mycelia are visible to the consumer. Visible mould growth may appear within just a few days. What one actually sees is a combination of vegetative cell growth, along with sporulating bodies. The latter is responsible for the characteristic blue-green or black color normally associated with mould growth. The ability of fungi to grow and which species predominate depend on several factors, including pH value and water activity, and storage temperature and atmosphere. Finally, it is important to recognize that some mould strains not only can grow on steamed bread, causing spoilage and economic loss, visible mould growth ordinarily precedes mycotoxin formation.

During storage, steamed bread may have obviously spoiled sensory character when the total bacteria reach 10^6 cfu/g. While stored with heat preservation measure, the total bacteria of steamed bread is under 10^4 cfu/g and the mildew number is below 60 cfu/g within 12 hr;

while storied at 4°C, steamed bread sensory character has no obvious changes within 5 days. After reheated, the total bacteria of cold steamed bread can be reduced to less than 10^3 cfu/g, if the total bacteria are less than 10^6 cfu/g. So the steamed bread stored at low temperature would be acceptable. The market investigation shows that the total bacteria of most commercial steamed bread is between 800~5000 cfu/g. If total bacteria standard of steamed bread is too low, it is difficult for manufacturer to carry out.

Several different approaches have been adopted to control mould growth. Steamed bread that is packaged while still warm is very susceptible to mould growth, due to localized areas of condensate that form within the package. So the package should be carried out after steamed bread being cooling drastically. The packaged products stored in 30, 20°C for less than 12 and 24 hr, respectively.

Given that biological spoilage of steamed bread is caused primarily by molds, it is not surprising that preservation strategies have focused on controlling fungi, both in the production environment and in the finished product. Mould and mould spores are present in flour and other raw materials and may be widespread in producing processing. Therefore, rigorous attention to plant design and sanitation is essential. The post-production environment should be separated from pre-production environment. Air handling systems should be designed such that airborne mould spores can not gain entry to the product side.

The addition of chemical additive is a common preservation method. Calcium propionate is especially effective against most of the moulds associated with bread, and is widely used in commercially products. With the additions of calcium propionate and sodium benzoate are 0.085 and 0.08% (based on flour weight), respectively, CSB could be stored for more than 48 hr in the temperature of 38°C. Due to interest in chemical additive-free products, alternative approaches for preservation have been considered. Steamed bread can be exposed to ultraviolet, infrared, or microwave radiation to inactive mould and mould spores. These method, however, are not widely used. Another indirect way to extend the shelf-life is via freezing. Many manufacturers freeze the steamed and packaged products as a means of preserving prior to delivery.

In addition, staling is associated with an increase in skin softness and a decrease in fresh flavor. In summer, CSB could not be kept for more than 24 hr. Even in winter, it only could be contained for just 2-3 days. The best containing method is freezing. The shelf-life of CSB depends not only on microbial activities, but also on physical-chemical changes. Specifically, the staling most frequently causes consumers to reject products. The staling refers to the increase in CSB inner firmness that makes the product undesirable to consumers. Staling is basically a starch structure and moisture migration problem. The reactions that eventually lead to staling actually start when the bread is steamed, as starch granules in the dough begin to absorb water, gelatinize, and swell. The amylose and amylopectin chains separate from one to another and become more soluble and less ordered. Then, when the steamed bread is cooled, these starch molecules, and the amylopectin, in particular, slowly begin to re-associate and re-crystallize. This process, called retrogradation, results in an increase in firmness due to the rigid structures that form. Amylose retrogrades rapidly upon cooling, while amylopectin retrogrades slowly. It is the slow retrogradation of amylopectin that is now thought to be primarily associated with staling. Furthermore, moisture migration from starch to gluten and from skin to inner makes the skin drier and firmer. Although staling is an inevitable process, a number of strategies have been adopted to delay these reactions and extend the shelf-life.

The freezing and prolonged frozen storage influence dough properties by reducing gas retention and yeast cell viability. Loss of viability of cells upon freezing has been attributed to intracellular freezing and increased internal solute concentrations causing effects such as low pH, dehydration, ionic toxicity, damaging of essential membrane processes and lowering of activities of glycolytic enzymes. Although some microorganisms are killed during freezing, many of them may survive depending on a number of factors such as: the type of organism, the rate of freezing and the composition of substrate being frozen. Bacterial spores are unaffected by freezing and, in general, Gram-positive rods and cocci are more resistant than Gram-negative bacteria. At conventional freezing, the viability of organisms is enhanced as the freezing rate increases probably due to the diminished contact time of the susceptible organisms with harmful high solute concentrations in the unfrozen water. At more rapid freezing, viability decreases probably due to the formation of internal ice crystals, which cause destruction of the cell membranes. Finally, at extremely fast freezing, ice crystal formation is reduced and replaced by vitrification. Furthermore, reports show that freezing dough at low rates (<2 °C/ min) is preferable in order to get high yeast survival and bread scores. In general, the effects of freezing on shelf life are additive. Exception exists in cases of wide fluctuations in temperature.

The major challenge for CSB is how to preserve aroma, taste and texture. In the past this limited the scale of production to the local community but as time has gone by the move has been to more widespread geographical distribution networks and extended shelf life.

The trend to limit "chemical" additives has raised the profile of the use of enzyme active materials as more "natural" additives. While the addition of many enzyme active materials does improve dough gas retention. Such materials do not perform the same protein cross-linking function of oxidizing materials. It is likely that the improving effect many enzyme-active materials contribute to their ability to change dough rheology and therefore the influence of the molding and processing operations. Since a significant proportion of bread quality derives from the cultivars of wheat that are used in the milling grist it is certain that greater attention will be paid to matching wheat flour quality with end use. This may entail selective breeding or greater attention to agronomic practices.

Further benefits to the steamed bread producing from the rapid developments in frozen dough technology may be expected. It is clear that there are still many potential lines of research that should be followed to minimize the adverse effects of dough freezing, especially with respect to improvers that can be included in the formulation to give superior baking properties and longer shelf life. The exploration of possible modifications to freezing and frozen storage technology to improve the baking properties of frozen dough will be worthwhile. In wheat bread made from non-frozen dough, the use of emulsifiers such as monodiglyceride (MDG) and hydrocolloids such as sodium alginate have been shown to produce significant beneficial effects. MDG was found to inhibit amylopectin retrogradation in bread significantly at all levels of usage, which in turn retarded the staling mechanism, and increased oven rise during baking. The effects of such improvers in frozen dough, however, have yet to be studied.

The use of alginate as bread improver has been reported. However, little research has been conducted on its inclusion in frozen dough formulations. In non-frozen dough, the addition of 1% alginate caused a reduction in pasting temperature, implying an earlier start of starch gelatinization and subsequently an increase in the availability of starch polymers as the dextrinization amylase substrate during production processing. Alginate also showed the highest level of water absorption compared to other hydrocolloids such as HPMC, k-carrageenan and xanthan gum, resulting from the extensive hydroxyl groups in its structure, which allow more water interactions through hydrogen bonding. As a result, the water content of the bread crumb is augmented. Furthermore, the addition of sodium alginate increases dough consistency, stability, and strength. The resistance of dough to deformation (tenacity), which is a predictor of the ability of dough to retain gas, is increased with the addition of sodium alginate. However, the dough height and specific volume of the final bread is lowered. Such conflicting effects would make it worthwhile to explore the effects of sodium alginate on frozen dough.

The conflicting results found in the literature on the effect of guar gum and green tea extract also indicate the need for further studies on the behaviour of hydrocolloids and other improvers in frozen dough systems. Synergistic effects of different types of improvers are also worthy of exploration.

REFERENCES

Ang, C. Y. W., Liu, K. and Huang, Y.W. (1999), *Asian Foods: Science and Technology.* Lancaster Press: 71-89.

Beranbaum, R. L. (2003), *The Bread Bible.* Norton Press, New York.

Chen Dongsheng, Zhang Yan1, He Zhonghu, Wang Desen1, Pena R. J. (2005), 'Effect of Water Addition on Northern Style Chinese Steamed Bread Processing'. *ACTA AGRONOMICA SINICA*, 31(6):730-735.

Dong B., Zheng X.L., Wang F.C. (2005), 'The relationship of wheat flour composition and steamed bread quality'. *Grain processing*, (4):39-42.

Ding C., Qi G.., Zhang J., Chen F. and Li L. (2007). 'Microbial analysis of traditional starter culture (Jiaotou) and its influence on the quality of Chinese steamed bread'. *Food Science* 28 (4): 69-74.

Gianou, V., Kessoglou, V. and Tzia, C. (2003), 'Quality and safety characteristics of bread made from frozen dough'. *Food science and technology*, 14: 99-108.

Goesaert, H., Brijs, K., Veraverbeke, W.S., Courtin, C.M., Gebruers, K., and Delcour, J.A. (2005). 'Wheat flour constituents: how they impact bread quality, and how to impact their functionality'. *Trends in Food Science & Technology*, 16: 12-30.

He, Z. H., Liu, A. H., Pena, R. J. and Rajaram, S. (2003). 'Suitability of Chinese wheat cultivars for production of northern style Chinese steamed bread'. *Euphytica.* 131: 155-163.

Huang, S., Betker, S., Quail, K. and Moss, R. (1993), 'An optimized processing procedure by response surface methodology (RSM) for northern-style Chinese steamed bread'. *Journal of Cereal Science*, 18: 89-102.

Huang, S., Yun, S., Quail, K. and Moss, R. (1996), 'Establishment of flour quality guidelines for northern style Chinese steamed bread'. *Journal of cereal Science*, 24: 179-185.

Huang, S., Qauil, K. and Moss, R. (1998), 'The optimization of a laboratory processing procedure for southern-style Chinese steamed bread'. *International Journal of Food Science and Technology*, 33: 345-357.

Hutkins, R.W. (2006). 'Microbiology and Technology of Fermented Foods'. *IFT Press, Ames*: 268-299.

Liu, J. (2004). 'Chinese Foods'. *China Intercontinental Press*, Beijing: 9-11.

Liu, C. H. (2005). 'Steamed Cereal Food Processing Technology'. *Chemical Industry Press*, Beijing.

Owens, G. (2000). *Cereals Processing Technology*. CRC Press, New York. pp 57, 131, 205-213.

Pomeranz, Y., Huang, M. and Rubenthaler, G.L. (1991). 'Steamed bread. III. Role of lipids'. *Cereal Chemistry*, 68 (4): 353-356.

Qin, P., Cheng, S. and Ma, C. (2007). 'Effect of waxy wheat flour blends on the quality of Chinese steamed bread'. *Agricultural Science in China*. 6(10): 1275-1282.

Riaz, M. N. (2006). *Soy Applications in Food*. CRC Press, New York: 70-77.

Su, D., Ding, C. Li, L. Su, D. and Zheng, X. (2005). 'Effect of endoxylanases on dough properties and making performance of Chinese steamed bread'. *European Food Research & Technology*, 220: 540-545.

Sun H, (2008). 'Explanation on state standard of Chinese steamed bread based on wheat flour'. *Cereal and Food Industry*, 15(2):1-2.

Veraverbeke, W.S., Delcour, J.A. (2002). 'Wheat protein composition and properties of wheat glutenin in relation to breadmaking functionality'. *Critical Reviews in Food Science and nutrition*, 42: 179-208.

Xu, J. Liu, C., Yang, Z. and Han, X. (2007). 'The usage of soy protein for Chinese steamed bread'. *Grain Oil Processing*. 1: 82-84.

Zhang Guoying. (2005) 'The effects of addition of water on the quality of CSB'. *Flourmilling*, 1:31-32

Zhang, X, Sun, J, and Li, Z. (2007). 'Effects of DATEM on dough rheological characteristics and qualities of CSB and bread'. *Cereal Chemistry*, 84 (2): 181-185.

Zhu, J., Huang, S., Khan, K. and O'Brien, L. (2001). 'Relationship of protein quantity, quality and dough properties with Chinese steamed bread quality'. *Journal of Cereal Science*, 33: 205-212.

Chapter 2

CHINESE NOODLES

1. HISTORY AND DEVELOPMENT OF NOODLES

The noodle is an old, traditional Chinese food with a long history, and which is of ancient origin and long development. In order to make noodles, firstly, dough is made of grain powder or bean powder mixed with water, and then pressed or rolled into sheets, or rubbed, drawn, pinched into strip-form (narrow or wide, flat or round) or into fragments, at last the noodle is finished through boiling, sautéing, stewing or frying. In 2002, Chinese archaeologists found that noodles were buried in the earthquake 4000 years ago that destroyed Lajia in Minhe County, Qinghai Province. They were 50cm long, 0.3cm wide and made of milled foxtail millet and foxtail millet (Figure 2-1). They are the most ancient noodles that have been found in the world up to now, and it proves that China is the first country that invented noodles (Ye, 2006).

Noodle products recorded in literal histories have been traced to the *Han Dynasty* more than 2000 years ago. In the early times, people made a cakey shape with dough that was cooked in a pot, which was considered as the predecessor of noodles. It was called "boiling cake" and "cake mixed with water", which was recorded in "Explaining Name •Explaining Food and Drink" written by Liu Xi in *Han Dynasty*. The noodles became strips in *Tang* (618–907), which were called "cold washing", "Butuo" (Figure 2-2) and renamed as noodles in the Song Dynasty (960–1279). The variety of noodles increased gradually, and developed into many kinds having local special features. Noodle manufacture developed quickly in the *Yuan Dynasty* (1271–1368) when dried noodles, which could be stored for long time, appeared. The drawing technique of noodles advanced significantly in the *Ming Dynasty* (1368-1644); there were still special sliced noodles which were made in Shanxi province. During the beginning of the *Qing Dynasty* (1368–1912), notable products in Chinese history, "Yifu" noodles, were made by mixing boiled or fried noodles with cooked food burning and stewing with a slow fire.

Chinese noodle-processing technology was introduced to Japan during the *Tang Dynasty*. The Japanese invented a rolling and pressing machine for producing flour in 1883. With the development of drying technology came the industrial manufacturing of noodles. Fine dried noodles are produced quickly nowadays, and fine dried noodles have even become a necessary food in China.

Figure 2-1. The archaeological and epigeous noodles found in 2002.

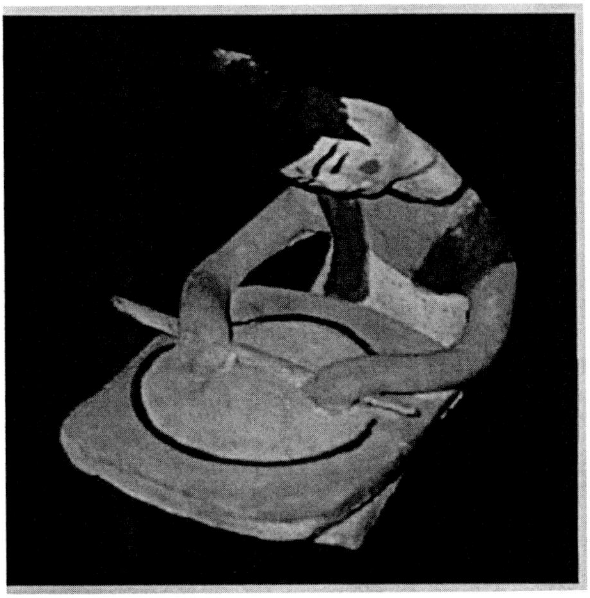

Figure 2-2. Maidservant who rolled dough in Tang Dynasty.

The development of instant noodles is based on the fine dried noodles' manufacture technology. Industrial manufacture of instant noodles began in Japan in 1958. Owing to its merits of being easier to eat, lower in price and so on, instant noodles quickly occupied Japanese markets and spread through China, Korea, Australia and other countries. Production technology of instant noodles was introduced into China from Japan around 1982. The market of instant noodle had become the largest in convenience foods.

Different places in China have different ways of processing or cooking noodles, different varieties and different ways of eating them. It is unique in the world that the flavors of noodles are diverse in different places of China. According to the character of noodles, they can be classified as wheat noodles, rice noodles and miscellaneous grain crop noodles. Based on processing methods, there are machine-made noodles and hand-made noodles. As for the flavors of noodles, they include soup noodles, cool noodles, bittern noodles, oil-doused noodles, salvage noodles, stewing noodles, braising noodles, sliced noodles, hollow noodles, drawing noodles and so on. By the shape, noodles can be wide, narrow, sliced, and silver silk thread and so on. By the method of cooking, there are cool mixing noodles, cooking salvage noodles, crisp frying noodles, braised noodles, and soup boiling noodles and so on.

Noodles play an important role in Chinese traditional foods. Because of their tasty flavor, excellent quality and reasonable price, consumers are very fond of noodles. Many people make noodles their staple food. Especially, noodles are closely related to China's customs and etiquette. It is an indispensable staple food in banquets for a birthday, nuptial, the day celebrated for giving birth to a baby, and other major feasts. For instance, when celebrating the birthday of older people, we must eat "longevity noodles", which was very popular in the Tang Dynasty. During a wedding ceremony, we eat "festive noodles". When a new baby is born, relatives and friends are invited to eat "jubilant noodles". The noodle is a kind of very popular food in many other countries, too. About 40% of the annual wheat harvested in all Asian countries is used for noodles manufacture.

2. The Processing Technology of Noodles

2.1. Raw and Supplemental Materials

2.1.1. Wheat Flour

2.1.1.1. Protein and Gluten

Gliadin and glutenin are the specific gluten protein of wheat. Wheat gluten forms a gluten network after the wheat flour absorbs water to form dough with certain elasticity, extensibility and plasticity properties. The dough is rolled and sheared into wet noodles. So the quality and quantity of gluten directly affects noodle quality, and have a great impact on the noodles' processing. Noodles made from flour with high gluten strength have a good processability. The flexibility and extension of wet noodles are strong so the amount of broken noodles decreases. Noodles made from flour with low gluten strength have a poor processability; the flexibility and extension of wet noodles are weak and the quality of noodles is affected. However, if gluten strength of wheat flour is too high, the flexibility and shrinkage rate would be increased, which is unsuitable for noodle making. It is said that wheat flour with medium gluten is a smart choice—in wet gluten content for shear noodles and fine dried noodles is about 28–32% and for instant noodles and wet noodles is about 32–36% (Shi et al.2001; Song et al.2005; Lei et al.2006).

2.1.1.2. Starch

Most of the carbohydrate in wheat flour is starch which accounts for 65% to 75%. Starch constitutes the main body of wheat flour and noodles, does not dissolve in water at normal temperature, but has some water expansion, which causes wheat flour to produce plasticity after absorbing water. So wheat flour is suppressed to make noodles after adding water. When wheat starch paste is heated to gelatinization temperature, the starch granules will be swelling, broken and pasting. These are the important processing performances in the process of wheat flour. The amylose accounted for about 24% in wheat flour, and amylopectin for about 76%. Starch with more amylopectin shows waxy endosperm quality. Noodles with more amylopectin are relatively soft and smooth in taste (Shi et al.2001; Song et al.2005; Lei et al.2006).

2.1.1.3. Crude Fiber

Crude fiber has adverse impacts in noodle processing. Water absorption of cellulose is 8 to 10 times its own weight in general, which is about 30 times higher than that of wheat starch, and about 5 times higher than that of wheat protein. When mixing flour and water, crude fiber absorbs the water firstly, and then affects the absorbency of wheat protein and starch. In addition, cellulose has no extension and plasticity. With more content, cellulose will lower the intensity of dough, which results in wet noodles breaking easily.

2.1.1.4. Fat

The fat in wheat flour is generally less than 2% and mainly in the wheat germ. The fat content of second flour is much higher than that of the first (or top) flour. Although, the fat occupies a small proportion of the wheat flour, it has a great relationship to the safety of storing wheat flour and noodles. The hydrolysis of fat into fat fatty acid results in the wheat flour and noodles recoming rancid and metamorphic.

2.1.1.5. Ash

Most of the ash is on the surface of wheat, the next highest amount is in the germ, and the least amount is in the endosperm. Generally, ash content is high in wheat flour with low-precision processing, and low with high-precision processing. The quantity of ash can decide wheat flour grade to a certain extent, and affects noodle quality. Ash is detrimental to noodle quality. Reducing the ash content in wheat flour is one of the important factors in improving noodle quality.

2.1.1.6. Pigment

The main pigments in the flour are carotene and flavonoids. Carotene is easy to oxidize; it can be oxidized by oxygen in the air. The increase of whiteness in aging white flour is the result of oxidized carotene by oxygen in the air. Adding oxidants such as peroxidation benzamide can also oxidize carotene. Flavonoids have no color in neutral conditions. In alkaline conditions, however, it appears yellow.

Industry standards for wheat flour used for noodles (SB/T10137-93) are shown in Table 2-1.

2.1.2. Water for Mixing Flour

Water is the auxiliary material for making noodles, and an indispensable, important material for the production of noodles.

Table 2-1. Industry Standard for Wheat Flour Used for Noodles

	Top flour	Second flour
Water % ≤	14.5	
Ash content (dry basis) % ≤	0.55	0.70
Granularity number screen	all passed	
CB42 sieve	the residue is no more than10.0%	
Wet gluten %	28	26
The stability time of farinograph min ≤	4.0	3.0
Falling number min ≥	200	
The sand content % ≤	0.02	
Magnetic Metals g/kg ≤	0.003	
Odor	normal smell	

2.1.2.1. The Role of Water in Making Noodles

The starch in wheat flour absorbs water and expands which causes dry flour to transform into wet dough with certain plasticity. The proteins in wheat flour absorb water, expand, and bond with each other to form wet gluten network, which endows dough with viscoelasticity and extensibility properties. Water not only can adjust dough humidity which gives dough an easy-to-roll coating, but also dissolves salt, alkali and other soluble materials. In the cooking of noodles, water can impel starch to paste after heated. Water is also the heat transfer medium when drying noodles.

2.1.2.2. The Effect of Water Quality on Noodle Quality

The pH of water has an influence on the production processing techniques and quality of noodles. If the pH were lower, gluten protein and starch would be decomposed in acidic conditions, which results in the decrease of dough processing. Generally, water alkalinity for preparing dough is required to be below 30 mg/kg flour. If the alkalinity were too high, gluten would be partly dissolved, dough flexibility and the processability would be reduced, and soluble substances in soup would increase when noodles were immersed in water.

Water hardness also impacts processing techniques and quality of noodles. Hard water will worsen the hydrophilic property of wheat flour, reduce water absorption speed, and prolong mixing time. Calcium and magnesium ions in hard water can combine starch in wheat flour, impact starch expansion during flour mixing and gelatinization during steaming noodles. Hard water also reduces dough viscosity and affects the dough processing property.

The recommended standards of processing water in making noodles are shown in Table 2-2.

2.1.3. Salt

Sodium chloride dissociates into sodium and chloride ions after dissolving in water which can accelerate water absorption speed, make water easy to distribute homogeneously after being added to flour. At the same time, sodium and chloride ions can fix water which is beneficial to form gluten by distributing around proteins. By the double media role of water molecules, sodium ions and chloride ions, the protein rapidly takes up water, expands and becomes more closely connected to each other, so that the flexibility and scalability of gluten is enhanced. In addition, there is a negative strengthening relation between sodium chloride and salt-soluble protein in wheat protein which regulates the role of the salt-soluble protein component in the formation of gluten; it improved the effect of dough results objectively. Salt is an important auxiliary material in the production of noodles; it not only plays the role of a seasoning but also can significantly improve the processing of dough. The salt ratio is small in noodle production, but has an influence on making noodles.

2.1.3.1. The Role of Salt in Making Noodles

Salt plays the role of convergence gluten, it can enhance the flexibility and scalability of wet gluten and improve the performance of the dough, so that the intrinsic quality of noodles can be enhanced. The effect of salt on the dough texture is in Table 2-3.

Since the salt water has a higher permeation, therefore, the flour absorbs water fast and uniformly, and the mix time of flour can be shortened, the dough is easy to mature and the quality of dough is improved. As the salt water reduces the pressure on the surface, so it has a certain moisture role. Preliminary processes can reduce losses of water and dough. To a certain extent, salt can suppress bacterium growth and activity of enzymes, and can prevent the dough from becoming sour very quickly on hot days. It also certainly has a function in blending flavors.

Table 2-2. The recommended standards of processing water in making noodles

pH value	5-7
Hardness degree (degree)	<2.0
Alkalinity(mg / kg)	<30.0
Fe content(mg / kg)	<0.1
Mn content(mg / kg)	<0.1
Odor	No
Else	Coincidence the tap water standards

Table 2-3. The effects of table salt on dough quality

Table salt (%)	Tensile force (g)	Tensile stretch (%)
1.0	15.0	31.75
1.2	14.7	32.12
2.0	13.0	36.25
3.0	11.5	53.41

Note: The wheat flour is the especially second-level powder, ash content 0.7%, moisture content 13.6%, wet gluten content 33.4%, protein content 10.58%.

2.1.3.2. Salting Quantity and Salting Method

The addition of table salt is decided according to the noodle variety. The mechanical system surface is below 3% generally, for example, while for fine dried noodles, it is about 2%~3%, and for instant noodles about 1%~2%. The addition of table salt in hand-made noodles is a little higher but must be less than 10% generally. Excessive table salt will reduce the quantity, the quality of wet gluten, and the elasticity, expansivity of the dough. At the same time, fine dried noodles are not easy to dry, and easily absorb moisture during the depositing period.

2.1.4. The Edible Alkali

The alkaline reagents can play a role in the protein and the starch in flour, enhance the dough biceps function, and enable dough with unique properties, such as toughness and elasticity. The alkaline reagents can also neutralize the free fatty acid of flour which is harmful to gluten, and extend dough formation time and the hydration time of dough protein. Simultaneously the alkaline reagents can increase dough stability time, strengthen gluten disulfide bond so as to increase ability to bear agitation and improve dough toughness. Moreover, suitable addition of alkaline reagents can improve fluidity properties and cohesion power of dough.

2.1.4.1. Function of Alkali in Noodle Production

The alkali has similar function as table salt on the gluten; it can restrain the gluten nature, improve toughness, elasticity and smoothness of dough.

Effect of edible alkali on dough quality is shown in Table 2-4.

2.1.4.2 Alkalizing Quantity and Alkalizing Method

The addition of edible alkali is about 0.1%~0.2% of wheat flour. Generally the sodium carbonate is dissolved in water to adjust to a good concentration, and then salt is added in the dough kneader.

2.1.5 Other Additives

Many other food additives can be used for quality amendment, nutrient supplement too. Compound phosphate, emulsifying agent, food gum, enzyme, nutritional additions such as protein, vitamin or amino acid are these kinds of materials usually used in noodle making.

Table 2-4. Effect of edible alkali on dough quality

Alkali addition (%)	Tensile force (g)	Extensibility (%)
0.2	18.6	32.3
0.3	16.0	36.5
0.4	15.0	43.0

Note: The wheat flour is the especially second-level powder, ash content 0.7%, moisture content 13.6%, wet gluten content 33.4%, protein content 10.58%.

Phosphate is one of the most widespread food additives. Sodium tripolyphosphate, sodium hexametaphosphate, sodium pyrophosphate, sodium phosphate, dibasic sodium phosphate, sodium dihydrogen phosphate, acid-form sodium pyrophosphate, and sodium acid pyrophosphate are authorized now in China.

Emulsifying agent is one kind of the urface active agents which have hydrophilic group (polar, oleophobic) and hydrophobic group (nonpolar, oleophylic). Because of its special molecular structure, the emulsifying agent can interact with macromolecule substance in flour such as starch, protein and fat so as to change noodle properties. Monoglyceride, phospholipid, sucrose esters, stearyl lactate, biacetyl tartaric acid monoglyceride are mainly used in noodle processing and they can adjust the relative layout between the network structure of protein and starch in dough, reduce the adhesion of dough, enhance the water-binding power and the moisture divergence in dough, promote the forming of stabile and matured dough, improve the gloss of the product, prevent the noodles aging.

Common edible gum include guar gum, sodium alginate, konjak powder, xanthan gum, CMC are used to improve dough rheological property and processability, increase water absorption and viscoelasticity of dough, and reduce oil absorption rate of fried instant noodles.

The commonly used enzyme preparation is as follows: xylanase, lipase, glucose oxidase, amylase and so on. Enzyme has a high degree of specificity, has influences on cooking quality, texture and color quality of noodles, and improves dough rheological properties and processing properties.

Soybean protein is the best quality product in vegetable protein and is added in flour for noodle making sometimes. It can complement the amino acid of flour while improving the texture of noodles.

2.2. Processing Technology

Noodle production includes manual operation to semi-mechanized and mechanized operations. With the constant improvement of social productive forces, the progress of technology, the production of noodles has become increasingly mechanized. The noodle modes of production have been transformed from the previous one and a small workshop into modern large-scale industries with mechanization and automation of high-performance, the production capacity has been increased greatly. The mechanization and industrialization of the production not only improve the noodle production but also increased significantly the types of noodles at the same time.

2.2.1. Hand-made Noodles

Hand-made noodles have a long history and have gotten the honor of "eating one kind of noodle in hundreds of methods" (Hou G-Q2001), because of their various varieties and highly elaborated methods. Common flour can be made into hundreds of multiform noodles by rolling, pulling, pressing, poking, gliding, grinding, drawing, picking, sipping, cutting, twisting and so on; the types of noodles are different in length, shape, hardness, crudeness and fineness. The materials used for noodles are extensive, including wheat flour, broomcorn flour, legumina flour, buckwheat flour, oat flour, rice flour, corn flour and so on. These miscellaneous grain crops can be made into different kinds of noodles which have their own unique feature and distinct flavor. They can be eaten in such many kinds of forms as boiling, frying, stewing, steaming, decocting, and braising.

Noodles are the staple food for residents living in the north of China. Rice noodles and river noodles (mentioned in another chapter) are two types of noodles, but there have been thousands of multifarious noodle types in various parts of China because of their different producing methods and different seasonings. The more distinguished noodles included dragon whiskers noodles in Beijing, Fusan ramen and gravy noodles in Shandong, braising noodles in Henan, Saozi noodles in Xi'an, Sliced noodles,in Shanxi, beef flavor noodles in Lanzhou, hot dried noodles in Wuhan, Dandan noodles, in Sichuan, Yangchun noodles in Shanghai, wonton noodles in Guangzhou, salvage noodles in Hong Kong, Taiwan noodles and so on (Wang and Ya,1999).

2.2.2. Machine-processed Noodles

With the development of production and processing technology, the manufacture of noodles has made a qualitative leap. The higher the level of noodle industrialization, the more kinds of noodles made by industrialization. The consumption of noodles has become more convenient. Different processing techniques have brought about different species. Figure 2-3 is the technological process flowchart of various machine-processed noodles.

2.2.2.1. Cutting Noodles

Cutting noodles are the simplest wet noodles which use common mechanized production. Production machines only include a dough kneading machine and calendaring cutting machine. As the wet noodles are difficult to store which restricts the development of production, they are produced by convenience grain shops in the community, collective canteen and central kitchen in large-scale catering enterprises where the shops are very near the factory. The targets for selling noodles are tens of thousands of residents, small and medium-sized restaurants (Xu, 1988; Liu et al., 2004; Liu, 2005).

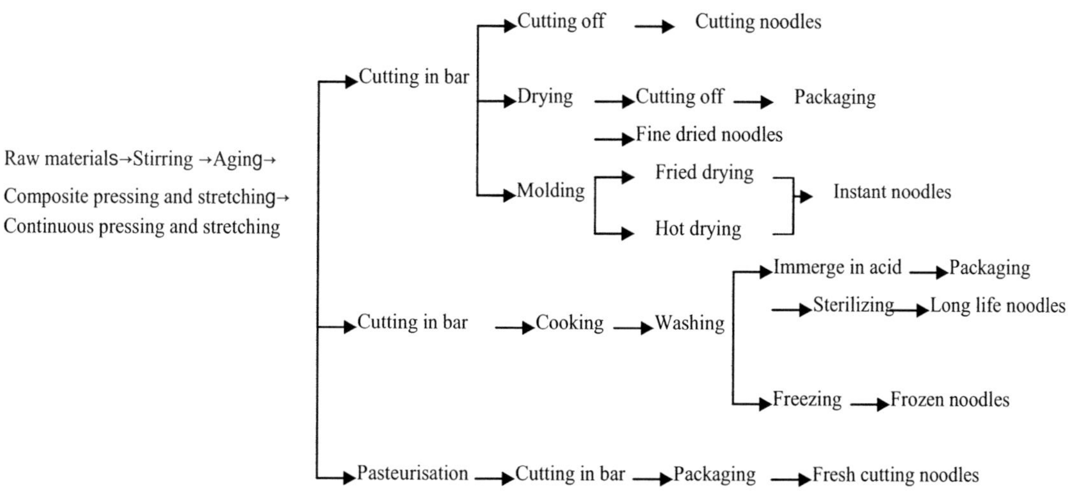

Figure 2-3. Technological process flowcharts of various machine-processed noodles.

2.2.2.2. Fine Dried Noodles

Fine dried noodles are made from wet noodles which are dried with hot air after being cut. According to the drying methods of fine dried noodles, they are divided: (1) drying slowly in low-temperature, the highest temperature of the drying area below 40°C, drying time is about 5~8 h and batch type equipment is often seen. (2) drying in the medium speed and wet, the maximum temperature of drying area is less than 45 °C, drying time is about 3~5 h, tunnel-type drying equipment is usually used. (3) high-temperature fast drying, the highest temperature of drying area is between 45 °C and 50 °C, the drying time is less than 3 h, tunnel-type drying equipment is usually used. As slow and low-temperature drying have a good affect on quality and stability of noodles, the majority of production lines use this method. At present, there are more than 2000 manufacturers of fine dried noodles in China and the annual yield is about 2 million tons.

Although there is little difference in fine dried noodles product line, the basic processing technology is almost the same. The low-temperature slow speed fine dried noodles product line is shown in Figure 2-4.

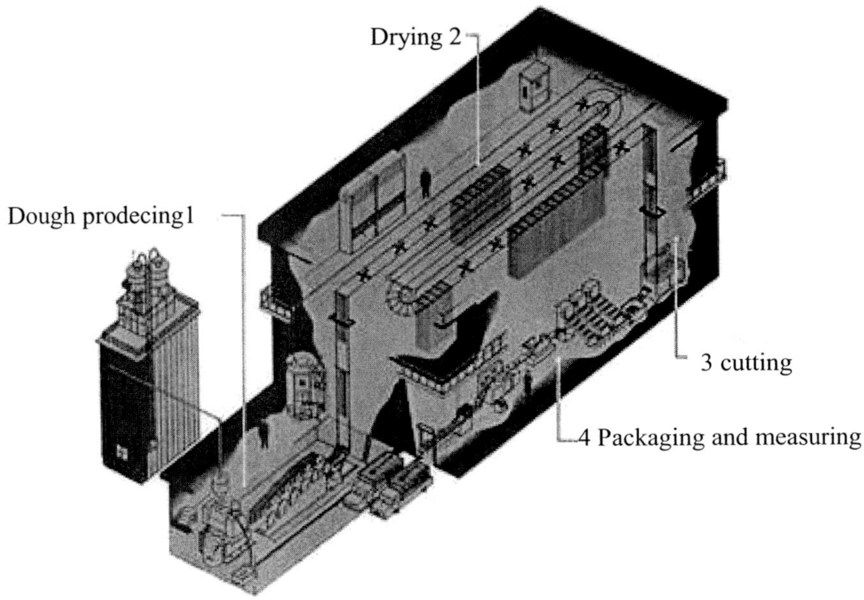

Figure 2-4. Low-temperature slow speed fine dried noodle product line.

2.2.3. Production Technique

2.2.3.1. Dough Kneading

Dough kneading is the first procedure in noodle making. It is also one of the key steps in insuring the product quality. Proper amount of flour, water and additives were added to the dough kneading machine, after certain period of stirring with suitable force, gliadin and glutenin in the wheat flour will absorb water and swell gradually, which will form a continuous membrane like matrix. These membrane like matrix cross binding mutually, they will form a stereo-gluten network system that has certain elasticity, extensibility, viscosity, and flexibility. The starch granule in the wheat flour does not dissolve in the water, but it also absorbs water and swells, then it will be surrounded by wet gluten system. This will makes the inflexible and loose wheat flour become wet dough with flexibility, extensibility and viscoelasticity, which is suitable for the condition of compound calendaring, cutting and molding.

There are three processes in dough kneading. First, unsolvable proteins including gliadin and glutenin, absorb water and swell, and then form the gluten. Secondly, starch granules swell. Thirdly, gluten networks form and extend. If there is no extension, the gluten will be irregular means rough, inelasticity, and poor-toughness of reticular tissue. After extension, the network tissue will be fine and smooth and have a larger elasticity and toughness when the starch granules filled the frame tissue after absorbing water and swelling. Water absorptivity of composition in flour is shown in Table 2-5.

Table 2-5. Water absorption of composition in flour

Name	Protein	Starch	Fiber
Water absorptivity (%)	200~300	30~40	800~1000

Generally, the production of fine dried noodles requires the wheat flour with wet gluten content 28%~32% and water addition with 25%~32% (flour weight), through water addition must be adjusted according to specific condition. Water addition can be a little high when protein is high. Kneading dough at about 30 °C will increase swelling ratio of wheat gluten. If room temperature is under 20 °C, hot water can be used in dough kneading. Dual axis dough kneading machine is mostly used and ideal rotating speed is 70~110 r/min. The dough kneading time is about 15 minutes.

2.2.3.2. Aging

"Aging" means maturing naturally. Wheat flour composed of protein particles and starch granules. Different particles have different diameters and water absorption speed. Because the time of dough kneading is short, part of water needs more time to infiltrate into the internal tissue of the wheat flour and be absorbed fully by protein and starch. During kneading processes, the dough is subjected to being hit with the stirring gear of the machine, so the gluten in the dough generates stress due to extrusion and extension. If noodles are processed by this kind of dough directly, the internal structure of the noodle will be unstable and easily deformed.

The main function of the aging procedure is to make the water permeate into the protein granule as much as possible, to make protein absorb water and swell fully, then adhere with each other to form gluten network tissue. By low speed stirring or standing, the internal stress of the dough will decrease, which makes the internal structure of dough become stable. This will promote the auto moisture regulation between protein and starch, cause the dough's homogenization, and have the function of tempering the particles.

According to the structure of the aging machine, it can be divided into horizontal aging machine, conveyer belt aging machine, suspender pole aging machine and plate aging machine. Plate aging machine is used mostly. Generally it is required to age for 15min. If the device conditions permit, longer aging time will be better. Aging temperature should be about 25°C and a temperature increase during aging is not desired. In the Chinese traditional manual noodles method, several hours of standing of dough is always used to improve the performance and taste of cooking.

2.2.3.3. Compound Calendering

After the kneading and aging of dough, the gliadin and glutenin have already finished water swelling and bond to each other to form the gluten, but the gluten network is still dispersed and loose, distribution of starch granules is still inhomogeneous. Because the particles in the dough haven't linked, plasticity, viscoelasticity and extensibility of dough are still weak. Through rolling dough from large rolls to small ones, the granular dough is rolled into a dough sheet under the external force, the gluten and starch scattered in the dough is assembled. Loose gluten is pressured into a tight network and distributed uniformly to surround the starch granules. The traditional method of making noodles in China is to repeatedly roll dough using a rolling-rod, which is to link up the wet gluten scattered in the dough by adding pressure to form the detailed gluten network which is used to surround the starch particles, and make them distribute uniformly in the dough sheet to improve their properties of processing and cooking.

The main function of compound calendaring is to make the loose dough form the thin dough sheet which is fine and to reach the required thickness. The sheet gets from compound calendaring the ability to tolerate more extrusion, which can benefit the formation of the gluten network.

Compound calendering machine is used in the compound calendaring process. In this process, the dough will be compressed into two dough sheets, and then be compounded into one. Then the sheet will go through the continuous squash machine, so the sheet will reach the required thickness gradually. According to the difference of the direction of calendering, it can be divided into mono-direction calendering and multi-direction calendering. Gluten will distribute along the pressing direction in the mono-direction calendering. Multi-direction calendering is an imitation of the manual operation which makes the gluten distribute and form gluten network well in each direction so multi-direction calendering can improve the rheological characteristics and eating quality of the noodle.

2.2.3.4. Slicing

Slicing is to let dough go through a series of rollers to form the sheet with a certain shape and then cut the dough sheet into wet noodles with certain length and width after rolling.

The technological requirements of slicing is to make the noodle have a smooth surface, have a uniform length and width, have no deckle-edged sheet and incorporative sheet, and the sheet should not be broken. Slicing equipment is composed of knife, cutter labyrinth and transmission part. Cutting and rolling machine parts can be equipped at the last roller, or cutting itself is a slicing equipment. Slicing specifications are shown in Table 2-6.

Table 2-6. Slicing specifications

The width of knife trough (mm)	1.0	1.5	2.0	3.0	6.0
Fine dried noodles varieties	Silver silk thread noodles	Thin noodles	Common Liner	Wide noodles	Yudai noodles

2.2.3.5. Drying

Drying is a process of dehydration and a key working procedure in noodle production, which not only relates to energy consumption, but also has great impact on the quality of products.

Surface water vaporization and internal moisture transfer are carried out simultaneously with different speed. When surface water vaporization is slower than the internal moisture transfer, the drying process of noodles depends on the surface vaporization. On the opposite, if the internal moisture transfer is slower than the surface vaporization, the surface dried first, then vaporization transfers from the surface to the inside. At this time, the moving speed of water plays a controlling role in the drying. When the internal moisture migration is much less than the speed of the surface water vaporization rate, there will be a large internal stress in some membrane on the surface resulting in a crisp noodle. Low temperature slow drying is usually used to reduce surface vaporization.

Noodle drying process can be divided into three stages: preparatory stage of drying, the main drying stage (divided into inner evaporation and the whole evaporation) and the final stages of drying.

The first is the preparatory stage of drying. The noodle with high moisture content is easily elongated and decreases cross-sectional area under the effect of noodle weight. The tensile stress will increase correspondingly and when fracture stress reaches its limitation, will lead to wet noodles being broken. Therefore, part of the surface water on the wet noodle must be removed so as to make noodles shift from plastic body to elastic-plastomer, make noodle tissue fixed, increase the strength of wet noodle and decrease noodle dropping during the early drying period. This drying process is known as the preparatory stage of drying. This stage may use the air flow with low temperature, usually 20°C ~30°C, to evaporate water off noodles' surface naturally.

Main drying stage is the main stage for drying wet noodles which is also the key stage. Main drying stage can be divided into inner evaporation stage and the whole evaporation stage.

Moving speed of water from inside to surface must be set as the reference for modulating temperature and moisture because the moving speed decides the drying speed. In this process, water moved from the internal to the surface would be evaporated, so it is named as evaporation stage. In this stage, the air humidity must be kept at a relatively high level (75%), water vapour pressure in the drying tunnel is equal or slightly higher than the vapor pressure of surface water in noodles.

Pervaporation is another part of the main drying stage. Internal moisture moves to the surface layer which makes moisture molecules accumulate in the surface layer, it is necessary to heat up and gradually reduce relative humidity of drying area air to evaporate the moisture in noodles fully and timely. This comprehensive evaporation process at high temperature and low humidity was known as the entire evaporation stage. In this stage, surface evaporation speed of noodles is fast in the whole process of dehydration by raising temperature and reducing relative humidity.

During the entire evaporation stage, most of the water in noodles has been taken off. In order to eliminate the noodles' internal stress resulting from drying shrinkage, to enhance the elasticity of noodles, and to avoid cracks and rupture for temperature imbalance, and at last to lower the moisture content of noodles to 12.5 ~ 14.5%, wind is applied for mediating the surplus temperature cycle to reduce noodles' temperature slowly. This drying process is called the final drying stage or complete drying stage.

The drying equipment for noodles consists of a heating system, ventilation system, drying tunnel and transportation machinery. The cableway drying method is used mostly by large-scale makers. Noodles can be dried at low temperature and high humidity for a long time, and the temperature of the air and quality of the noodles are easy to control.

2.2.3.6. Cutting

Noodles are cut off by using the relative movement between noodles and the cutter including the shear or cutting role of the cutter. The length of most noodles in China is 240 mm or 200 mm, the permissible variation of length is about 10mm. In the process of cutting, there are no great effects on the intrinsic quality, but the procedure is greatly related with appearance quality of noodles and the quantity of dry-head. The end breaking (dry-head) should be controlled below 6%~7% of the yield. The mechanical cutter includes a disc noodle cutting machine and a reciprocating noodle cutting machine.

2.2.4. Weight Measurement and Packaging

2.2.4.1. Weight Measurement

Weight measurement is an important procedure before noodles packaging. The accuracy of measurement relate to the interests between consumer and manufacturer. The "noodle quality standards" (SB/T10068-92) regulates that deviation of net weight should be controlled within 2.0%.

2.2.4.2. Packaging

The basic requirements for noodle packaging are that it should be orderly and beautiful-looking, clear in design, integrated marker, and safe for the health of consumers. Most of the noodle factories in China are still packaging manually, and only a few makers use automatic packaging machines.

2.3. Instant Noodles

According to the drying methods of instant noodles, it includes fried instant noodles and hot air dried instant noodles. The biggest difference in the production technique between hot air dried and fried instant noodles is that the former uses hot air for dehydration of cooking dough, while the later uses oil. Fried instant noodles accounted for more than 90 percent, non-fried instant noodles less than 10 percent of products on the China market. Though hot air dried instant noodles are still in its infancy state, due to low fat content, the product has a strong market potential. Instant noodles have become a best-selling convenience food in China, because of characteristics such as convenience, fast preparation, appeal to diverse tastes and ease in storage. Now, there are more than 3,000 instant noodle production lines in domestic factories with an annual production of about 360 million tons. A fried noodle production line is shown in Figure 2-5.

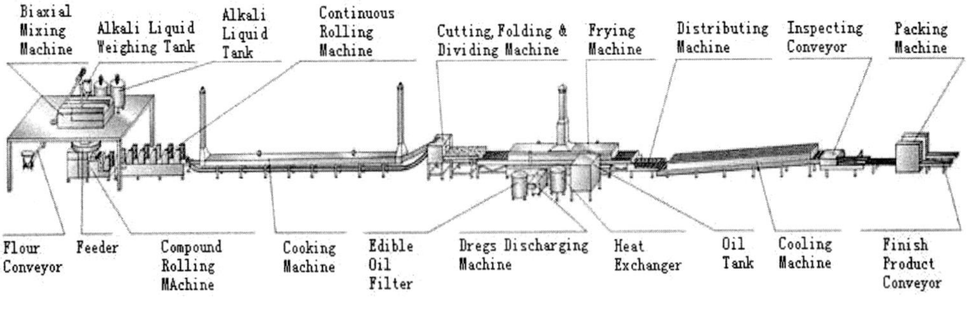

Figure 2-5. Fried noodle production line.

2.3.1. Dough Modulation

Dough modulation in instant noodle production and fine dried noodles are almost the same. The flour with 32%~36% wet gluten content is suitable to instant noodle making. Water, salt and alkali addition are usually 33%~36%, 1%~2% and 0.1%~0.2% of the wheat flour, respectively.

2.3.2. Aging and Compound Calendaring

Aging and compound calendaring processes are almost the same with fine dried noodle.

2.3.3. Cutting, Folding and Shaping

The basic principle of slicing and folding for instant noodles is that flour belts become vertical noodles cut by intermeshing from the two sides of many concave-convex slots with equal intervals. Two symmetric cuprum combs are installed at the lower part of a toothed roll, clinging to the groove of the toothed toll, which ensures the slicing runs continuously and clears the sheared noodles so as to not adhere to the toothed roll. The noodles sheared by the cutter have some swaying property, and then go through the shaper which has a specially designed section with a flat rectangle. Under the shaper, a pore mesh-belt weaved with stainless steel wire is installed. The line speed of mesh-belt is less than that of the noodles. Because of this differential speed, noodles passed by the shaper swing backward and forward as they are subjected to a certain resistance, then they twist and pack into a waveform surface with erected wave crests and back-to-back wave crests. As the speed of the same belt is speedier than that of noodles in the shaper, noodles are elongated gradually to form a waved pattern. When noodles are transported into the steaming machine, the waved pattern is fixed by steaming and boiling. The device schematic diagram of cutting, folding and shaping is shown in Figure 2-6.

The noodles gotten from cutting, folding and shaping must have a smooth surface, a proper density, equal branching, and have no bonding between noodles.

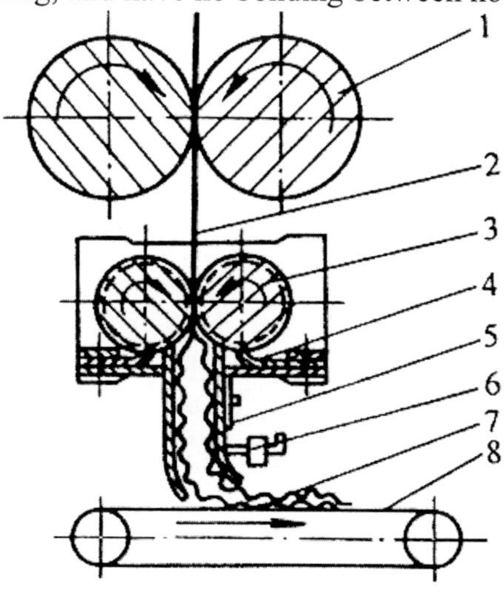

1. Roller 2.Flour tape 3.Knife 4. Hackle comb 5. Molding box 6. Pressure regulating hammer 7. Bellows-block 8. Speed controlled stainless steel net-belt.

Figure 2-6. The device schematic diagram of cutting, folding and shaping.

2.3.4. Dough Steaming

Dough steaming is one important step in instant noodle production. Corrugated noodles were heated in a saturated humidity environment, and the starch in noodles was gelatinized while the protein was denaturized thermally. Dough steaming has the function of fixing noodles pattern shape, too. During the steaming process, β-starch changes into α-starch. β-starch is raw starch and its molecules are in a crystallized state under certain arrangement. It is difficult for enzymes to get inside the molecular pattern of β-starch so it is hard to digest and has a bad taste. After absorbing water and being heated, β-starch will change into α-starch which has a confused molecular arrangement and it is easier to digest, decompose and has a better taste. The gelatinization degree for hot air dried instant noodles should be above 80% and for fried instant noodles above 85% so as to improve the rehydration of noodles.

A continuous steaming machine is usually used in the step. The inlet temperature is 60~70°C and the outlet temperature is 95~100°C. Steaming time should be controlled in 90 ~ 120 s, the length of the steaming box is 16–20 m. In order to improve the gelatinization degree, it is necessary to increase water absorption as high as possible in mixing flour, and a moistener is often located at the entrance of the steaming box to spray a film of water on the surface of noodles to increase water content.

2.3.5. Quantitative Cutting

The process of quantitative cutting is to cut ripe noodles from the steaming box at a certain length by a couple of cutters and rollers with relative rotation. At the same time as cutting, the folding board, is reciprocating and intercalating in the interspaces of noodles which are been cut. Conduplicate noodles are sent to the next dry process by a guided roller and conveyor belt. The device schematic diagram of quantitative cutting off process is shown in Figure 2-7.

The basic requirements of the quantitative process are accurate quantities, folded neatly, and accuracy when noodles get into the hot air dry machine or automatic fry drop-box. It is a unique multi-functional process in the instant noodle production line.

2.3.6 Drying and Dehydration

Fry drying is the process which puts noodles gotten from quantitative cutting into the automatic frying machine. Noodles are surrounded by high-temperature oil while going through the high-temperature oil tank. Noodle temperature will increase rapidly, and the water in noodles will evaporate quickly. Porous structures in noodles resulting from evaporation will further improve the gelatinization degree. When noodles are been soaked, it is easier for hot water to get into noodles' micropores which let noodles have a very good rehydration. Due to the rapid drying during frying, the state of starch gelatinization is fixed after cooking, and aging rate is greatly reduced in the storage period.

Palm oil with the freezing point of 20~36°C is used in fried noodle drying. The oil surface should be above the top of the fried box about 30~60 mm. The reasonable dried temperature is 130~150°C, while the dried time is 70~80 second. The water content after frying should be less than 8%.

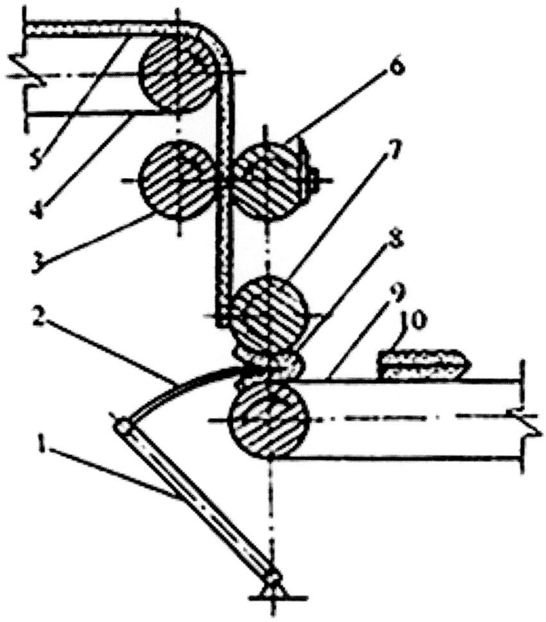

1. Linkage; 2. Folding board; 3. Cutter roller; 4. Feed conveyor belt; 5. Steamed noodle belt; 6. Cutter; 7. Folding guide roller; 8. Noodle under folding; 9. Transmission net-belt; 10. Noodle shaped

Figure 2-7. The device schematic diagram of quantitative cutting off process.

Frying equipment includes the main host, oil heating installations, oil supply system and electronic control system. Spiral plate heat exchanger and tube array heat exchanger are used in the oil heat installations. Net-belt scrubbing filter or scraper grille filter (also named automatic slag discharge machine) and net field filter are often used in the oil filter box. The oil circulation pump will pump out oil from the fry pan, send it to the heat exchanger, and then send it back to the fry pan which plays a complementary role of heat loss, making the oil cycle between fry pan and heat exchanger. Vertical hot oil pipeline pumps are often used as the circulation pumps.

Hot-air drying is the process of putting the quantitative cut noodles into a high temperature environment. Hot air transfers a large amount of heat to the blocks rapidly, and boils the water in noodles in a very short time (20~30 s). The steam will enter into the inside of noodles and form a large amount of stomata which increase the starch gelatinization degree.

Drying temperature for hot air drying instant noodles is 140~160°C, when hot air speed increases to 25~30 m/s. The block will be dried within 180~200 seconds and the water content of noodles is below 12%. Hot air drying equipment includes hot air dryers, hot air drying host, hot air circulation system and electronic control system as shown in Figure 2-8.

2.3.7. Cooling

After drying, instant noodles still have a high temperature of 60~100°C when they are sent to the cooling machine. If these blocks are packed without cooling, aqueous vapor produced in the package will cause mildew. There are two kinds of cooling method including natural cooling and forced cooling. Forced cooling uses air cooling equipment usually to enhance the air flow. The objects being cooled go through the cooling tunnel, in which the heat will dissipate rapidly under the function of cool wind. The cooling process requires that the blocks after cooling are close to the room temperature or above the room temperature about 5°C. The common cooling machine is net-belt cooling fan and centrifugal multi-storey air-cooled equipment.

2.3.8. Packaging

Packaging is an important process in instant noodle production. It's mainly composed of weight testing, finishing, distribution, transportation and soup delivery, packaging and packing processes.

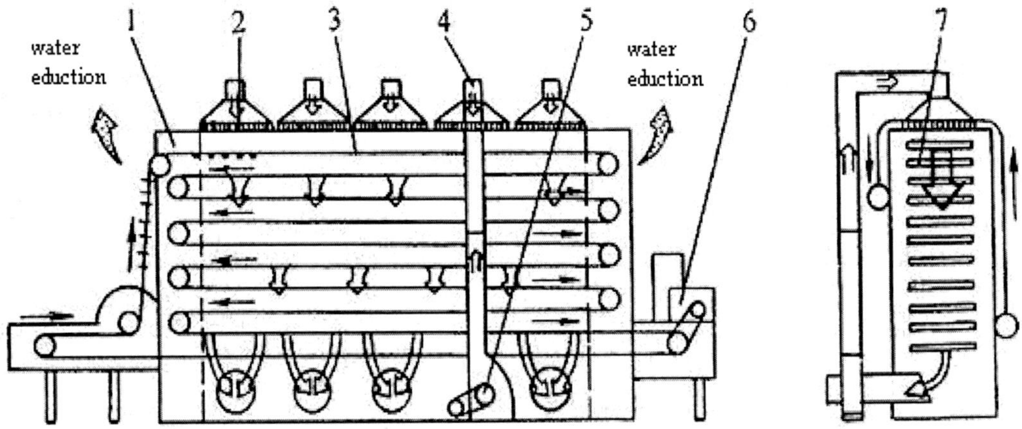

1.Frame 2.Heat exchanger 3.Chain 4.Wind pipe 5.Hot air cycle blower. 6.Stepless speed change transmission 7. Stainless steel-box.

Figure 2-8. Hot-air drying machine.

2.4. Long Life Noodles

As Huang (2006)) reported, Long Life noodles (LL noodles) are called ready-to-eat instant noodles by cooking, washing, pH adjusting, germicidal treating, which can be stored for more than six months. The LL noodles have a crystal appearance and short dehydrating time, good chewing character and continuous strip. It has high moisture and good taste like fresh noodles, and can be directly consumed after dehydration. The product was developed in Japan in the early 1990s, and the LL noodles in China are just in the initial stage, and have only a 0.5% market share. In Japan it has accounted for about 10% of the market.

2.4.1. Kneading flour

The kneading of LL noodles is the same as that of the common noodles; it makes protein in flour form wet gluten by absorbing water. And starch is enveloped by wet gluten to form dough suitable for noodle processing. The suitable content of wet gluten for LL noodle processing is generally 32~36%. It is better to knead flour in a vacuum kneader. In a Vacuum state, injected water is very easy to atomize which ensures water uniformity, and the air in the flour is aspirated to form a vacuum state. The water can easily infiltrate the internal parts of flour particles so it can increase the water content of flour, which leads to formation of compact dough and a network structure, at last it enhances the elastic strength of dough. The water addition to dough is generally 30% to 34% in the ordinary flour mixers. Water content can be increased to 35 ~ 40% in the vacuum mixers, which is favorable to increase the quantity and quality of wet gluten. At the same time, it is suggested to use high-speed mixing at the beginning stage, because it is favorable for the uniform contact of water and flour, while low-speed mixing in the last stage can avoid damaging of the formed gluten network.

2.4.2. Aging

Flour aging and flour belt aging are generally used in LL noodles processing. The dough is generally aged in a belt aging machine with fragment devices. The conveyor runs slowly, which plays the role of static aging in the process of transportation. Flour belt aging is operated with compound squash and continuous squash. Flour belt aging device is a closed box, and the aging time is normally 30 ~ 90 min and temperature is about 25°C.

Figure 2-9. The schematic diagram of Multi-direction Sheeting.

2.4.3. Compound and Continuous Squash

The production technology of LL noodles is almost the same as that of the conventional noodles. The difference is that the former uses a unique corrugated roll-squash in the first squash because the dough sheet is bearded with great longitudinal force which decreases the transverse strength of the formed gluten network compared to squashing with a smooth roll. However, multi-direction sheeting has better imitation with manual kneading in dough state because it can promote the formation of gluten network and can improve the quality of dough, so it is beneficial to improving the gluten product (Chen et al., 2006). The schematic diagram of multi-direction Sheeting is shown in Figure 2-9.

2.4.4. Slicing and Quantitative Cutting

The slicing and quantitative cutting are operating in the last roller after continuous squashing.

2.4.5. Boiled (Steamed), Washed and Acid Leaching

The Water boiling (steaming) is to heat and cook sliced noodles so that starch in the noodles gelatinizes, protein denaturates, and noodles absorb water fully which helps to form a good taste. Water boiling is better than steaming for noodles. First, the effects of contact between noodles and water are good. Secondly, as the noodles absorb moisture in water, it is beneficial for gelatinization of starch. Thus the time of gelatinization is just about 3 ~ 5 min. The weight of noodles after being boiled is about 2.3 times that before being boiled.

After water boiling and gelatinization, noodles should be washed with cold-water in order to remove the starch paste and other adhesion materials on the surface of noodles. By rapidly cooling, the surface of noodles cools and shrinks while the viscoelasticity is enhanced. Noodles after washing undergo acid pickling to adjust the pH of noodles below 4.5. It can effectively inhibit the breeding of bacteria and ensure the noodles can be stored for a long time at room temperature. The additives which are used to adjust pH value include lactic, malic acid and citric acid.

After cooking noodles, using the low temperature water, the surface of the noodles shrink rapidly. Adhesion between noodles can be prevented by washing away mucus and lowering the temperature.

2.4.6. Sterilization and Packaging

The product packaging is divided into two parts, the washed noodles are packaged after removing water in the automatic packaging machines. Though noodles are pickled, bags of packaging and air in workshop which contain many bacteria will still affect the shelf life. After packaging, noodles are sterilized with steam in 93 ~ 97 °C for 30 ~ 45 min so as to kill the bacteria in the packaging bags completely. After the sterilized bags are cooled in the cold water, they can be preserved for 7 to 14 days at room temperature.

2.5. Frozen Noodles

Frozen noodle is a kind of noodle which is stored below -18°C after it is cooked, shaped and frozen. Frozen noodles can not only maintain the organizational structure, taste, smell of fresh noodles, but also have a better rehydration, and can be preserved for a long period without adding any preservatives. The production technology of the frozen noodles is the same as the LL noodles in the same aspects including kneading the flour, aging, compound and continuous squash, slicing and quantitative cutting, boiling with water, water washing. The other technology is as follows.

2.5.1. Freezing

Put the washed noodles into the instant freezer; it is frozen quickly under the -38 °C. The quick freezing method makes the free water in the frozen noodles into small ice crystals. It keeps the virgin color and luster, taste and the nutritional ingredients of the natural food to a maximum extent.

2.5.2. Packaging and Cold Preservation

The frozen noodles are put into different flavor seasonings for packaging, then stored and transported under the conditions of -18 °C or lower temperature. As most microorganisms can not breed when food is frozen to -18 °C or even lower, the enzyme activity will be greatly restricted, the biochemical reactions in food will become very slow. Due to the free water of frozen food becoming small ice crystals, water creates the necessary conditions for the role of microbial growth and enzyme, microbial growth, so that the frozen noodles can be stored more than six months.

3. RESEARCH ON NOODLE PROCESSING

Noodles have a long history in China, and come in various varieties. In recent years, with the development of noodle industrialization, the consumption of noodles increased quickly. Requests on the quality of noodles becomes higher and there are many researches on noodle processing.

3.1. The Impact of Wheat Quality on Noodle Quality

Many researches showed wheat protein, starch and kernel hardness have significant effects on noodle quality (He, et al. 2004; Zhang, et al. 2000; Lei, et al. 2003; Lan and Wang, 2006; Song et al. 2005).

3.1.1. Protein

Protein content in wheat flour is 9 ~ 17 % generally in China. Although protein content is much lower than starch content in wheat, it plays an important role on noodle quality. Protein absorbs water, expands, and forms the gluten network structure which can provide space for uniform distribution of starch and other ingredients. Gliadin and glutenin in the endosperm of wheat play a major role in forming the network; the content and ratio of these two kinds of protein determine the type and quality of wheat to a great extent. If protein content is too high or gluten is too strong, it will cause flour to be difficult to process. As a result, the dough surface becomes rough and white color degree of noodles decreases. Dried noodles are prone to bend and uniformity of noodles may be affected. But if protein content is too low, dough strength and boiling resistance of noodles may be weak which give noodles bad taste and lack of toughness and flexibility (Kovacs et al. 2004). It shows significant negative correlation between protein content and water absorption. There is significant negative correlation between wet gluten content, settlement value and the loss of solids, too. So it is better if the protein content of flour is 12~13% for noodle processing. Research results showed that protein quality is even more important than protein content. It showed significant or much significant positive correlation between the toughness, hardness, flexibility of noodles and quality value of flour such as settlement, dough stability time, extension, the biggest anti-extension resistance, stretching area. Appearances of noodles (such as color, apparent situation) are negatively correlated with the value of settlement and the time of formation. The luster of noodles and the content of flour protein are in inverse proportion.

3.1.2. Starch

Starch, one of the important components, takes up about 54~ 72% of the dry weight in wheat. Starch includes amylose and amylopectin. The ratio of amylase and pasting properties of starch determine starch quality. Studies found that viscosity trait and swelling volume have very high correlation coefficient with the noodles' quality. The majority of the research agreed that the gelatinization characteristic of wheat flour, including peak viscosity, revitalization value and an increased ability to dissolve have an influence on the surface texture of noodles. Boiling loss rate of noodles, peak viscosity and the gelatinization temperature is extremely positively correlated with starch. Boiling absorption rate of noodles and the peak viscosity have an extremely positive correlation. Usually, the higher the viscosity of starch pasting peak, the better quality of the noodles in smoothness and flexibility. That is to say, noodle quality can be improved by increasing starch pasting properties, and wheat flour (or starch) viscosity can be used as an indicator for noodle quality evaluation. The composition of starch has a very important impact on noodles quality. High amylose content of wheat flour may affect ductile property and total quality of noodles while low or middle content of amylose and high content of amylopectin may improve toughness and quality of noodles. Some noodles with lower paste temperature taste better (Zhang et al. 2000; Lei et al. 2003; Lan and Wang, 2006; Song et al. 2005).

3.1.3. Kernel Hardness and Noodle Quality

In a certain range, with the increase of wheat hardness, the quality of noodles becomes better. However, if kernel hardness is too high, it would have a negative effect on noodle quality. There was significant negative correlation between kernel hardness of wheat and color, appearance, smoothness and taste of noodles. Kernel hardness has a significant effect on the texture of noodles, and wheat with medium hardness is suitable for Chinese noodles.

3.2. The Research and Development of Noodle Production Technology and Equipment

There are many researches on the development of noodle processing technology (Yang et al. 2003; Yu et al. 2003; Liu 2005; Chen et al.2006). In order to adapt to the demands of noodles processing, the equipment and technologies of noodles have been researched and developed. The grinding technology in cryogenic temperature can make finer particle size, good gluten strength and good lubricities of wheat flour. The particles of wheat flour are stirred under negative pressure, absorbing water fast and equably, and then form protein network structure which is beneficial to starch gelatinization when vacuum mixer is used. In addition, compared with conventional mixer, the water absorption of dough increased about 10% and the quality of product improved. The ripple roller technology, which imitates artificially kneading dough conditions, can improve the formation of gluten network. The pressed dough sheet can bear obvious higher intensity of pulling. Use of static ripening installment can increase ripening time and enhance ripening effect. The use of a flour mixer with double-shaft and two-speeds results in atomized water and can enhance the uniformity of water absorption of dough. The pressing shell with special wire rope in drying chamber transmission realizes the slung-load driving in T-track, the operation is stable which successfully solved the phenomenon of dropping bar. Hot air drying equipment for instant noodles has been researched and developed, too. Through high temperature and rapid dehydration, small holes appear on the surface of noodles that have the same quality as fried instant noodles but lower fat content. Due to the application of airstreams blowing and absorbing degreasing technology, nutrition fortification and spraying technique, ABE automatically selected and arranged device, automatic pouch dispenser, automatic pouch layer, bowl noodles pouch dispenser, carton packer and so on, the production lines of instant noodles showed great progress in China. The level of electromechanical integration has been improved in noodle product machinery. The mechanical device has a strong adaptability to adjust the processing technique during product processing, and the automatic control of various parameters has made great contributions to improve noodle quality, saving energy, improving labor productivity and reducing cost.

3.3. Research and Application of Flour Quality Improver

There are many researches on the use of a flour improver in noodle processing (Sun et al.2003; Huang et al. 2004; Guo and Zhou 1997; Wu and Li 1998; Liu and Liu 1999; Feng 2000; Tian et al. 2001; Peng et al. 2002; Qiu 2002; Shao et al. 2002). Because there are many kinds of wheat in China, the quality of wheat flour is much different. Tailored flour is still in development and the application is still not wide in China. To improve this situation, chemical additives are used to improve noodle texture, workability and the nutritive peculiarity. Additives used in noodle processing mainly include modified starch, thickener, emulsifier, preservative, enzyme preparation, compound additives and so on.

The results showed that monoglyceride, lecithin or salad can be used as an emulsifier type of noodle improver. Fine dried noodles are improved obviously by adding a noodle improver (emulsifier) if quality and content of gluten are high. While for the soft nature flour, quality of fine dried noodles may be improved by adding few inorganic salts chemical additive or proteins. Emulsifier can improve surface conditions and reduce the loss of steam boiling. Adding stearyl-2-lactic acid into flour can enhance the gluten intensity and reduce the starch outflow in the steam boiling process.

The glucose oxydase may improve intensity and elasticity of dough, apparent condition and cooking tolerance of noodles obviously. Glucose oxidases and 1% gluten (based on flour) showed similar improving effects on fresh noodles. Glucose oxydases and lipase used together showed better effect on strengthening of dough to bear agitation than solely added glucose oxydases. Settling time of the dough can be extended from 3min to 10min. If using oxidases, fungal amylase, ascorbic acid, wheat gluten emulsifier at the same time, settling time of dough may be prolonged to 20min. MTGase can strengthen protein network architecture and improve physical properties of noodles.

In studying the use of natural plant gum, results showed that colloid may improve dough rheological characteristics, increase the caking property of dough, reduce the loss of noodles in steam boiling, increase the tensile strength, enhance vigor and improve the surface condition (Charles et al. 2007). The linseed gum may improve noodle processing and edible performance properties, and make noodles flexible, lively and palatable. After adding the edible gum, noodle hardness was increased, cohesiveness was cut down, section joint and dissolving of solids and the turbidity of soup after boiling were decreased. Noodles are translucent, feel slippery in the mouth, and don't easily stick to the teeth. By studying the effect of guar gum, sucrose esters, wheat gluten, and modified starch on the single screw rod extrusion noodles quality, gelatum and the hydrophilic of gluey gum were found to have the functions to improve noodle quality after being rehydrated. The compound hydrophilic colloid and the pre-gelatinization craft may form the developed network structure in the corn dough and reinforce the biceps of dough.

It is possible to improve the dough characteristic obviously and enhance noodles intensity and cooking quality by using the compound chemical additives which are matched with the compound alkali, the compound phosphate and the compound thickening agent. Results indicated that the compound phosphate can obviously enhance dough settling time and noodle quality, especially strength, viscoelasticity and toughness of noodles, making noodles endure longer boiling time (Wang et al.2000).

The compound chemical additives of wheat gluten, odium polymannuronate, sodium chloride and compound alkali can obviously improve the rheological characteristic of dough, the intensity and cooking quality of noodles. Studying the influence of wheat gluten, starch, emulsifier, and natural plant gum on the rheological characteristic of noodles indicated that these additives can cut down the cooking loss, increase the tensile strength, reinforce the vigor and improve the surface condition of noodles. The anti-staling agent including propylene glycol, sorbitol and alcohol may increase the shelf (storage) life of fresh noodles.

3.4. Evaluation Technology of Noodle Quality

The quality of noodles is evaluated mainly by sensory evaluation including the color, apparent state, edibility (hardness and softness), toughness (elasticity and stickiness), viscoscity of noodles suggested by a professional standard (SB/T 10137-93). But it is difficult for noodles must be boiled before sensory evaluation and boiling itself affected the quality of noodles significantly. So even evaluated by a well-trained panel group, the deviation of evaluation values would be large. Some researchers tried to evaluate the quality of noodles by instruments such as texture analyzer, color measurer (Li et al. 2006; Zhang et al. 2007; Huang, 2006). Parameters include color of raw noodles, drawing force, cooking quality (water absorption and cooking loss), surface hardness, chewing taste, color and luster of cooked noodles can be determined directly and correctly. The micro-quality of noodles can be detected by gel chromatography technology, polarizing microscope, electronic scanning technology and so on. But the problem is that the relationship between these parameters and sensory quality of cooked noodles is still not clear.

As a conclusion, there are many popular traditional foods of noodles with reasonable nutrition which is accordant with Chinese eating habits in China. It embodies the wisdom of the Chinese Nation for thousands of years. Various kinds and many different manufacture methods are an important part of the cultural heritage of Chinese foods, but the development is still not enough and much research still has to be done.

REFERENCES

A.L. Charles, T.C. Huang, P.Y. Lai, C.C. Chen, P.P. Leeand Y.H. Chang (2007). 'Study of wheat flour–cassava starch composite mix and the function of cassava mucilage in Chinese noodles'. *Food Hydrocolloids*, 21(3), 368-378.

Chen Haifeng, Zheng Xueling,Wang Fengcheng (2005). 'The Development of Noodles at Home and Abroad'. *Grain Processing*, 1:39-42.

Chen Jie, Lu Junchang, Zhu Zhiguang, Wangchun (2006). The Effect of Sheeting on Dough Lateral- axial Stress Characteristics. *Journal of the Chinese Cereals and Oils Association*, 21(3):44-47.

Feng Wenrui (2000).'Application of Glucose oxidase in Noodle Processing'. *Science and Technology of Food Industry*, 6: 67-68.

Guo Dachun, Zhou Huiming (1997). 'Effect of additive on Noodle Rheological Properties'. *Journal of the Chinese Cereals and Oils Association*, 3: 14-18.

He Z.H., Yang J., Zhang Y., Quail K.J. & R.J.Pefla (2004). 'Pan bread and dry white Chinese noodle quality in Chinese winter wheats'. *Euphytica*, 139: 257-267.

Hou Guoquan (2001). 'Oriental noodles'. *Advances in Food and Nutrition Research*, 43:141-193.

Huang Deming, Zhao Guohua, Kan Jianquan (2004). `Studies on the Interaction of Emulsifier and Starch`. *China Food Additives*, 6:59-63.

Huang Weifeng (2006). 'Exploitation of Type LL-4200 Product Line of Humid and Instant Noodle'. *Food Science and Technology*, 7:162-166.

Jiang Zhihong, Wu Ying (2003). 'The Present Situation and Development of Noodle products '. *Journal of Cereals & Oils*, 3:16-19.

Lan Jin, Wang Lekai (2006). 'Relation between Wheat Protein, Starch and Noodle Character'. *Grain Processing*, 4: 58-60.

Lei Changgui, Lu Daxin, Chen Jingpin (2006). Properties of wheat protein and noodle quality. *Food and Drug*, 2006, 8(6):27-30.

Lei Ji, Liu Zhongqi, Qin Wen (2003). 'Effects of protein, starch, hardness and color on wheat noodle quality'. *Southwest China Journal of Agricultural Sciences*, 16(4):122-125.

Li Lite (2000). 'Problems and Prospects of Noodle Processing Technology in China'. *Journal of Zhengzhou Institute of Technology*, 21(3):12-14.

Li Mengqin, Zhang Jian, Feng Zhiqiang, Lei Na, Chang Zhiwei (2007). 'Study on the Evaluation Indexes and Methods of Noodle Quality'. *Journal of Triticeae Crops*, 27(4):625-629.

Liu Minnian Liu Donghong (1999). 'Application of Composite Emulsifier to Fine Dried Noodles for Quality Improvement'. *Transactions of the Chinese Society of Agricultural Engineering*, 4: 264-268.

Liu Xingxin (2005). 'Noodles Industrial Profile and Trends in China'. *Science and Technology of Cereals, Oils and Food*, 13(6):24-25 Liu Yuanyuan, Lei Zhengsheng, Ling Zuoyi, Wang Ruifang, Yang Huiming, Lai Qingru (2004). 'Effect of Wheat Quality to Noodle Quality'. *Journal of the Chinese Cereals and Oils Association*,19(3):18-20.

Li Zhuowa, Wang Chun, Chen Jie (2006). The Application of Texture-tensile Test in Flour Quality Evaluation. *Grain Processing*, 2006(4):90-91.

Lu Qiyu (2005). 'Processing Technology of Cereals, Oils and Food'.Beijing : China Light Industry Press, pp 34-65.

M. I. P. Kovacs, B. X. Fu, S. M. Woods and K. Khan (2004). 'Thermal stability of wheat gluten protein: its effect on dough properties and noodle texture'. *Journal of Cereal Science*, 39(1):9-19.

Peng Zhiying, Zhao Mouming, Hu Kun (2002). 'Development of The Complex Additive for Noodles'. *Cereal & Feed Industry*, 9: 42-44.

Qiu Weifen (2002). 'Application of Enzymes to Improve the Flour Qulity'. *Food Science and Technology*, 3: 28-31.

Shao Xiuzhi, Yu Xiaoxia, Bian Jianhua (2002). 'Study on Improveing Flour Quality with Compound Enzyme Products'. *Science and Technology of Cereals, Oils and Food*, 2: 1-2

Shen Zaichun, Shen Qun (2004). ' New R&D and the application in production and development trend in instant noodles and vermicelli'. *Global Food Industry Information*, 11:26-28.

Shi Junling, Wei Yiming, Zhang Guoquan (2001). 'Study on The Effect of Protein and on noodles quality'. *Journal of Northwest A&F University(Natural Science Edition)*, 22(1):32-35.

Song Yazhen, Yan Jinting, Hu Xinzhong (2005). 'Wheat Flour Viscosity and the Relationship with Fresh Noodle and Cooked Noodle Texture Character'*Journal of the Chinese Cereals and Oils Association*, 6,12-14.

Sun Xiaodong, Shi Fengshan, Du Ping, Zhou Shuohen (2003). 'The Application of Linseed Gum in Noodle'. *China Food Additives*, 3: 64-66.

Tian Jiahua, Huang Kerong, Xiao Kai, Lei Min, Lu Xiaoli (2001). 'Application of Food Gum and Pregelatinization Technique in Processing Corn Noodle'. *Food Science*, 5: 34-37.

Wang Changfei, Ya Fei (1999). 'Shanxi Pasta'. Shanxi Science & Technology Press, pp23-45.

Wang HaiHong, Sun DaWen, Zeng Qingxiao and Lu Yinquan (2000). 'Effect of pH, corn starch and phosphates on the pasting properties of rice flour'. *Journal of Food Engineering*, 46(2), 133-138.

Wang Renxiang (2006). The Age of Noodle. *China Cultural Heritage*, 2006(1):75-79.

Wu Xuehui, Li Ling (1998). 'Improving Effects of Compound Phosphate on Noodle Quality'. *Cereal & Feed Industry*, 12: 43-44.

Xu Qiushui (1988).*Production Technology of Various Types of Noodle*. Beijing: China Food Press, pp 55~59.

Yang Mingzhe, Yu Yali (2002). 'Study on mode of noodle production and noodle characteristics'. *Journal of Harbin University of Commerce (Natural Sciences Edition)*, 18(6):657-660.

Yang Mingzhe, Yu Yali, Gao Feng, Jia Qingsheng (2003).' The Evolvement of Noodle Study in China and Foreign Country'. *Journal of the Chinese Cereals and Oils Association*, 18(1):1-4.

Ye Maoling (2006). Crack the Mystery of The Millennium Noodles. *Encyclopedic Knowledge*, 2006(4):46-49.

Yuki Sawatari, Hisashi Sugiyama, Yasuyuki Suzuki, Akihiro Hanaoka, Katsuichi Saito, Hiroaki Yamauchi, Sanae Okada and Atsushi Yokota (2005). 'Development of fermented instant Chinese noodle using Lactobacillus plantarum'. *Food Microbiology*, 22(6), 539-546.

Zhang Guoquan, Wei Yiming (1999). 'Study on The Relationship Between Flour Milling Quality and Noodle Quality'. *China Western Cereals & Oils Technology*, 24(4):39-41.

Zhang Guoquan, Wei Yiming, Ou Yangshaohui (2000). 'Study on The Relationship Between Wheat quality and noodle quality'. *Journal of the Chinese Cereals and Oils Association*, 15 (3): 5-8.

Zhou Xiuqing (1999). 'New Noodle Exploitation at Home and Abroad'. *Food Science and Technology*, 6:45-46.

Zhang Y, Nagamine T, He Z H, et al. (2005). 'Variation in qualitytraits in common wheat as related to Chinese fresh white noodle quality'. *Euphytica*, 141: 113-120.

Zhang Yan, YanJun, H. Yoshida, Wang Desen, Chen Dongsheng, T. Nagamine, Liu Jian-jun, He Zhong-hu (2007). 'Standardization of Laboratory Processing of Chinese White Salted Noodle and Its Sensory Evaluation System'. *Journal of Triticeae Crops*, 27(1):158-165.

Chapter 3

CHINESE RICE NOODLES

1. INSTRUCTION

Production and consumption of rice in China is the largest when compared to all countries in the world. It is reported that rice production and consumption respectively accounted for ~ 35% and 85% of the total amount in the world. Normally, most of the rice is consumed just after primary processing such as dehulling, milling and polishing, and only 6% of the consumption is after intensive processing. Rice products appearing on the market in China can be summarized as follows:

1. *Milled rice:* milled rice is produced by dehulling of paddy to brown rice, followed with milling of brown rice. Except for normal milled rice, there are also some specific products included, such as water-mated milled rice, cleaning free rice, colored nutritional rice, parboiled rice, embryo maintained rice, and so on.
2. *Processed rice products:* brown or milled rice is used as material for processing. Products such as germinated brown rice, rice crackers, fresh wet rice noodles, instant rice noodles, and dried rice noodles are included in this group.
3. *Intensive processing rice products:* starch and protein separated from rice belong to this group.

Rice noodle is a kind of striped or filamentary food made of rice with the processing procedures of soaking, steaming and molding, etc (Wang et al., 2004).With changes of consumption habits and an increase in comprehensive utilization of rice, rice noodles have become more and more important in recent years, because of its convenience and high utilization of material. Rice noodles are a kind of nutritious and tasty food. Because of advantages such as convenience, nutrition, and diverse flavors, rice noodles have been very popular both at home and in restaurants in southern China, Hong Kong and Macao for a long time. And now, with the development of the rice noodle industry and distribution of food, people could enjoy rice noodles all around the world. Rice noodles could utilize rice completely, and during rice noodle production, both intact rice kernels and broken rice could be well used for desired products. It was reported that in 1995, production of rice noodle reached 100 thousand tons in China (Deng et al., 2000).

1.1. Origin and History of Rice Noodles

The origin of rice noodles could be traced back to the Yuan Dynasty of China. It was recorded that when people from northern China invaded the south, they were not accustomed to eating rice. Since their preference was noodles made from wheat flour, they tried to make noodles with rice. This was the beginning of the rice noodle.

Now, there are various types of rice noodles on markets, and the most famous products are from southern China in the regions of Jiangxi, Fujian, Guangdong and Guangxi. Rice noodles from Jiangxi are famous for their specific flavor produced by fermentation. Guilin, Guangxi is proud of its traditional culture-related rice noodle known as Guoqiao Mixian. Fujian and Guangdong also are the regions where the history of the rice noodle originated. In Taiwan, the rice noodle from Xinzhu is the most well-known.

After hundreds of years' development, rice noodles can serve as both a staple food and a side dish, consumed as breakfast, lunch, dinner or side dishes. Edible qualities, such as taste, appearance, noodle-shape and convenience of consumption, also changed with the development of technology and science. Rice noodles thin as a silver filament or shaped like macaroni, nutritious rice noodles and instant rice noodles, which could be eaten directly just after a simple heating in boiled water, have all appeared on the market in recent years.

1.2. Classifications of Rice Noodles

Many criteria, such as processing technology, cultural history stories and names of regions are used for names and classification of rice noodles. Some criteria and their classification are summarized below.

1.2.1. Classification Based on Molding Methods

According to molding methods, rice noodles are grouped into cutting noodles and squeezing noodles (Table 3-1).

1.2.2. Classification Based on Production Region

Consumption of rice noodles is also affected by production regions, since there's some special processing, or cultural or historical background behind. Classification (or names) of the main rice noodles followed by regions is illustrated in table 3-2.

1.2.3. Classification Based on Shapes

Rice noodles can also be classified based on their shape (Table 3-3).

1.2.4. Classification by Eating Style

Rice noodles can be eaten directly after soaking in boiled water for several minutes, or after cooking. Table 3-4 showed the classification of rice noodles by eating style.

Table 3-1. Rice noodles classified by molding methods (Li et al., 2000)

Molding	Shapes of transect	Description of noodles	Moisture content* %
Cutting	rectangle	Thickness: 1 mm Width: 4-6 mm Length: 200 mm	High: 40-65 Low: 15-25
Squeezing	circle	Diameter: 1-2 mm Length: 200-400 mm Diameter: ~3 mm Length: 200-400 mm Diameter: 2 cm Length: 500 mm Diameter: 2-3mm Length: 200-400 mm (Light acidic because of fermentation)	High: 40-50 Low: 15-25 High: 40-50 Low: 15-25 40-50 40-50

* High or low used to distinguish whether the noodles are dehydrated or not.

Table 3-2. Rice noodle named by production regions (Liu et al., 2007)

Names	Production Regions
Guoqiao Mixian	Yunnan
Chencun Mifen	Shunde, Guangdong
Guilin rice noodle	Guangxi
Shatian rice noodle	Fujian
Rice noodles	Sichuan

Table 3-3. Shapes of rice noodles (Xing et al., 2007)

Groups	Shapes of final products	Processing
Wave-like	waves block	Mechanical processing
Pai rice noodle	Strip, rectangle and circle	By hand
Instant rice noodle	Stripes with rectangle transect; block	By hand

Table 3-4. Rice noodles classified by edible quality (Cen et al., 2007)

	Dry instant products	Normal dry products	Wet rice noodles
Samples	Instant Guoqiao Mixian; instant Hefen; instant water-washed rice noodle	Pai rice noodle; straight-strip noodle; wave-like rice noodle	Fresh Shahe noodle; Guilin rice noodle; Luoxiu rice noodle

1.2.5. Classification by Moisture Content

Moisture content affects the shelf life of the products, and consumption style as well. According to moisture content, rice noodles are classified into two groups—dehydrated and instant wet rice noodles.

In addition, there are still many other criteria, such as fermented rice noodles and unfermented rice noodles according to whether there is a process of fermentation.

Table 3-5. Types of rice noodles by moisture content (Deng et al., 2000)

Type	Description of products
Dehydrated rice noodles	Noodles dehydrated for prolong shelf life, which could be as long as 2 years. This product has good water rehabilitation and toughness.
Instant wet rice noodles	Fresh noodles without dehydration, which taste smooth and delicate, convenient for consumption.

2. MATERIALS FOR RICE NOODLES

2.1. Raw Rice

Rice is the most important material for rice noodles. Proper rice has to be selected for certain kinds of rice noodles. For example, rice with protein content > 7.1% is fit for instant rice noodle. Many components in rice, such as starch, protein, fat, water and minerals affect the quality of rice noodles. The relations between contents of chemical components in rice and sensory quality of rice noodles were systematically studied. Suitability and quality properties of 21 varieties of rice widely cultivated in Hubei and 14 *Xian* rice cultivated in Hunan to instant rice noodles were studied and the results indicated that amylose content, protein content and fat content affected the sensory quality of instant rice noodles (Liu et al., 2008; Zhao et al., 2002). Types have a significant effect on quality of instant rice noodles. Compared to Jing (japonica) rice, Xian (indica) rice is more suitable for rice noodle making [10]. The content of starch is positively correlated to color, taste, mouth feel and comprehensive evaluation of instant rice noodles. The content of amylose is also positively correlated to taste, while negatively correlated to color. The sensory quality of instant rice noodles can be well predicted by their content of starch, protein and fat since correlation between these indexes was well fitted to an exponential model (Liu et al. 2008).

Starch has the greatest impact, although other components also affect the quality of final products. Starch accounts for 70–80% of the total weight of rice (Li, 2005), which is the main component for the formation of a gel structure during processing of rice noodles. The content, solubility, swelling power, gel consistency, paste properties of amylose, and staling value of rice starch have different effects on the quality of rice noodles.

2.1.1. Content of Amylose

The formation and major structure of rice noodles are supported by the gel, where amylose plays an important role in the net when it absorbs water, swells and becomes sticky during the gelatinization of starch in rice [1]. Researchers indicated that molecules of amylose are feasible for processing and for the quality of final products (Zhao et al., 2002). It is reported that the content of amylose in rice is positively correlated with facilitation of processing ($r = 0.9552$), while significantly negatively correlated to the sensory quality of rice noodles ($r = -0.9149$) (Zhang et al., 2003). A higher content of amylose correlated with a lower rate of broken rice noodles. That result can be attributed to the amylose content. The changes of state of starch, from α state to β state during noodle formation, in turn enhances the strength of the rice gel and decreases the rate of breakage. Besides, with an increase in amylose content, the yield of rice paste and its maximum strain will also be improved, while the retro gradation rate decreases. Amylose content of rice was significantly correlated with the mouth-feeling attributes of rice noodles, which means that the amylose content can be an important evaluation index for rice noodles (Cheng et al., 2000). It was also reported that amylose content, solubility, expansibility and aging value of rice starch had remarkable influence on cooking loss, dehydration time, breakage rate and taste, and amylose content had the most significant influence on processing (Zhang et al., 2003).

The content of amylose in rice is mainly affected by the genes of the rice varieties, as well as growing conditions including the temperature, light, altitude, irrigation, and so on (Wang et al., 2004; Zhang et al., 2003). The content of amylose should be at the level of 20–25% for rice noodles, since this level will keep the yield of rice paste, stretch force and retro gradation rate at the desired level for a good and stable quality of rice noodles [11]. Other researchers proposed that a proper content of amylose for rice noodle making ranged from 23–28% (Ding et al., 2004).

2.1.2. Solubility and Swelling Power of Starch

Solubility of starch is the percentage of starch—dissolved in water under a certain temperature—to total starch. Swelling power is defined as the quantity of water absorbed by 1 g of dry starch under a certain temperature (Liu et al., 2008; Zhang et al., 2003). Solubility and swelling power under 25 ºC can be determined as follows.

Make a starch mixture (2%) using water and shake for 30 min.; then, 50 mL of the mixture is taken and centrifuged at 3000 rpm for 15 min. The supernatant is evaporated in a water bath and dried for analysis of solubility of starch. At the same time, sediment is also collected for swelling power.

$$S\,(\%) = (A/W)*100\% \qquad (3\text{-}1)$$

$$B = P*100/W\,(100-S) \qquad (3\text{-}2)$$

Where S is solubility, A is soluble starch (g, dry matter), W is total starch content (g), B is swelling power, P is dry matter of sediment. In general, high solubility and swelling power always relates to poor quality, large cooking loss, high breakage rate, turbidity of cooking water and tough sensory quality of rice noodles. It was reported that proper swelling power for rice noodles ranged from 8 to 9 (Ding et al. 2004)

2.1.3. Gel Consistency and Gelatinization Temperature

Gel Consistency is a characteristic of rice starch colloid, which indicates the staling tendency of rice starch in the process of heating (gelatinization) and cooling (Li, 2005; Cheng, 2000; Sun et al., 2004). It is expressed by the ductility of cooling rice endosperm starch colloid under the concentration of 4%, so it is a simple but exact indicator to reflect the characteristics of gelatinization of rice. Gel consistency is mainly related to contents of amylose and effect of integration between amylose and amyl pectin.

Gel consistency is determined as follows:

- Prepare rice powder by grinding the polished rice (mainly rice endosperm) till it fits a 0.15 mm sieve followed with adjusting moisture content to ~12%;
- Take ~ 0.1 g (accuracy 0.0001 g) powder in tube and mix well with 0.2 mL indicator (thymol-sulfonphthalein), then add exactly 2.0 mL 0.2 M potassium hydroxide and mix well again;
- Cover the tube with glass ball and heat it with boiling water bath for 8 min, during which keep the height of starch gel as 2/3 of tube length;
- Cool the tube under ambient for 5 min, then cool in icy water bath for 20 min, followed with laying the tube on horizontal platform in 25±2°C for 1 h;
- Gel consistency is the length (mm) of gel at the bottom of tube.

Gel consistency less than 40 mm is named a tough gel consistency rice variety, higher than 60 mm is a soft variety, in the range of 40–60 mm is a medium variety [18]. Proper gel consistency for rice noodles is in the range of 35–55 mm, neither too high nor too low (Ding et al. 2004). High gel consistency corresponds to better fluid and weaker strength of the rice slurry, which will cause a higher breakage rate of noodles. Low gel consistency relates to weak toughness and an unsmooth taste of rice noodles. Gel consistency is significantly correlated to low viscosity, final viscosity and setback values. Lower gel consistency is relate to higher contents of amylose, higher low viscosity, final viscosity and setback values, which means better processing quality.

Gelatinization is the response, such as swelling and forming an even paste of starch, of moisture and high temperature. Gelatinization temperature of rice starch is significantly related to cooking quality of rice.

Gelatinization temperature can be determined according to the standard method (NY-147, China Ministry of Agriculture) as follows:

- Put 6 mature, full and polished rice kernels in a box and add 10.0 mL 1.70% (m / V) of potassium hydroxide;
- Evenly distribute the kernels in the box with a glass rod and cover the box;
- Carefully move the box into an incubator and keep it at 30 ± 2°C for 23 h;
- Record and grade the decomposition of rice endosperm of each kernel. Gelatinization temperature is related to this alkali spreading value (Table 3-6).

Table 3-6. Grade of alkali spreading value of rice

Grade	Kernels Appearance	Clearness Kernel
1	No visible change	White central
2	Swelling	White central; chalk ring
3	Swelling; half-circled or thin ring around kernel	White central; cloudy ring
4	Swelling; complete round and wide ring around kernel	Milk-white central; cloudy ring
5	Crazing; complete round and wide ring around kernel	Milk-white central; clear ring
6	Partly dissolved kernel mixes with ring	Cloud-white central; ring disappeared
7	Whole kernel dispersed completely	Central and ring disappeared

Table 3-7. Gelatinization temperature of rice

Group	Grade of alkali spreading value	Gelatinization temperature m(°C)
High	1~3	> 74
Medium	4~5	70-74
Low	6~7	< 70

The alkali spreading value of rice is expressed as

Alkali spreading value = $\sum$ (G*N) /6 (3-3)

Where G is the level of each rice kernel, N is the numbers of rice kernels in the same grade.

When measuring the gelatinization temperature, standard (including high, medium, and low gelatinization temperature materials) should also be prepared for inner mark, grade differences between the measured and the standard value must be lower than 0.5.

According to alkali spreading value, the gelatinization temperature is divided into three groups—high, medium and low (Table 3-7).

It is clear from table 7 that rice starch gelatinizes at lower temperatures when alkali spreading value is at higher levels, which relates to strong stickiness of rice, and will deduce poor cooking properties and high breakage rate of noodles. It was reported that the proper range of alkali spreading value of rice for noodles is 3.0–4.5 (Wu et al., 2005).

Correlation of the content of amylose with mechanical properties of rice were also studied (Table 3-8) (Ding et al. 2004).

Amylose content, gel consistency and swelling power are important in choosing materials for rice noodles. However, the content of amylose should be considered first [14].

Table 3-8. Correlation of amylose content and mechanical properties

	Amylose	*Gel consistency*	*Swelling power*	*Shear Force*
Amylose	1			
Gel consistency	-0.690*	1		
Swelling power	-0.699**	0.700*	1	
Shear Force	0.899**	-0.805**	-0.822**	1

** significant at 0.01.
* significant at 0.05.

2.1.4. Staling

The phenomenon of gelatinized α-asylum becoming opaque, and even precipitation occurring when stored at room temperature or below, is known as staling (starch aging). This can be interpreted that at lower temperature, the gelatinized molecules of starch will automatically rearrange in sequence as before gelatinization, hydrogen bonds among adjacent molecules gradually occur, and highly dense and crystallized starch molecules will also form as a result. Staling of starch is determined by degree of aging, which is expressed by the weight of separated water from contracted and dehydrated starch gel after centrifuging (Li, 2005; Ding et al., 2004; Jin et al., 2004).

A detailed measurement procedure is as follows:

- Heating 100 mL starch slurry (6%, take 6 g starch and fill to 100 mL) in boiling water bath for 20 min (during heating, keep the concentration at 6%);
- Keeping a certain quantity of heated slurry in the refrigerator 2°C for 24 h;
- Centrifuging stored slurry by 3000 rpm for 15 min;
- Value of aging is calculated with water isolated from centrifuged slurry.
- High aging value of rice is related with high cooking loss, high breakage rate and viscous texture of final rice noodles.

2.2. Additives and Modifiers

In rice noodle production, some additives and modifiers are also applied to improve the sensory quality of products, extend the preservation period, or facilitate the processing. Various kinds of additives and modifiers can be applied, and the methods and ratios of addition should be varied with the nature of the raw rice, processing conditions, and properties of the additive itself. At present, additives relatively widely used are listed below.

2.2.1. Salt

Salt, of which sodium chloride (NaCl) is the main ingredient, is the most commonly used additive in the production of rice noodles. It is used by some manufacturers for its positive effect on dehydration of noodles, although the mechanism is still unclear. Addition of salt should be about 1% (Jiang et al., 2001). Salt could be mixed with ground rice as a solid, or as a solution. Proper addition ranges from 0.1–0.5% with adaption of seasons. Excessive addition will make noodles crispy, or easily moistened in humid seasons (Fu, 2000; Peng, 2000).

2.2.2. Compounded Phosphates

Main compounded phosphates applied in rice noodle production are sodium hydrogen phosphate (Na_2HPO_4) or sodium pyrophosphate. Both chemicals are a white powder, easily dissolved in water, and can be used as nutrient enhancers. Compounded phosphates strengthen the connection and integration between starch molecules by the promotion of soluble starch leakage with increase of temperature. And at the same time, phosphate ions could make bigger starch and protein molecules by its chelating capacity, which will improve the toughness of rice noodles (Jiang, 2001; Fu, 2000; Peng, 2000; Fu, 2000). Compounded phosphates can improve the quality of rice noodles by increasing exquisiteness and toughness, reducing breakage rate, and possibly increasing the luster. Compounded phosphates could be added by 0.1–0.4% mixing with rice powder after dissolving in cold water. Excessive addition will result in a light yellow or yellow color of rice noodles.

2.2.3 Distilled Monoglycerides (DMG)

Distilled monoglycerides, light yellow wax-like solid, is a common emulsifier. It does not dissolve in water, but could mix evenly with hot water after strong oscillations. It exists as crystal form B, which is inactive and difficult to change to active state (crystal form A) at ambient temperature. However, after heating, DMG could change to form A, combine with starch and protein, and achieve improvement of mechanical and sensory quality of the rice noodles. It is reported that addition of DMG could bring about the production of an even emulsified layer around the granules, which effectively reduces the dissolving of starch, decreases stickiness by rapidly close passage for water absorption, and prevents water penetration into starch. Besides, the formation of complexes of monoglycerides and amylose is an irreversible reaction, which is significantly helpful in prevention of retrogradation of instant rice noodles and shortening the rehydration time (Fu, 2000; Peng, 2000; Deng et al., 2001; Fu, 2000; Fu et al., 2003).

DMG could be applied as follows: steep DMG with cold water till swelling after enough water absorption, and then heat the mixture to paste followed by mixing with ground rice. The proper addition is 0.3–0.6% because excessive addition will make the noodles yellow and result in poor exquisiteness.

2.2.4. Starch from other Sources and Modified Starch

Starch from other sources—potato, cassava, maize, sweet potato, modified starch, etc.—are also used in rice noodle processing for the facilitation of processing and improvement of the edible quality of final products. It was reported that mixing *Xian* rice with starch from cassava (ratios as 3:1) could decrease the content of amylose from 23.68% (in Xian rice) to 18.19% (in mixture), and consequently improve toughness and smoothness of the final rice noodles (Zou et al. 1999). Jiang and Liu (1999) compared the effects of starch from different sources on the quality of rice noodles (Figure 3-1). They concluded that addition of starch from maize could facilitate rice processing by an increase in the content of amylose in the slurry, while addition of starch from cassava would not affect the processing since amylose in cassava was at a similar level as rice. However, starch from cassava could improve sensory quality including mouth feeling and appearance (Jiang and Liu, 1999).

2.2.5. Sulfite

Sodium pyrosulfite and sodium sulfite, known as the sulfite, are white granular powders with smells of sulfur dioxide. They are the main chemicals used in rice noodles for whitening final products [21], because these chemicals could release sulfur dioxide under the acidic conditions and have the effect of bleach. Sulfite could be added in the soaking water (0.5%, sulfite), or mixed with the rice slurry (0.5%, sulfite); both methods should adjust pH to acidic levels with acetic acid. After sulfating, kernels or noodles should be treated with clean water to make the remaining sulfur dioxide less than 20 mg/kg.

2.2.6. Acetine (Acetic Acid)

Acetine (containing 5–30% acetic acid) was mainly used to adjust the pH value during processing and loosen the structure of noodles (Fu, 2000). It was reported that staling of noodles would be accelerated with the increased acidity.

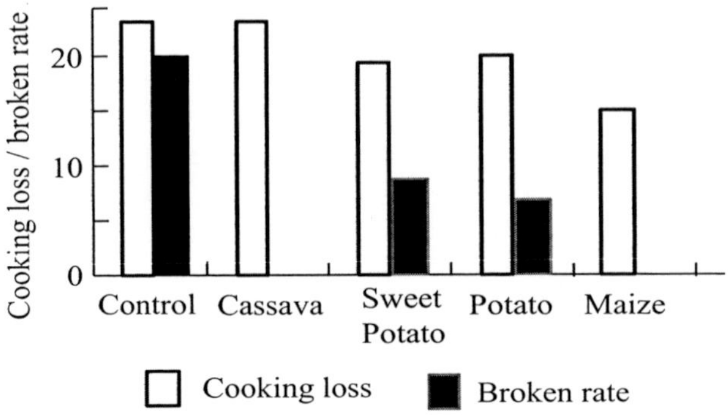

Figure 3-1. Effect of different starch on cooking loss and broken rate of rice noodle (Jiang and Liu, 1999).

2.2.7. Other Additives

Other additives such as wheat gluten, propylene glycol, konjaku flour (Cen, 2007) are also used in rice noodle processing.

3. Processing Procedures

In China, rice noodles are mainly produced in small-scale factories and workshops. Pasting of rice slurry and molding of noodles are often done by machines in factories, while traditionally, they were done by hand or semi-mechanization in workshops. Generally, procedures of rice noodle making consist of selection of raw materials, ingredients adjustment, cleaning, soaking, pasting, molding, steaming/boiling, cooling and drying. Sequences and conditions of each step vary with the noodle styles and quality necessary for final products.

For different types of rice noodles, processing procedures are also changed. Procedures for instant wet rice noodles and dried rice noodles are given. Instant fresh wet rice noodles becomes more and more popular because of their smooth taste, convenient consumption and relatively long shelf life. Dried strip rice noodles are consumed world wide for their easy transportation and long shelf life; they can be stored for longer than two years at room temperature.

In Figure 3-2 and Figure 3-3, it is clear whether the final products are dried or wet, or whether they are from cutting or squeezing. The procedures include material selection, cleaning, soaking, slurry adjusting, molding, steaming and cooling or dehydration.

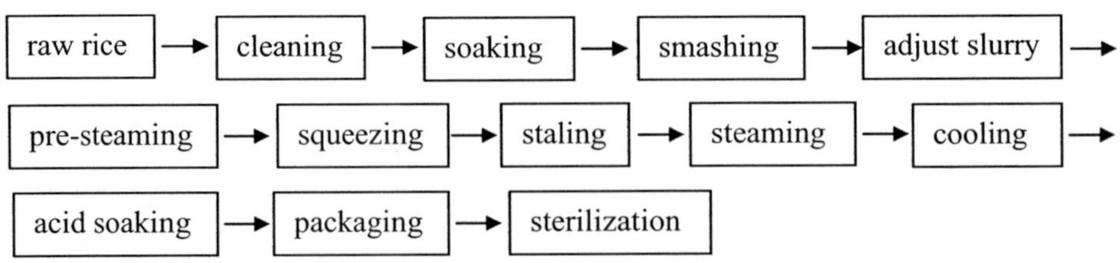

Figure 3-2. diagram of instant fresh wet rice processing (Ye et al., 2005).

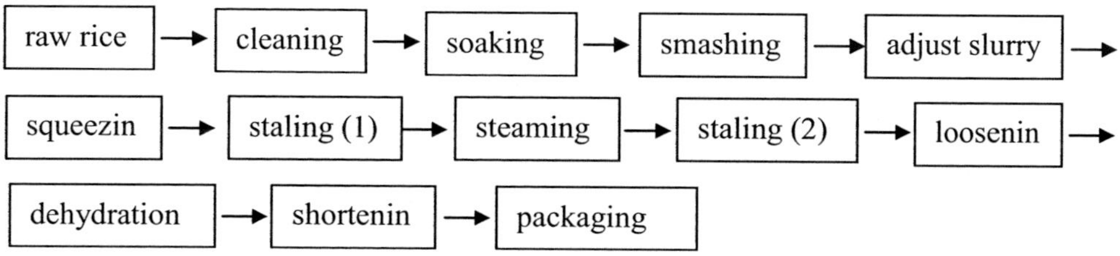

Figure 3-3. diagram of dried strip rice noodle processing (Pan and Deng, 2002).

3.1. Selection and Pre-treatment of Raw Rice

3.1.1. The Selection of Raw Material

Components and characteristics of raw materials significantly affect feasibility of processing and quality of the final rice noodles. Studies on the effect of rice on quality of rice noodles have shown that good quality could be obtained from raw material with high amylose content and high protein content [9]. According to harvest time, *Xian* rice can be grouped as early-harvested and late-harvested. Compared to early-harvested rice, late-harvested rice is more expensive and stickier, and therefore more difficult to process, and this leads to a low yield of noodle production, but gives better a edible quality [24]. In order to achieve an economic purchase of materials and assure the quality of the final rice noodles, a mix of early- and late-harvested *Xian* rice are normally used. Researchers proposed different ratios of the mixture, such as ranges from 1:0.25–0.67 (Pan and Deng, 2002), at 1:1.5 (Fu, 2000), ratio was 7:3 (Chen, 2002), determined by comprehensive consideration of price, processing and sensory quality of noodles.

Take the fermented rice noodle as an example. Yield of slurry, viscosity of preheated paste, rehydration rate of dried noodles and taste on toughness of noodles are the main factors considered for ratios of early- and late-harvested *Xian* rice. It is reported that stickiness of paste with high proportion early-harvested rice is suitable for processing, while high proportion of late-harvested rice will lead to high rehydration rate and soft toughness of noodles.

3.1.2. Cleaning

In order to achieve the whiteness and lucence of the final products, bran powder, dust adhered on the kernels, and other foreign matter should be removed by cleaning, which is commonly carried out by the combination of desanding and water-jet washing machine. At present, a three-step water-jet washing machine is widely used, and four-step water-jet washing machine is also introduced to improve the efficacy of cleaning (Pan and Deng, 2002)·

A fluid bed cleaning system also can be used (Figure 3-4).

Cleaning water can be recycled in this system and some ingredients, such as starch, could be collected in the sediment tank.

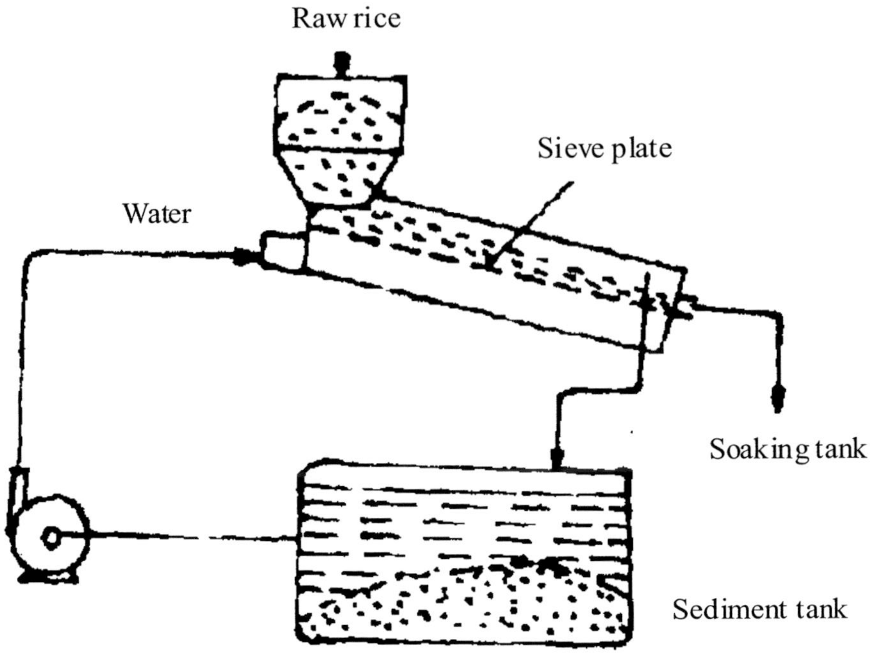

Figure 3-4. Fluid bed cleaning system (Wu and Li, 1998).

3.2. Soaking (Steeping)

3.2.1. Soaking

Soaking enables water to penetrate into the central part of the kernel, resulting in even distribution of moisture and loose structure of rice kernels, which will affect starch gelatinization and processing feasibility (Cen, 2007). Water absorption depends on soaking procedure and tightness of rice kernel. Soaking procedure should be adjusted with treating and environmental conditions. Effect of soaking time (at 40°C) on mechanic quality of rice noodles is in table 3-9.

Normally, when soaked at room temperature, rice should be soaked for 2–3 h in summer and 4–5 h in winter till the moisture content of rice reaches 26–30% (Cen, 2007; Fu, 2000; Ma and Lin, 2003). After soaking, rice should be kept in the air for 30 min to dry the surface of the kernel.

Table 3-9. Soaking time on stretch properties and hardness of instant fresh wet rice noodle (Ye et al., 2005)

Soaking time/h	0	4	8	12	16	18	24
Stretch properties/cm	7.3	7.6	7.9	8.2	7.9	7.5	7.3
Hardness/kfg	89	75	69	66	61	56	51

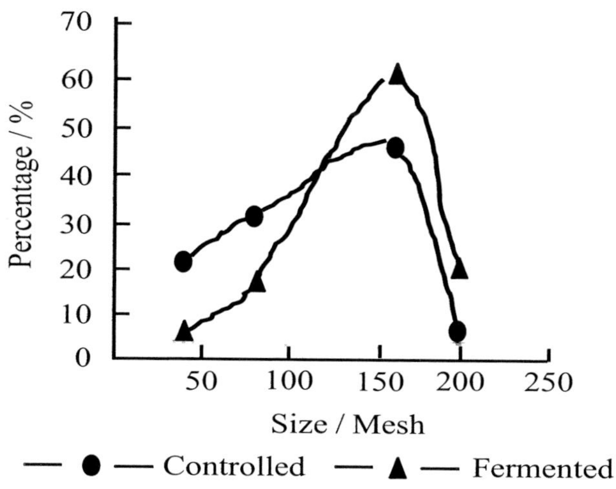

Figure 3-5. Distribution of size of granules after smashing (Ding and Wang, 2005).

3.2.2. Traditional Soaking (Natural Fermentation)

Traditionally, soaking is essential for rice noodles. It is not only a procedure for water absorption, but also behaves as a process of natural fermentation. It will take 2–3 d in summer and 4–5 d in winter, respectively. Soaking (fermentation) will soften and loosen the kernel structures, consequently, kernels will be fully saturated and could be smashed by hand, and the percentage of smaller size granules in fermented rice apparently is higher than the controlled after the same smashing procedure (Ding and Wang, 2005). Distribution of size of granules with or without soaking at room temperature for 4 d after the same smashing procedure is illustrated in Figure 3-5.

Meanwhile, with progress of the natural fermentation (mainly lactic acid fermentation), the acidity of the soaking liquid increases apparently, with the pH value decreasing from neutral (~7) at the beginning to a strong acidic condition (~ 3.5).

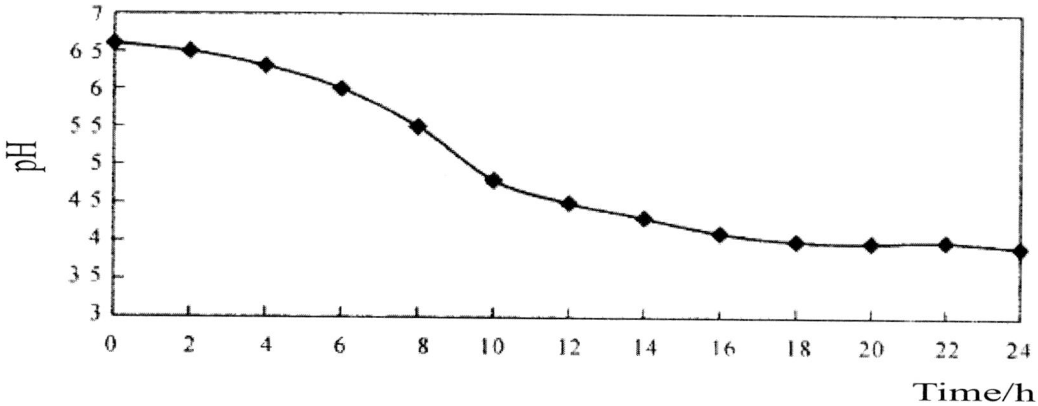

Figure 3-6. Changes of pH during fermentation (He and Sun, 2005).

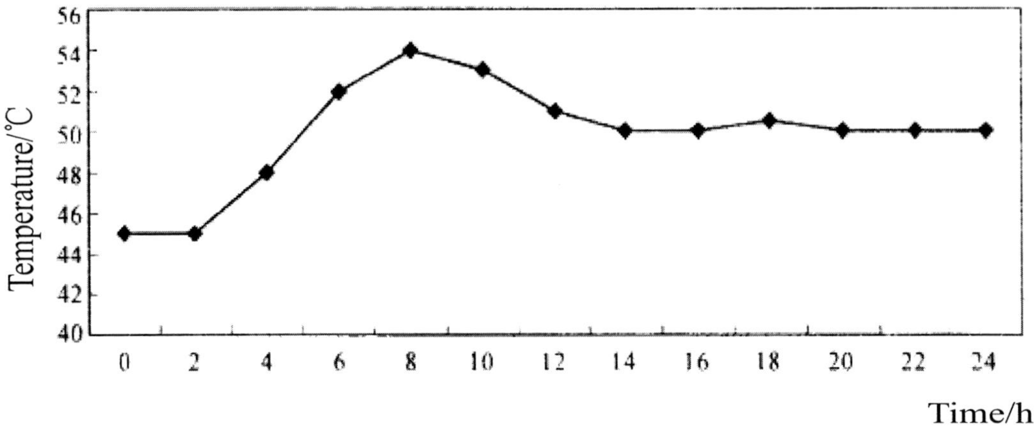

Figure 3-7. Change of temperature during fermentation (He and Sun, 2005).

In order to make full use of the advantages of natural fermentation, the natural fermentation was optimized to accelerate the natural fermentation, in which rice kernels are soaked at 45°C for 1 d. Changes of pH value and temperature of soaking medium during accelerated fermentation are presented in figure 6 and figure 7 respectively.

According to the results in figure 1, whole fermentation can be divided into three phases: 1) from beginning to 4 h, pH values remain stable at the level above 6; 2) from 4 h to 16 h, pH values decreased rapidly from 6.3 to 4.1; 3) from 16 h to the end (24 h), pH values remain stable at ~ 4.

Similar to changes of pH, changes in temperature also can be divided into three phases: 1) from beginning to 8 h, the temperature of the system increased rapidly from 45°C to 54°C (the highest point of the whole process); 2) in the duration of 8–16 h, temperature of the soaking system goes down gradually to 50°C; 3) from 16 h to the end (24 h), temperature of the system remained stable at about 50°C.

Here, soaking is also a process of natural fermentation, which is dominated by lactic acid bacteria existing in raw rice, water and/or the environment. Changes of temperature indicated that natural fermentation is a procedure that releases energy. Lactic acid fermentation and accumulation of energy during soaking (fermentation) are the main reasons for microbial safety of noodles and stability of temperature during soaking in traditional rice noodle processing. From figure 6 and 7, we can see that in accelerated fermentation, at the beginning 10 h, since lactic acid bacteria grow well under suitable temperature and abundant nutrients in soaking medium, heat and metabolic substances accumulate rapidly, which leads to increase of temperature and decrease of pH value. However, higher temperature and acidity will in turn inhibit the growth of microorganisms. Balance of accumulation of acid and energy will finally maintain temperature and acidity at stable levels. Furthermore, organic acids produced and accumulated in fermentation play an important role in the latter procedures and in the preservation of rice noodles (He and Sun, 2005).

Except for the changes in temperature and in the pH of the soaking medium, the physical and chemical properties of rice also changed with natural fermentation (Table 3-10).

Table 3-10. Physical and chemical changes after natural fermentation (4 d) (Ding and Wang, 2005)

	Controlled sample	Natural fermented sample
Protein contents (%)	8.26	6.58
Starch content (%)	89.2	91.0
Amylose in rice noodle (%)	28.2	29.5
Amylose in rice starch (%)	32.6	33.0
Soluble sugar (%)	0.42	0.12
Swelling power	8.27	8.55
Gel consistency	34	25

From the results, the changes of chemicals in rice were small after fermentation (decrease of protein mainly caused by dissolving). However, gel consistency, which means gel characteristics of starch, changed significantly after the treatment of natural fermentation. Lu et al. also studied the effect of natural fermentation on chemical components of rice (table 3-11) and concluded that fermentation did not have significant effect on degradation of starch (Lu et al., 2002a; Lu et al., 2002b).

Table 3-11. Chemical components of the samples before and after natural fermentation (Lu et al., 2002b)

	Fermented (27h)	Control (3h)
pH	4.0	7.7
Total titrate acid (g 1000mL-1, lactic acid)	1.062	0.004
Reducing sugar (mg L-1, in medium)	460.8	4.3
Reducing sugar (%, in rice kerne)l	2.48	0.76
Apparent amylose (%)	20.9	21.6
Total starch (%)	87.1	87.3
Protein (%)	5.2	7.3

3.3. Smashing

Soaked rice with appropriate moisture content goes into a smashing (some for grinding) procedure. According to technology, there are two different types of smashing.

Table 3-12. Size of ground rice on quality of rice noodle (Jiang and Liu, 1999)

Size (sieve mesh)	Cooking loss (%)	Broken rate (%)
55	Difficult to be molded	
70	25.1	20.0
80	22.4	13.3
100	21.1	6.7
120	19.8	0

3.3.1. Dry Grinding

Rice soaked with a normal procedure needs dry grinding, where the hammer milling machine is widely used. The size of ground rice powder is positively correlated to the quality of the final noodles. However, the yield of rice noodles also needs to be taken into consideration. The yield of rice noodle is negatively correlated to the size of the ground rice; the finer the ground size, the lower the yield of final rice noodles. Proper processing (considering both yield and quality of rice noodles) is grinding till smashed rice fits the sieve of 60 or 80 mesh (Jiang and Liu, 1999; Pan and Deng, 2002). Moisture content of soaked rice should be controlled in the range of 26–28%, since higher moisture may lead to agglomeration of ground rice, or to plugging of the pores of the sieve. Jiang et al. studied the effect of size of grinding rice on cooking loss and broken rate of rice noodle (table 3-12).

3.3.2. Wet Smashing

Rice soaked with a traditional procedure, or accelerated fermentation, will be treated with wet smashing (slurry making), during which water will be added and a wheeled-grinder is widely used. Moisture content of slurry made from naturally fermented rice after wet smashing should be controlled at ~45%. For some processing with separated fermentation, strains will be added in the press filtration of slurry after wet crashing and the moisture content in filtration should be kept at ~ 42%.

3.4. Mixing and Moisture Adjustment

Rice powder or slurry will be obtained after grinding or smashing of soaked rice kernels. Other ingredients for different demands, such as modified starch, DMG and salts, will be mixed. And moisture content in the material obtained from dry grinding are often lower than what's proper. Mixing procedure, namely stirring the adjusted materials with a horizontal dough mixer for more than 30 min till all materials could be a lump by hand, but not too tight. Proper moisture content of the mixture is 38–40% (Cen, 2007; Zou et al., 1999).

In order to improve the sensory quality of rice noodles, some ingredients are mixed with smashed rice. Several additives and their effect on physical characteristics and sensory quality of rice noodle are listed in table 3-13.

Table 3-13. Effect of some ingredients on sensory quality of noodles (Cen, 2007)

Additive	Broken rate %	Yield of slurry %	Sensory score (total 100)
Phosphated distarch phosphate (PDSP)	7	5	95
Carboxymethyl cellulose sodium (CMC-Na)	9	8.7	80
Starch (maize)	8.5	7.5	85
Control	-	-	70

Ingredients affect physical characteristics and sensory quality of rice noodle significantly. Addition of PDSP apparently make noodles smooth and tough like the typical flavor and physical quality of traditional rice noodles because of its effect of strong connection.

3.5. Fermentation

Ground rice will be mixed with added strains and fermented with controlled speed and extent. Temperature and time will be controlled during fermentation because excessive fermentation will lead to higher acidity and viscosity of slurry, which may cause the process to become out of control. However, insufficient fermentation will make a low yield of slurry, lead to noodles being sensitive to cooking, and have a sensory quality not as smooth as after fermentation.

3.6. Molding of Rice Noodles

Molding of rice noodle is categorized into two types—squeezing and cutting.

3.6.1. Squeezing Rice Noodles

Squeezing is an important method of rice noodle molding. A certain shape of rice noodle is obtained when molded by squeezing machinery after a mixed rice slurry is pre-steamed till it forms dough having a certain viscosity and hardness. Mixed materials and ingredients are pre-steamed for the starch's proper temperature and gelatinization suitable for squeezing. Mixed ingredients containing fermented rice is pressed into blocks (*Pizi*) of 30 mm thick and pre-steamed in a tunnel continuously for 4–5 min at 90°C. Normally, *pizi* is kneaded several times in the duration of pre-steaming for prevention of uniformity of heating. It is also very important to control the extent of pre-steaming since excessive pre-steaming will cause the noodles to be connected together, and insufficient pre-steaming will lead to a high breakage rate of the noodles.

At present, the ideal squeezer has the function of self-cooking with a dual-drill. Materials feed into the machine evenly and continuously, after heating properly in the drills, they enter the spiral and are squeezed into the noodles. Size and shape of rice noodles can be adjusted with demands.

3.6.2. Cutting Rice Noodles

Molding by cutting is mainly applied for instant fresh rice noodles. Cutting noodles are obtained as follows: mixture of ground rice and added ingredients is steamed for a sheet of a certain thickness, and then the sheet will be cut into noodles with certain blades following the needs of consumers.

3.7. Steaming / Boiling

The role of steaming (boiling) is to make rice noodles completely gelatinized to the α-state in rice noodles, and as a result decrease the leakage of dry mass during cooking. According to working pressure in the cabinet, steaming can be divided into two types—common-pressure and high-pressure (higher than 0.05–0.06 MPa) steaming (Pan, 2007). Steaming normally takes place in a steaming cabinet in batches. Working pressure in the cabinet of common-pressure steaming is the same as air pressure, so it takes some time to fulfill the effect of steaming. High-pressure steaming has higher pressure and temperature (above 100°C) in the cabinet so it can shorten the steaming period and the treatment is more efficacious (Pan and Deng, 2002). Steaming also can be continuously carried out in a tunnel by being heated directly with steam for 3 min.

Boiling also could achieve the same role as steaming for gelatinization of starch. During boiling, temperature and time must be strictly controlled to avoid the excessive gelatinization. Boiling can be treated at 98°C for 1–2min, together with the addition of salt and additives, which are used to eliminate foams.

3.8.Cooling

Steamed or boiled rice noodles go into a cooling water (0–10°C) to decrease their temperature rapidly to 24–26°C for constringing and tough gel. And at the same time, removing starch granules on the surface of rice noodles will make noodles separate well, and reach a smooth and slippery sensory quality. Normally, cooling will take 1.5–2.5 min. Cooled noodles can be sold in retail as instant rice noodles, or be packaged for long-shelf life fresh wet noodles.

3.9. Drying and Packaging

Rice noodles are dried with a procedure using high temperature and high moisture at the beginning phase, and then gradually going to low temperature and low moisture until the proper moisture content is reached. Presently, two different types of drying equipment—continuous drying cabinet and drying room—are widely used.

Indirect steam is used for the heat exchange system in the continuous drying cabinet, and heat exchange will happen between the heating system and the air in the cabinet. Air in the cabinet is heated via radiant heat exchange and cycled in the cabinet by fan. In order to achieve satisfactory quality, rice noodles will be overturned one or two times manually during drying. A continuous drying cabinet consists of at least two separate cabinets (Figure 3-8), and in between there is the transport chain open to facilitate the overturning manually.

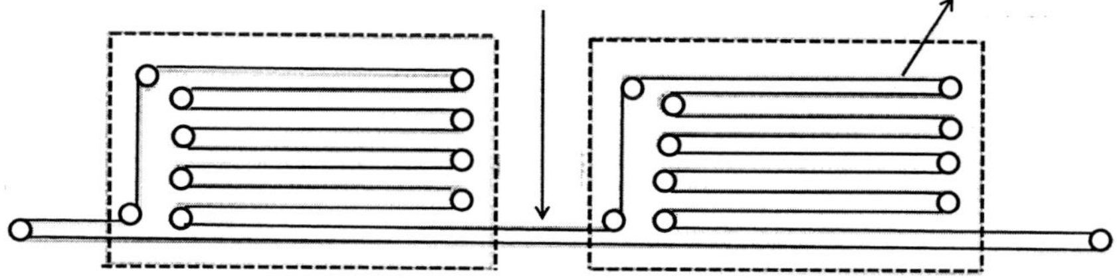

Figure 3-8. Sketch of continuous drying cabinet.

The drying room is another widely used drying system for rice noodles. Gradient of temperature and moisture in the room should fit the necessary technological requirements of drying of rice noodle throughout the entire processing. Rice noodles produced by a self-cooked squeezing machine tend to crisp after drying because of its dense structure (Pan and Deng, 2002). Only drying with a low-temperature, long-time procedure—drying under 50°C for more than 8 h—could noodles reach a good quality, including white color, strong transparence, high uniformity and no apparent crisp.

It was also reported that rice noodles could be dried with hot air for removal of water in the outer layer of noodles and determination of shape of the noodles, followed by dehydration with microwaves. After dehydration with a combination of hot air and microwaves, the breakage rate and cracks in noodles would decrease, while rehydration would improve (Zao et al., 1999).

After drying, the noodles will be cooled to room temperature in time. Then rice noodles will be shortened and packaged to reduce the evaporation and to prevent crisping. In order to reduce the cut powder debris, a round disc with small saw tooth and a thin blade is generally used. After shortening, rice noodles will be packaged using plastic bags or paper containers.

3.10. Packaging of Fresh Wet Rice Noodles

Fresh wet rice noodles tend to deteriorate at ambient temperatures because of their high moisture content (50–80%) (Cen, 2007). Most products with high moisture content are preserved by application of preservatives, which consumers would be aware of. Some researchers studied the effects of application of organic acid, thermal treatments and a combination of two methods on storage of fresh wet rice noodles.

3.10.1. Pretreatment and Package

The cooled fresh wet rice noodles will be soaked in acid to decrease the pH value (4.2–4.3) and to extend the shelf-life. Depending on the form and texture, rice noodles will be soaked in acidic medium with acidity ranged 0.3–2% (by lactic acid, pH 3.8–4.0) under 25–30°C for 1.5–3 min. After soaking, the noodles will be separated from the soaking medium and free liquids on the surface removed to prevent water absorption and texture swelling in the following high-temperature processing. Normally, the noodles should be filtered for 8–10 min to keep the moisture content in the final product at 65–68%. Rice noodles will be low-vacuum packed together with 3–4 drops of plant oil, which can prevent the connection and sticking of noodles.

3.10.2. Sterilization and Prevention of Microbial Growth

Packed fresh wet rice noodles need sterilization or antimicrobial treatments for long-period preservation. Treatments of thermal sterilization, chemical anti-corrosion and combined treatments are different in principles and processing technology.

1) Thermal Sterilization

Thermal sterilization is different in heating medium as in Table 3-14.

Table 3-14. Comparison of different sterilization (He and Sun, 2005)

Methods	Condition	Shelf-life
High pressure & temperature (steam)	$\frac{10\text{ min - }20\text{ s - }10\text{ min}}{121}$	>3 months
Boiling	Boiling for 60 min	>3 months
Pasteurization	Water heated, 95°C 35-40min	~30 days
	Water heated, 95°C 35-40min	~2 months

The temperature of the cold point in the package should reach 92°C and keep for 10 min in thermal sterilization, followed with cooling down to room temperature rapidly with a fan.

2) Antisepsis with Organic Acid

In order to preserve rice noodles with controlled environment, pH of rice noodles can be adjusted with organic acids, such as acetic acid, lactic acid, ascorbic acid and their relevant salts. Sensory quality of rice noodles preserved at 37°C for a certain time after treatment with acid and salt for 1 min are presented in Table 3-15.

Table 3-15. Sensory quality of rice noodles treated with acid and salt (He and Sun, 2005)

Organic acid	*Concentration (v/v, %)*	*Store time (d)*	*Sensory quality*
Acetic acid	1	4	Strong smell and sourness
Lactic acid	1	3	Apparent sourness, mild smell
Ascorbic acid	1	2	Apparent sourness, dark color by oxidation
Lactic acid + sodium lactate	1	15	Apparent sourness
		15	Apparent sourness
		12	Sourness
		7	Weak sourness
Lactic acid + sodium acetate	1	15	Apparent sourness
		14	Apparent sourness
		13	Sourness
		8	Weak sourness

The preservative effect of organic acids orders from strong to weak as: acetate> lactate> ascorbic acid. Effect strengthens when acid applied together with relevant salts. Lower pH is related to better preservation and worse edible quality. For soaking medium of lactic acid-sodium lactate, preservation could prolong up to 12 d with little noodle sourness when pH is adjusted to 4.12. In addition, effect of lactic acid-sodium acetate system is better than that of lactic acid-sodium lactate system.

Table 3-16. Effect of different combination (He and Sun, 2005)

Sterilization		*Bacteria growth*	*Mold growth*
control		+++	+++
Lactic acid + sodium lactate 1% (V/V) 95° C 35 min	pH4.5	++	++
	pH4.2	-	-
	pH4.0	-	-
Lactic acid + sodium acetate 1% (V/V) 95° C 35 min	pH4.5	-	++
	pH4.2	-	-
	pH4.0	-	-

3) Combination of Acid Treatment and Thermal Sterilization

Effect of combined treatments of acid and thermal sterilization (95°C, 30 min, pasteurization) on growth of microorganisms after incubating at 37°C for 3 months is presented in Table 3-16.

It is apparent that a combination of pasteurization and acid can effectively inhibit the growth of microorganisms, and the fresh wet rice noodle could be preserved for longer than 3 months with acceptable sensory quality.

Fresh wet rice noodles also can be preserved for more than 3 months after treatment and sterilization (Table 3-17).

Table 3-17. Comparison of procedures for rice noodle preservation

Pretreatment	Sterilization	Reference
Soaking in acidic solution for 1 min	at 90°C for 90 min	7
Soaking with lactic acid solution (1.5%) for 5 min	at 100°C for 5min	29
Soaking in acidic solution for 1 min	at 90°C for 60 min	23
Soaking in acidic solution (pH 4.0-4.5) for 2-3 min	at 80-90°C for 20-40 min	39
Soaking with lactic acid solution (1%, other organic acid such as nitric acid also can be used) for 35-40 sec	at 90-95°C for 40-45 min	33

4) Other Treatments

In addition to thermal sterilization and organic acid antisepsis mentioned above, it is also reported that natamax (active ingredients such as natamycin) and sodium diacetate are also effective in preservation of fresh wet rice noodles. It was reported that optimum addition of natamax was 50–200 mg/kg, while addition of sodium diacetate was 0.5–3%. While having the same effect of preservation, addition of natamax was only one tenth that of sodium diacetate [39]. Ma et al. studied the effects of several natural preservation agents, such as Nisin, Natamycin and Lysozme on the extent of shelf-life of fresh wet rice noodles. It was reported that the complex made of these agents could significantly inhibit microbiological contamination and extend shelf-life of fresh wet rice noodles even under neutral conditions [33].

4. Quality Evaluations of Rice Noodles

4.1. Sensory Properties and Evaluation

The quality of rice noodles is comprised of sensory and physicochemical properties, and there are some correlations between sensory quality and physicochemical properties (Zhao et al., 2002). High quality of rice noodles means a smooth and soft sensory quality, certain elastic properties, low broken rate and cooking loss and proper cooking time. Its quality also differed with types of rice noodles (Cheng et al., 2000).

4.1.1. Quality Evaluation of Fermented Wet Rice Noodles

In the sensory aspect, fermented rice noodles should have the proper color and the specific fermented smell, no mold infection, foreign matter, adhesion and apparent broken noodles; also, the thickness and softness of the noodle strips should be at a similar level. Rice noodles should be smooth, tough, and no sandy sense after rehydration by steaming, boiling or soaking in boiled water (He and Sun, 2005).

In the physicochemical aspect, fermented rice noodles must meet the following requirements:

Moisture content: 65–70 %; broken rate: $\leq$ 20%; titrate acidity: $\leq$ 1.5 (NaOH mL/10 g rice noodle); hardness: 0.15–0.3 kgf; rehydration duration: $\leq$ 3–5 min (He and Sun, 2005).

In addition, for the safety of the rice noodles, food security, concentration of chemical and metals, existence of microorganisms also should be controlled as follows: aflatoxin $B_1 \leq$ 5 μ g kg^{-1}, lead (pb) $\leq$ 0.25 mg kg^{-1}, arsenic (As) $\leq$ 0.28 mg kg^{-1}, the total number of bacteria $\leq$ 1 000 cfu g^{-1}, coliform $\leq$ 30 cfu g^{-1}, pathogens: not be detected (He and Sun, 2005).

4.1.2. Non-fermented Rice Noodles

Some researchers proposed that the quality of non-fermented rice noodles can be evaluated with criteria consisting of six indexes—aroma (smell), luster, stickiness, hardness, toughness and integrated evaluation. Grades and evaluation criteria are shown in Table 3-18.

4.1.3. Other Sensory Properties

Besides the sensory properties mentioned above, cooking loss and broken rate are also used for quality evaluation (Cheng et al., 2000; Wu et al., 2005).

1) Cooking Loss

Cooking loss stands for the dry matter loss in the cooking. It is expressed as percentage of loss mass to total weight of rice noodles.

Cooking loss = (G0-G1)/G0*100% (3-4)

Where G0 was dry matter content of samples before being cooked (g); G1 was dry matter content of samples after being cooked (g).

2) Broken Rate

Raise noodles with chopsticks and hold for 10 s, then count broken strips with the naked eye for calculation of broken rate, which is expressed as the percentage of broken to the original strips. The assessment of breaking rate should complete in 2–4 min after cooking.

Table 3-18. Sensory evaluation criteria of rice noodles (Cheng et al., 2000)

Aroma	Luster	Stickiness	Hardness	Toughness	Integrated evaluation	Score
Strong light scent	White and bright	A little teeth-stick	Difficult to chew	Apparent	good	2
Weak light scent	Mild brightness	Teeth-sticky	mild hard	A little	better	1
Difficult to judge	Difficult to judge	Difficult to judge	Difficult to judge	Difficult to judge	ordinary	0
Weak off-flavor	gloom	No stick feel	soft	Lack	bad	-1
apparent off-flavor	no luster	smooth	sodden	Apparent lack	worse	-2

4.2. Physical Properties and Evaluation

Apart from the Sensory quality mentioned above, the taste of rice noodle also relates to mechanical properties, which can be measured by the corresponding equipment, such as the rheological meter and stretching instrument, etc.

4.2.1. Mechanic Properties

Mechanic properties of rice noodles include hardness, shear force, maximum strain, etc. Significant correlations have been found between sensory quality and mechanic properties in non-fermented rice noodles, hardness and stickiness, hardness and shear stress, shear force and chewiness, maximum strain and chewiness, maximum strain and general sensory evaluation (Cheng et al., 2000; Ding et al., 2004; Li et al., 2003).

Min et al. (2003) found that fermentation by lactic acid bacteria had a significant effect on mechanic properties of rice noodles made from early *Indica* rice. The rice starch molecular profile could be divided into two distinct fractions as amylopectin and amylose. During fermentation, the number of average molecular polymerization of amylopectin decreased from 12676.3 to 11500.4 glucose, while the average molecular polymerization of amylose increased from 2975.3 to 3563.2 glucose. The average chain length of amylopectin decreased from 6112 to 4510 glucose and amylose increased from 2317 to 2814 glucose. Amylose content increased from 12.33 % to 17.37 % and protein content decreased from 6.89 % to 4.2 %. With the changes of starch and protein, the maximum strain, the ratio of extension rate to the section shrinking rate and elastic of rice noodle were 39.7%, 81.78% and 3.5% higher than the controlled, while hardness was 4.5% lower. Consequently, sensory evaluation of the rice noodle after lactic acid bacteria fermentation was also improved, which was smoother, softer and tougher than the non-fermented ones (Min et al. 2003a, 2003b, 2005).

Li et al. (2003) studied the effect of natural fermentation on mechanical properties of rice noodles. It was revealed that the hardness, maximum breaking strain and plasticity of rice noodles tended to increase, however the yield intensity tended to decrease with the increase of fermentation period. Young modulus and the maximum breaking stress achieved peak value after 4 d fermentation and then decreased. 5% paste made of fermented rice dough was non-Newtonian, pseudoplastic fluid. The shear stress and apparent viscosity of rice dough fermented for 7 d were higher than in other treatments.

4.2.2. Retrogradation of Wet Rice Noodles

Except for preservation of fresh rice noodles, prevention of retrogradation (starch aging) is another hot research point. Rice contains 75–80% starch, which presented as minicrystals (known as β-state) unprocessed. During processing of rice noodles, especially pre-steaming, steaming or boiling, minicrystals would be destroyed and gelatinized to the α-state. However, gelatinized starch will be retrograded and crystallized spontaneously forming a high degree of minicrystals (β-state) and will become dissoluble when stored at room temperature. A study on retrogradation characteristics of instant wet rice noodles during storage by X-ray polycrystal diffractometer and texturemeter showed that there was no microcrystalline structure in fresh instant wet rice noodles and a few microcrystal structures were formed after storing at 4°C for 30 d. With changes of texture, toughness, flexibility and shearing force of noodles will increase, while adhesive property and rehydration ratio will decrease [39, 44].

Table 3-19. Effect of additives on hardness of rice paste (Xie et al., 2006)

Additives	Hardness/(kgf·cm-2)	Relative hardness %
Control	4.32	100
Plant oil	2.39	55.32
Konjac fined power	2.77	64.12
Potato starch	2.79	64.58
DMG	3.34	77.31
Sucrose ester	2.85	65.97
Xanthan gum	2.38	55.09
Guar gum	2.05	47.45
Wheat gluten	2.42	56.02
Potato modified starch	2.95	68.29
Cassava modified starch	3.47	80.32
MC	3.09	71.53
Maltodextrin	1.88	43.52

* Stored at 4°C in refrigerator for 7 days.

After staling, rice noodles become brittle and easily broken, taste rough and hard, do not have quality of fresh noodles even after soaking in hot water (80°C) for 3–5 min. Only when heated in boiling water for several minutes, staled noodles could have similar quality as the fresh. Staling is one of the key issues in production of fresh rice noodles industrialization [28].

At present, substances such as oil, emulsifier, starch substitutes, edible hydrocolloid, and enzymes, are being studied for prevention of staling of rice noodles as in Table 3-19.

Studies on wet rice noodles indicated that natural fermentation will speed the retrogradation of starch, although it could improve other edible qualities, and application of a modifier, such as modified starch and DMG, and controlling of moisture content in noodles, temperature of storage and size of smashed rice could effectively prevent retrogradation of wet rice (Ding and Wang, 2005; Tu et al., 2003).

4.3. Main Factors Affecting Quality of Instant Rice Noodles

4.3.1. Content of Amylose in Rice

The content of amylose significantly affects the quality of rice noodles. The operation will be easily controlled and nice sensory qualities on toughness, certain elastics, good color and luster will result when the content of amylose in *Xian (indica)* ranged from 18–24%. Amylose content higher than 25% related to better sensory qualities of rice noodles, together with more difficult gelatinization. While a content lower than 18%, satisfactory sensory qualities of the rice noodles may not be reached and the steamed paste might not be molded because of its high stickiness. Even if rice noodles could be formed, cooking loss would be at a high level (Zhang et al., 2003).

4.3.2. The Size of Ground (Smashed) Rice

The size of the smashed rice contributes a lot to the quality of rice noodles (Table 3-20).

Rice noodles are difficult to be molded, or have a high broken rate (close to 100%) even if it formed, when the size of the smashed rice is bigger than a 40 mesh. A large smashed size also leads to the problems of incomplete gelatinization and high cooking loss, which make the cooking medium turbid. Only when the smashed rice can fit the sieve of 60 mesh can rice noodles be made with normal processing procedures. Better integration of paste, lower broken rate and cooking loss, shorter rehydration time will be obtained with a smaller smashed size; especially when the size is finer than 120 mesh, there will be no broken strips. However, if the grain is smashed too small, the yield of rice noodles will decrease, resulting in the increase of the production cost. According to processing experience, 80 mesh is the appropriate size for both edible quality and economic production expense (Zhang et al., 2003).

4.3.3. Moisture Content of Rice Slurry

Moisture content of smashed rice slurry also has a great influence on the quality of rice noodles. To ensure the complete gelatinization of starch, the moisture content must be higher than 30%. The effect of different contents on the quality of noodles and processing is showed in Table 3-21 (Zhang et al., 2003).

Study and producing experience revealed that 38% of slurry moisture content is the best, which will facilitate the molding processing and make noodles smooth and tough. If the content is lower than 30%, it will result in uneven and incomplete gelatinization of starch, cracks on dried noodles, and a tendency to a high broken rate. If the water content is too high, the high fluidity of the slurry will decrease the squeeze pressure of molding and lead to difficulty in molding. Even if noodles are formed, the adhesion of strips, lower strength and toughness of noodles, and a dark color of the final product will occur.

Table 3-20. Effect of size of rice noodles on their quality (Zhang et al., 2003)

Sieve size/mesh	Cooking loss /%	Rehydration time/min	Broken rate/%
60	27.9	6.5	26. 7
80	21.4	5.5	22.2
100	21.1	5.0	6.7
120	18.8	4.0	0

Table 3-21. Effect of moisture content of slurry on quality of rice noodles (Zhang et al., 2003)

Moisture content %	Sensory evaluation	Physical property	Score
30	White surface, cracks on noodle	High broken rate	4.6
32	White surface, rough, bad taste	High broken rate	5.9
34	Smooth surface, improved taste	Easy to mold	6.7
36	Tough, smooth	Easy to mold	7.9
38	Sheer surface, smooth	Easy to mold	8.5
40	Sticky skin, dark colored	Hard to mold	5.0

Up till now, the rice noodle industry has developed on larger and larger scales since more and more people use the noodles for breakfast or side dishes at night. It is reported that in recent years, the output of rice noodles was more than 2 billion RMB per year. Take Guangdong province as an example, there are more than 30 large- and medium-scaled companies or factories for rice noodle preparation in the Dongguan region, with the production about 240 thousand tons and 20 thousand tons exported every year. In regions of Fanyu and Foshan, there are about 20 companies or factories larger than medium-scaled producing Pai rice noodles, with production about 20 thousand tons and half is exported. Production of wave-like rice noodles, representative of Heyuan, and Pai rice noodles, produced in western and eastern Guangdong, is about 60 thousand tons (Xing et al, 2007). As an important product for daily life and for export, the improvement in processing of rice noodles should be a goal now.

REFERENCES

Chen An. (2002). 'New technology and equipment of rice noodle processing'. *Storage and Processing.* 11:31-32.

Cen Junjian. (2007), 'The prospect of rice noodles industry, its technology and equipment'. *Science and Technology of Cereals, Oil and Foods.* 15(2):4-7.

Cheng Minghua, Li Lite,(2000) 'The evaluation systems of non-fermented rice noodle'. *Journal of the Chinese Cereals and Oil Association.* 15(5):14-18.

Cheng Minghua. (2000). Study on the evaluation systems and Processing technology of rice noodle [Doctoral Dissertation]. China Agricultural University.

Deng Danwen, Zheng Gonguyuan, (2000). 'The processing technique of rice vermicelli'. *Food Science & Technology.* 5:15-16.

Deng Danwen, Zheng Gonguyuan (2000). 'Comparison of processing technology of rice noodle'. *Western Cereal and Oil Science and Technology*, 25(3):29.

Deng Wendan, Zheng Gongyuan, (2001), 'Study on preservation of fresh rice noodle'. *Food Science.* 22(4):70-72.

Ding Wenping, (2004). 'Establishment of parameters of Rice Selected for producing Rice Noodles'. *Food Technology.* 10: 24-26.

Ding Wenping, Wang Yuehui,(2004). 'Establishment of instrumental evaluation system for rice noodle'. *Science and Technology of Food Industry.* 25(1):135-137.

Ding Wenping, Wang Yuehui, (2004). 'Determination of index for noodle making rice'. *Food Industry.* 5:16-18.

Ding Wenping, Wang Yuehui. (2005), 'Effect of soaking and fermentation on physiochemical properties of rice noodle'. *Food Industry.* 3:5-7.

Fu Xiaoru. (2000), 'Fresh wet rice noodle and processing technology'. *Food Industry.* 5: 9-10.

Fu Xiaoru.(2000). 'Utilization of additive and ingredients in the production of rice noodles'. *Science and Technology of Sichuan Grain and Oil.* 4: 44-45.

He Xinyi, Sun Qingjie. (2005), 'Research on sterilization and fermentation of fresh rice noodle'. *Food Science & Technology.* 12(5):23-26.

Jiang Fuguo, Gu Guangyu (2001). 'Study on optimum processing technology of high quality of strip rice noodle'. *Food Research and Innovation.* 22(6):32-33.

Jiang Zhengqiang, Liu Xiaoyan. (1999). 'The processing technology and quality improvement of a new type rice noodles'. *Journal of the Chinese Cereals and Oils Association.* 14(3):40-43.

Jin Maoguo (1995). 'The characteristics of rice noodles starch and its relationship with the quality of rice noodles'. *Journal of Wuxi university of light industry.* 14(4): 307-312.

Li Hongbin. (2005), 'The relation of amylase content in rice slurry and quality of instant rice noodle'. *Food Science & Technology.* 4: 29-31.

Li Lite, Cheng Minghua, (2000). 'Production and research status of rice noodle'. *Food & Machinery,* 3:10-11.

Li Lite, Min Hongwei, et al. (2003), 'Effect of natural fermentation on mechanical properties of rice noodle'. *Food Science.* 24(4):45-49.

Lin Jing, Lin Xia, et al. (2000), 'Inspection of rice noodle sanitation in Guangxi'. *Guangxi Journal of Preventive Medicine.* 6(1): 12-14.

Li Xiugui, Lin Xia, et al. (2001), 'Survey of microorganism pollution of wet rice noodles in Nanning'. *Hubei Journal of Preventive Medicine,* 12(2): 46-47.

Liu Zhaomin, Ma Lin, et al. (2005), 'Inspection of sanitary condition of wet rice noodle and analysis of relative factors'. *Chinese Journal of Public Health Management.* 21(2): 161-163.

Liu Xingxin.(2007), 'Reconstruction of rice noodle industry and technical innovation in china'. *Cereal and Feed Industry,* 14:1-3.

Liu Youming, Tan Rucheng, (2008). 'Study on raw material selection for instant rice noodle'. *Food Science & Technology.* 3:133-136.

Lu Zhanhui. (2005) Starch modification in rice roodles and it gelatinization mechanism induced by biological fermentation process [Doctoral Dissertation]. China Agricultural University.

Lu Zhanhui, Li Lite, (2002). 'Effect of natural fermentation on rehological properties of rice noodle'. *Journal of Chinese Institute of Food and Technology.* 6(2):8-12.

Ma Xiaoming, Lin Yu. (2003), 'Study on the technology of productional and function to Fresh and Wet Rice Line'. *China Food Additives*, 6: 96-99.

Min Hongwei, Li Lite et al. (2003), 'Mechanism of improving eating quality of rice noodles by lactic acid bacteria fermentation'. *Food and Fermentation Industries.* 29(6):5-9.

Min Hongwei. (2003). Study on the Mechanism of improving eating quality of rice noodles by lactic acid bacteria fermentation [Doctoral Dissertation]. China Agricultural University.

Min Hongwei, Li Lite, et al. (2005). 'Study on physical property of rice noodle by biological fermentation'. *Food Science.* 26(1):97-99.

Pan Shaopeng. (2007). 'The technology of fermented instant rice noodles'. *Jiangxi Food Industry*, 2:42-43.

Pan Shaopeng, Deng Danwen. (2002). 'High quality straight strip rice noodle making in Jiangxi'. *Food Science & Technology.* 1:11.

Peng Shanshan. (2000). 'The application of additives in production of instant rice noodles'. *Guangzhou Chemical.* 28(4): 28-30.

Sun Qingjie, Ding Wenping (2004). 'Studies on the Standard for Rice Noodle Material'. *Journal of the Chinese Cereals and Oils Association.* 19(1):12-15.

The origin of rice Noodles. http://zhidao.baidu.com/question/19669690.html.

Tu Zongcai, Liu Chengmei, (2003). 'Prevention of staling and qulity improvement on rice noodle'. *Food Industry.* 5:7-8.

Wu Tao, Wu Hui, (2004). 'Study on preservative technology of fresh rice vermicelli'. *Cereals & Oil.* 4:24-26.

Wang Xuehua, Li Hesong (2004). 'Characteristics of rice for rice noodles and processing technology'. *Crop Research.* 5:304-307.

Wu Weiguo, Li Xianguang.(1998). 'Problems and solved strategies or rice noodle processing'. *Food Science*, 19 (6): 49-51.

Wu Weiguo, Zhang Yu, (2005). 'The relationship between characteristics of rice raw material and quality of rice noodle'. *Cereal & Feed Industry.* 9:21-14.

Xie Ding, Liu Yongle, (2006). 'Investigation of anti-retrogradation of fresh instant rice noodles'. *Food & Machinery*, 22(2): 8-10.

Xing Fang, Liang Lanlan (2007), 'Situation of safety and stfandard of rice noodle in China'. *Modern Food Science and Technology.* 23(8):68-70.

Ye Min, Zhang Yan, (2005). 'A study on processing technology for production of instant wet rice noodle'. *Cereal & Feed industry*. 11:15-17.

Zhang Yu, Yang Miquan, (2003). 'Study on relationship between characteristics of starch and quality of rice noodle'. *Food Science*, 6: 35-38.

Zhao Siming, Liu Youming, (2002). 'Study on the processing suitability and quality properties of instant rice noodles'. *Cereal & Feed Industry*. 6:37-39.

Zou Guoyao, Yao Huiyuan (1999). 'The Research of Instant Rice Vermicelli'. *Cereal & Feed Industry*. 5:45-46.

Chapter 4

STARCH NOODLES

1. INTRODUCTION

1.1. Definition

Starch noodles, produced from purified starch from various plant sources, are a major category of Asian noodles. They are produced by the following steps: 1) mixing dry and gelatinized starch to form a slurry or dough, 2) extruding it directly into boiling water to cook, 3) cooling the formed noodles in cold tap water, 4) holding at refrigerated or freezing temperature, 5) warming in cold tap water, and then drying (Galvez et al., 1994). Starch noodles are obviously different from other types of noodles, such as pasta and wheat flour, since it is made from gluten-free starch. Thus, starch itself plays an essential role in both the production of starch noodles and the final quality of starch noodles. Excellent starch noodles would be clear or transparent and have fine threads, high tensile strength, and low cooking loss even with prolonged cooking (Collado et al., 2001; Purwani et al., 2006).

1.2. Naming

The use of starch from various sources to manufacture products in noodle shapes has been practiced for centuries in China and subsequently spread to neighboring countries. These products are called "starch noodles" or "cellophane noodles" because of their translucent or transparent appearance of pre- or post- cooking. This transparent appearance is due to the gelatinization of starch in the manufacturing process. After drying, the product shows a translucent appearance; thus, Westerners call them "cellophane noodles" (Hui, 2007).

Nomenclature of Chinese starch noodles is difficult due to the vast spectrum available and the many dialects of Chinese being used to name them. Mung bean threads were probably the first starch noodles manufactured (Hui, 2007). In Chinese, therefore, starch noodles are called lǜ dòu miàn ("mung bean noodles"; literally "green bean noodles"), fěn sī ("soft white noodle"), or dōng fěn ("winter white noodle"). In China, the primary site of production of starch noodles is in the city of Longkou, Shandong province in China, and thus the noodles are also marketed as Longkou fensi.

Table 4-1. Names of various starchy foods in different countries or areas (http://en.wikipedia.org/wiki/Chinese_noodles)

Common English name	Winter noodle	Bean threads	Mung bean sheets	Liang pi
Characters	冬粉	粉丝	粉皮	凉皮
Pinyin	dōngfěn	fěnsī	fěn pí	líang pí
Cantonese	dung fun	fun sze	fan pei	---
Hokkien	dang hun	---	---	---
Thai	---	Wun sen (วุ้นเส้น)	---	---
Western equivalent	Thin mung bean vermicelli	Mung bean vermicelli	---	---
Description	Very thin mung bean starch noodles	Thin cellophane-like noodles	Wide, clear noodles made from mung bean starch	Translucent noodles made from wheat starch left from producing gluten

The Korean sweet potato vermicelli ("*dang myun*") is a product similar to mung bean threads.It is not purely colorless, but a kind of light brownish-green. It also has excellent *al dente* properties, which remain upon re-heating. They are produced as long and thick noodles (Hui, 2007).

In Japanese cuisine, they are called harusame, lit. "spring rain." Japanese harusame is a similar starch noodle product, as it is also made from potato, sweet potato, rice, or mung bean starches (Hui, 2007). In Vietnamese they are called *bun tau* or *bun tao*. In Thailand they are called *wún-sên*, an almost clear noodle made from mung-bean starch and water. While in the Philippines, rice flour, maize starch (MS), and mung bean starch are made into starch noodles locally known as *bihon* for those with thin strands and *pancit malabon* for those with thick strands (Collado et al., 2001).

1.3. History

Starch noodles have been a Chinese favorite food for at least 1,400 years (Zhang and Chi, 2001). Before that age, most grains with abundant starch content were simply boiled whole, as people now cook rice. The first written account of starch noodles dates from the North Wei Dynasty. Jia Sixie, in the latter part of Wei, the North Dynasty, recorded the production of starch and starch noodles detailedly in his famous book "Qi Min Yao Shu". Although its processing method was simple, it is the same principle as produced by machine nowadays (Zhang and Chi, 2001).

The production of mung bean starch noodles and the market were also mentioned in other books, Ben Xin Zhai Shu Shi Pu and Dong Jing Meng Hua Lu, in Song Dynasty. (Zhang and Chi, 2001).

Although an old legend said that Sun Bin invented starch noodles, there is no record of it (Zhang and Chi, 2001). The origin of noodles remains an unsolved historical question.

Longkou starch noodles, the first export production from Longkou haven, are produced around Zhaoyuan county, Shandong province. Historical sources said that a small quantity of mung bean starch noodles, which were packed with bulrush and contained 120 kilogram per bundle, was transferred from Zhaoyuan county to Longkou county, and then exported abroad in 1860. Large numbers of starch noodles, produced in Zhaoyuan county, were transported abroad from Longkou haven by ship in 1881. Therefore, these starch noodles were called "Longkou fěnsī" , which was translated into "Longkou starch noodles" (Zhang and Chi, 2001). In order to keep the high quality, the manufacture of starch noodles is seasonal, produced only in spring and autumn. The temperature is good for steeping, not too cold or too hot. A shining, clean, slightly windy air is good for drying (Chang, 1993).

The starch noodles' manufacture used to be a family business, employing primitive equipment. It was manually operated and controlled by experienced workers. Some machines such as rotary screens, mechanical mills, kneading machines, etc., are adopted but the principle of manufacture remained practically the same. It is a so-called "traditional acidic" method, because the steeping of beans, coagulation of milled pulp and soaking of silks are all carried out in an acidic liquor (Chang, 1993).

Starch noodles are a familiar product and food in Hubei province, Hunan province, Hebei province, Shanghai city, Beijing city, Tianjin city, and so on, around and after 1949 (Zhang and Chi, 2001).

Starch noodles, which have been an important part of the Chinese diet from ancient times, like other great Chinese concepts, spread to the surrounding Asian cultures and beyond. Nowadays, from the hawker stalls in Singapore's markets to the buy-a-ticket fast food cafeterias of Japan, starch noodles are available.

1.4. Eating Methods

Starch noodles can be used as a major ingredient for making a popular Chinese hot pot, fondue, starch noodles and vegetable salads, and many stir-fried dishes. Starch noodles come in a great variety of shapes and sizes, and can be cooked many ways. They can be boiled or stir-fried, then eaten either hot or cold, with toppings or with dipping sauces—like the spicy dandan mien, or tasty sweet potato starch noodles tossed with peppery sauce, a specialty of Sichuan Province. This is a Cantonese style of cooking in that cooks can coil steamed vermicelli into thick pancakes, fry them on both sides until crisp, and then top them with a stew. Cooks soak them in hot water until soft, and then use them in stir-fries, soups, and salads. The soaked starch noodles have a rather slippery texture and a glassy look (Figure 4-1).

1.5. Categories

Starch noodles can be classified according to different parameters such as the type of raw materials, the size of starch noodle strands, the manufacturing method, producing area and the form of the product on the market.

Figure 4-1. Cooking methods for starch noodles. (A) cold and dressed with sauce; (B) boiled to soup with flavoring; (C) stir-fried.

Starch noodles can be classified into mung bean starch noodles and coarse grain starches noodles according to the type of raw materials used in their manufacture. Traditionally, mung bean starch is used as a main and excellent ingredient in starch noodle making. Coarse grain starches noodles are made from various legume starches such as broad bean, pea, cowpea, bean, and various tuber or root starches such as potato, sweet potato, cassava, and a variety of grain starches such as maize, wheat, sorghum. But the qualities of of mung bean starch noodles are the best (Zhang and Chi, 2001).

There are three kinds of starch noodles, namely thin starch noodles, thick starch noodles (vermicelli) and flat starch noodles (broad strips), according to its shape or width. The thin starch noodle is the most common one for it is easy to cook (Zhang and Chi, 2001).

There also are three kinds of starch noodles according to the type of manufacturing method used. They are produced by dropping, extruding, cutting, respectively.

The dropping method which uses gravity as the extrusion force is the most traditional method in China. Part of the starch, approximately 5% is first gelatinised in excess hot water to yield a viscous paste. In this paste, the remainder of the starch is mixed at approximately

50 °C yielding a dough that can be extruded using gravity to filaments of partly gelatinised starch. To this end, the filaments are cooked in boiling water for approximately 10 seconds and subsequently run through cold water. The filaments are then cut and the noodle threads are hung on rods to drip and drain. The noodles are frozen either in winter in the open air or in special freezing rooms, thawed and dried in drying cabinets or in the open air, except that noodles made from legume starches can be dried without a freezing step.

The extruding method involves this process for producing starch noodles, in which the starch is first blended with water and then partly gelatinized in a single or twin screw extruder. The partly gelatinized starch is directly extruded at the second step. The resulting noodle threads are cut or shaped and then hung on rods, dried in the air or oven.

The cutting method is used when noodles are manufactured by gelatinizing starch into a sheet and then linearly cutting and drying (Pueraria thunbergiana Benth) (US Patent 5916616).

There are a great variety of starch noodles according to the producing areas in China, such as Zhaoyuan starch noodles, Hankou starch noodles, Hebei starch noodles, Hunan starch noodles, Shanghai starch noodles, Yunnan starch noodles (Zhang and Chi, 2001).

Although starch noodle texture and size—round, square, flat, or thick—may vary somewhat from area to area, the styles are still recognizable. Of course, the big difference comes when cooks prepare them and add seasoning and other ingredients.

1.6. Consumption

Chinese starch noodles are an important part of most regional cuisines within China (mainland, Hongkong, Taiwan), as well as in Singapore and other Southeast Asian nations such as Vietnam, Thailand, and Cambodia with sizable overseas Chinese populations. It has also entered the cuisines of neighboring East Asian countries such as Japan and Korea.

The most popular starch noodle is LongKou fěnsī produced in Shandong province. According to statistic data, in Shandong province, there were about 140 and 40 of factories for producing starch noodles in Zhaoyuan county and Longkou city, Shandong province, respectively, in 2001. There are more than 200 thousand tons of output of starch noodles, over 80 thousand tons for export per year in Longkou city. Longkou starch noodles are sold well in 56 countries and areas such as Japan, England, America. (Yang et al., 2007)

The starch noodles market in China is continuously growing for starch noodles can be easily affordable for people in the low-income bracket and a dish of oriental starch noodles can be a delicious and healthy meal.

2. Materials for Starch Noodles

Starch is a primary material in the production of starch noodles because it is the main ingredient of starch noodles. Owing to the absence of gluten as compared with wheat flour, physicochemical, thermal, rheological properties of starch affect the quality of starch noodles. Meanwhile, the influence of starch functionality on starch noodle quality would be noticed as a result of the retrogradation step included in the manufacturing process (Chang et al., 2006). Mung bean starch is the best raw material to produce high quality starch noodles for its high amylose content, restricted swelling of starch during gelatinization and the high shear resistance of its paste (Li and Chang, 1981).

Starches of various legumes such as broad bean, pea, cowpea, bean, and various tuber or root starches such as potato, sweet potato, cassava, and other grains, are competitive with the mung bean starch on a cost basis.

2.1. Mung Bean Starch

Mung bean (*Vigna radiata* (L.) Wilczek) or green gram, is native to the northeastern India-Burma (Myanmar) region of Asia. It is primarily grown in Asia, Africa, South and North America, and Australia principally for its protein-rich edible seeds (Liu and Shen, 2007). Mung bean is similar in composition to other members of the legume family, with 24% protein, 1% fat, 63% carbohydrate and 16% dietary fiber (US Department of Agriculture, 2001). It lacks sulfur-containing amino acids and uncooked beans contain trypsin inhibitors (Khader and Rao, 1996).

The people value the mung bean, because of the medicinal effect in addition to nutrition. It can provide health protection against hot summer weather, to cure some kinds of heat and toxicity in China. The most famous great treatise on herb medicines "Pen Chao Kung Mao" by *Li Shih-Chen*, 1578, Ming Dynasty, highly recommended mung bean as "The key food ingredient", "Good grains for the benefit of the society". Many foods are made from mung bean, such as drinks, soups, porridge, and cake. They are well consumed by the people, especially during hot summer days (Chang, 1993).

Mung bean starch is used in the production of vermicelli or glass noodles. These noodles are made from a mixture of mung bean starch and potato starch or other legume starch, and when boiled, become gelatinous in texture and transparent (Liu and Shen, 2007). Traditionally, mung bean starch is considered to be the most suitable raw material for starch noodle making, and the mung bean starch noodle is, consequently, regarded as the best of all kinds of starch noodles (Muhammad et al., 1999; Kasemsuwan et al., 1998).

2.1.1. Isolation of Mung Bean Starch

The method using sour liquid to extrude the starch is a traditional way in China. Selected sound mung beans are first steeped several hours in acidic liquor, which has been used once in the milled pulp coagulation step. The resulting steep liquor is siphoned off as a by-product, mixed with some cold water and steeped again. The resulting steep liquor is siphoned off and used again in the later step. Well-steeped beans are milled to a fine pulp, coagulated with acidic liquor to separate the starch from other materials. The natural fermentation occurring in the steeping operation is quite complex, but mainly lactic acid, which is effective in softening the beans, destroying the cell walls, and freeing the starch granules (Chang, 1993).

Well-steeped mung beans swell in volume, and are easily milled. The milled bean pulp is a homogeneous slurry. When a proper amount of acidic liquor is admixed, coagulation takes place immediately, forming three distinct layers. The heavy starch granules settle to the bottom, while the protein and fibrous materials remain at the top. Separation is quite fast, complete in about one minute. After a few mins, siphon off the top and middle layers, which are recovered as by-product. Admix some fresh water to wash the starch, settle for 1–2 h, siphon off the supernatant acidic liquor which would be re-used in the steeping operation. Pass the starch through a fine screen, mix again with some fresh water, settle for about ten hours, siphon off the supernatant liquor. The pure starch granules settle into a solid cake at the bottom. It is cut into pieces, put in cloths, hung up to drain off some water, sun-dried to a moisture content of 20–30%, then is ready for noodles making (Chang, 1993).

Other isolation methods are centrifugation and centrifugal cyclone separator processing. Slurry was centrifuged at 3000g for 10 min, and then the starch was collected. Liu and Shen (2007) compared the physicochemical properties of mung bean starches from sour liquid processing and centrifugation.

2.1.2. Morphological Property of Mung Bean Starch

Mung bean starch granules are small, smooth, and either spherical or elliptical (Figure 4-2). Large granules were kidney-shaped or oval, small granules were spherical, and some of them had internal fissures. It seemed that large granules had more internal fissures than small granules. The granules which had no internal fissures showed that the granules were broken or not grown yet (Liu and Shen, 2007).

Mung bean starches have polarization cross and obvious concentric circles from electron microscopy (Figure 4-3). The long axis looked like strands of rope twisted to the right. At the end of the "rope", it was opened like a "Y". Some granules showed two long axes slit. The short axis was not clear and sometime even could not be seen. Other granules had no horizontal polarization (Liu and Shen, 2007).

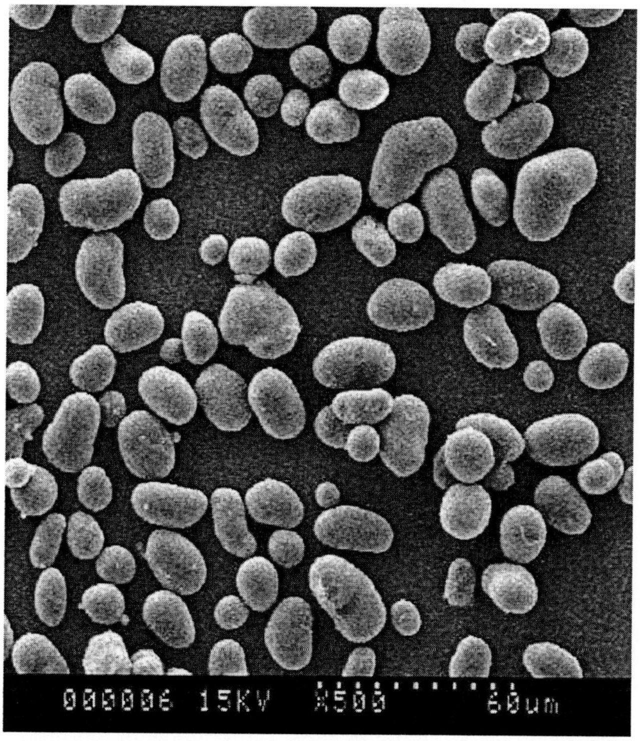

Figure 4-2. Micrograph of the granules of mung bean starch (500×) (Liu and Shen, 2007).

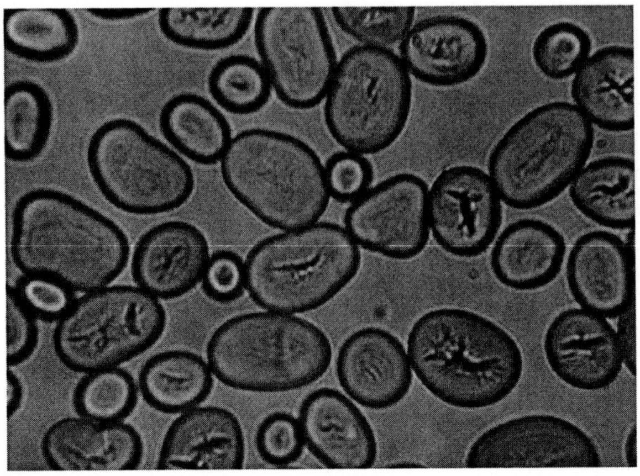

Figure 4-3. Polarization cross of mung bean starch granules (640×) (Tan et al, 2007).

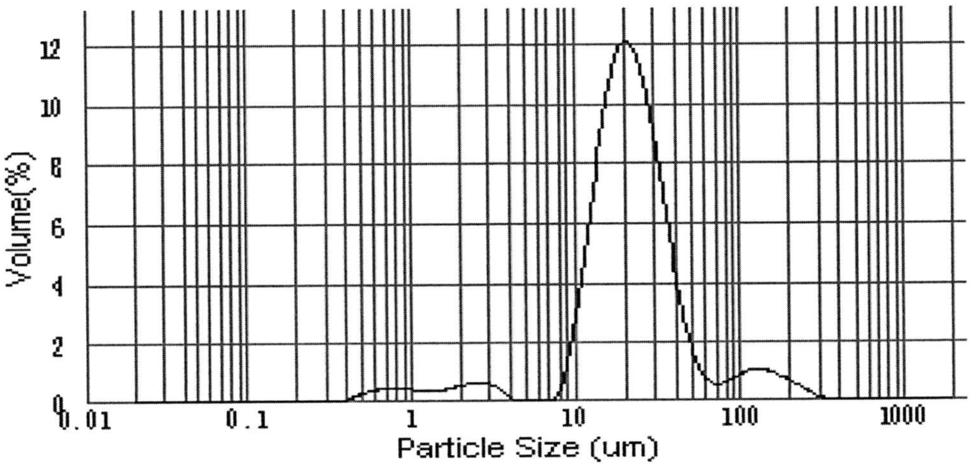

Figure 4-4. Granule size distribution of mung bean starch granules (Tan et al., 2007).

The granular size of mung bean starch ranged from 6.5 to 43.4 μm (Figure 4-4) (Liu and Shen, 2007; Tan et al., 2007). They were 14~15 μm in width, 18-21 μm in length with oblong or kidney-like shapes (Liu and Shen, 2007).

2.1.3. Chemical Property of Mung Bean Starch

Excluding ash, the components of mung bean starches from different isolating methods have significant differences (Table 4-2). Crude protein and lipid contents of starches isolated by different method range from 0.07%~1.34%, 0.05%~0.74% respectively. Streptococcus lactic which was the main microorganism isolated from sour liquid could break protein down into peptides, and separate protein from starch granules, so the starch granules deposited quickly, while the slurry including mung bean protein and peptides was thrown away. Part of the streptococcus lactic was deposited with the starches which caused the protein and fat of the starch of sour liquid processing to be higher than that of centrifugation (Liu and Shen, 2007).

Table 4-2. Chemical composition of mung bean starches from different isolated methods (% dry basis)

Mung bean starch	Moisture	Fat	Protein	Ash	Amylose	Reference
Sour liquid processing	12.93	0.13	1.34	0.15	34.3	Liu and Shen, 2007
Centrifugation	10.74	0.05	0.32	0.12	32.7	Liu and Shen, 2007
Tap water	8.99±0.13	0.74±0.07	0.68±0.08	0.14±0.04	33.7±0.41	Tan, 2007
LFS	——	0.20 ±0.03	0.08 ±0.02	——	30.9 ± 0.1	Chang et al., 2006
NaOH	——	0.16 ±0.03	0.08 ±0.01	——	30.9 ± 0.1	Chang et al., 2006
Na_2SO_3	——	0.19 ±0.02	0.09 ±0.00	——	31.0 ± 0.2	Chang et al., 2006
Distilled water	——	0.19 ±0.01	0.07 ±0.02	——	31.1 ± 0.2	Chang et al., 2006

LFS, Lactic acid fermentation solution. Many values were expressed as the mean±standard deviation.

Amylose content of starches isolated by different methods range from 30.9% to 34.3%, and far more than sweet potato starch (Tan et al., 2006). Amylose is the most important factor affecting the starch gel strength because of its prompt association and retrogradation and its interaction with lipids to form a helical complex and with amylopectin to give strong gel networks (Jane and Chen, 1992). Since amylose from sour liquid starch is higher than that from centrifugation, the transmissivity value of noodle is higher too.

Any pigmentation in the starch would be carried over to the noodles. This reduces the quality and the acceptability of the noodle (Galvez and Resurreccion, 1992). A low value of chroma and a high value of lightness are desired for the starches. Liu and Shen (2007) reported that starch from centrifugation was darker (L*), greener (A*), and more yellow (B*) than starch from sour liquid processing (Table 4-3).

2.1.4. Physical Property of Mung Bean Starch

2.1.4.1. Solubility and Swelling Power

The solubility and swelling power of starch were correlated in a direct way with the temperature. With increasing temperature, the solubility and swelling power of starch increased. The values of solubility and swelling power of mung bean starch from centrifugation were higher than that from sour liquid processing (Figures 4-5 and 4-6) (Liu and Shen, 2007).

Table 4-3. Brightness of mung bean starches from sour liquid processing and centrifugation (Liu and Shen, 2007)

	L*	A*	B*
Sour liquid processing	93.80 ± 0.4	8.05 ± 0.0	1.19 ± 0.02
Centrifugation	94.90 ± 0.04	8.18 ± 0.02	1.07 ± 0.01

Each value was expressed as the mean±standard deviation.

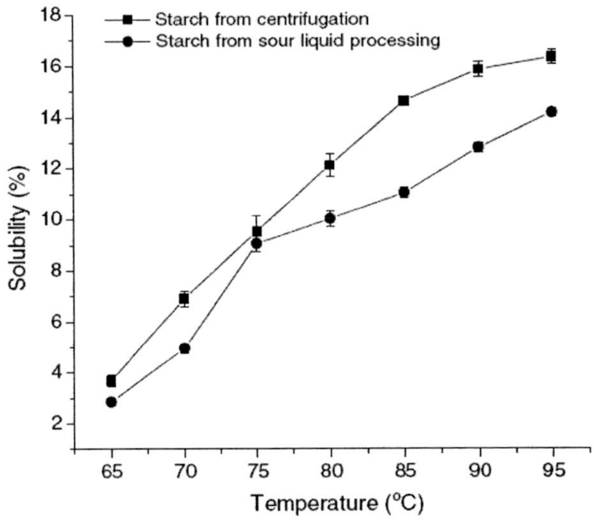

Figure 4-5. Solubility curve of mung bean starches from sour liquid processing and centrifugation (Liu and Shen, 2007).

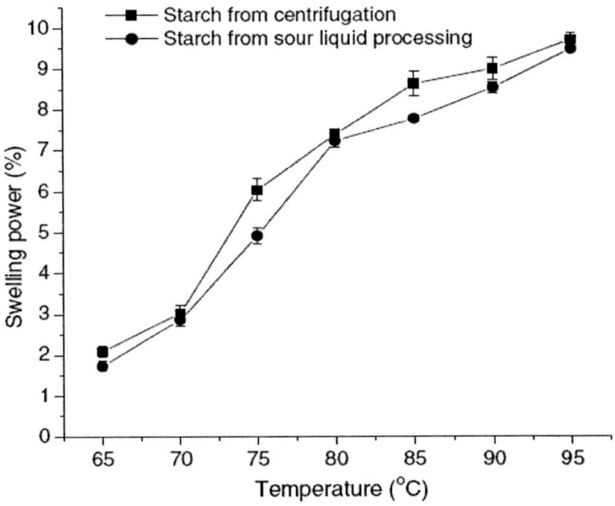

Figure 4-6. Swelling patterns of mung bean starches from sour liquid processing and centrifugation (Liu and Shen, 2007).

2.1.4.2. Viscosity

The trend curve of viscosity showed that if temperature is lower than 65 °C, the viscosity of the mung bean starch is at a low value (Figure 4-7). From 65 °C to 95 °C, the viscosity of mung bean starch from centrifugation increased gradually, but that of sour liquid processing starch increased sharply from 75 °C to 95 °C and was significantly lower than that from centrifugation at 65 °C~75 °C, while significantly higher if the temperature was over 75 °C (Liu and Shen, 2007).

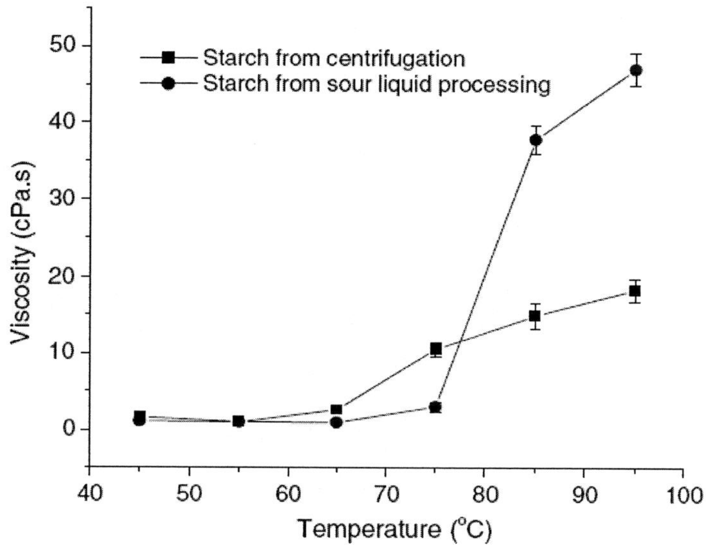

Figure 4-7. Viscosity curve of mung bean starches from sour liquid processing and centrifugation (Liu and Shen, 2007).

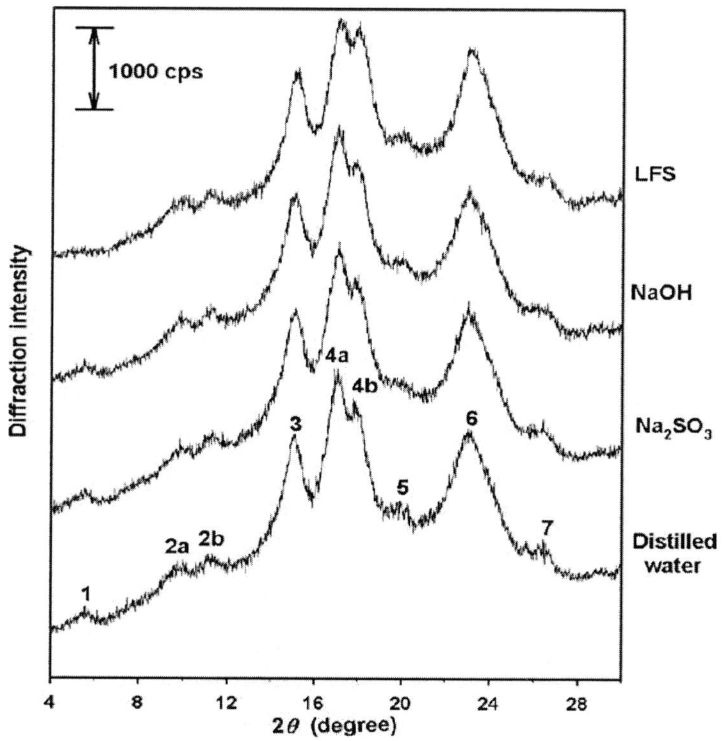

Figure 4-8. X-ray diffraction patterns of mung bean starches isolated using different steeping liquors (Chang et al., 2006).

The amylose portion of the starch likewise affected its swelling and hot-paste viscosity. Schoch and Maywald (1968) stated that as the amylose content increased, the swelling tended to be restricted and the hot-paste viscosity tended to be stabilized. Higher amylose contents are desired for the manufacture of starch noodles. Thus the vermicelli made from mung bean starch from sour liquid processing had a better quality (Liu and Shen, 2007).

2.1.4.3. X-ray Diffraction Pattern

The mung bean starch preparations had typical C-type X-ray diffraction patterns (Figure 4-8). The diffraction intensity of peak 1 and the intensity difference between peak 4a and 4b is for a lactic acid fermentation solution (LFS).

2.1.5. Thermal Property of Mung Bean Starch

2.1.5.1. Pasting Properties

The profile of mung bean starch measured by the RVA was similar to that of type C starches (Figure 4-9), i.e., without an apparent pasting peak during cooking and an obvious breakdown of hot paste, obtained by a Brabender Viscoamylograph (Schoch & Maywald, 1968). The LFS-isolated starch had significantly lower values of peak, hot paste and final viscosity than those of other starches (Table 4-4). However, no significant difference in pasting properties was found among starches isolated by NaOH, Na_2SO_3 and distilled water. (Chang et al., 2006)

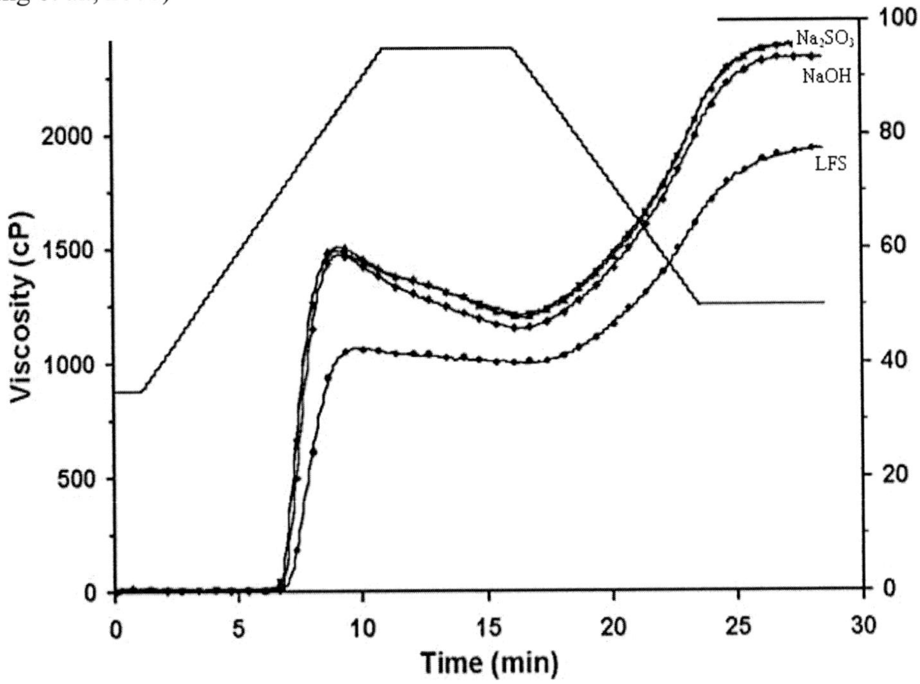

Figure 4-9. Pasting profiles of mung bean starches isolated by LFS (•), NaOH(□), Na2SO3 (▲), and distilled water (▼), respectively. (Chang et al., 2006).

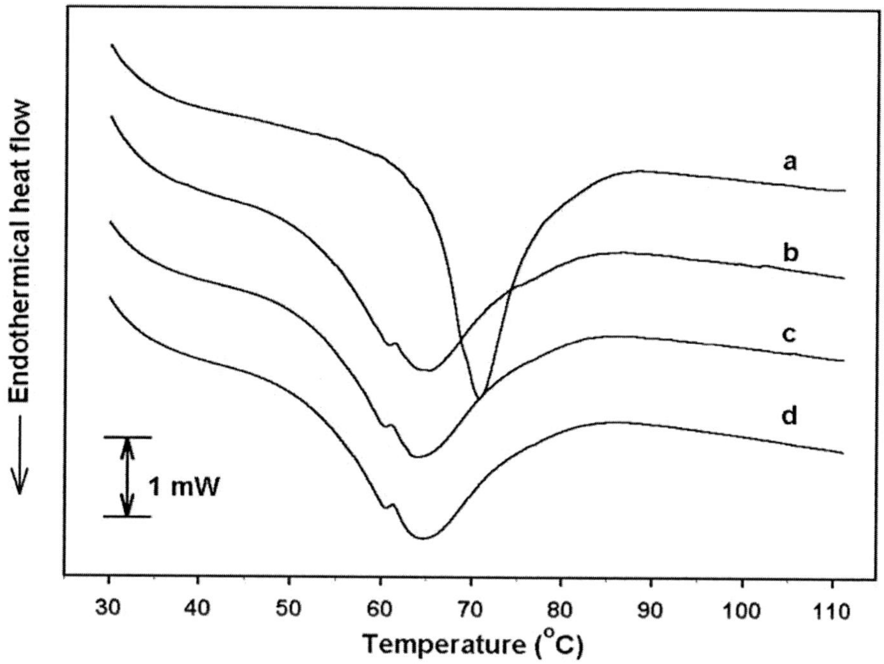

Figure 4-10. Differential scanning calorimeter thermograms at scanning rate of 1.2°C/min of mung bean starches isolated by LFS (a), NaOH (b), Na2SO3(c), and distilled water (d), respectively. (Chang et al., 2006).

The pasting properties of starch were considered to be affected by its amylose content and chain-length distribution of amylopectin, a larger proportion of long chains resulting in a lower peak viscosity if the starches had similar amylose contents (Jane et al., 1999). Moreover, the LFS-isolated starch had a lower setback than the other starches, which might relate to the lower degree of polymerization of amylose fraction (Jane and Chen, 1992).

2.1.5.2. Gelatinization Thermal Properties

The thermal transition profile of LFS-isolated starch exhibited a narrow, mono-modal distribution (Figure 4-10). The onset (T_o) and peak (T_P) temperatures of pasting of LFS-isolated starch were significantly higher than those of the other starches (Table 4-5).

Table 4-4. Pasting properties of mung bean starches isolated using different steeping liquors (Chang et al., 2006)

Steeping liquor	Peak viscosity (cP)	Hot paste viscosity (cP)	Final viscosity (cP)	Breakdown viscosity (cP)	Setback viscosity (cP)
LFS	1003 ± 34 b[A]	932 ± 30 b	1875 ± 51 b	71 ± 10 b	943 ± 26 b
NaOH	1411 ± 18 a	1082 ± 38 a	2273 ± 36 a	330 ± 43 a	1191 ± 08 a
Na_2SO_3	1446 ± 37 a	1129 ± 36 a	2327 ± 46 a	317 ± 10 a	1198 ± 14 a
Distilled water	1424 ± 45 a	1136 ± 68 a	2326 ± 66 a	288 ± 31 a	1190 ± 19 a

[A] Means within a column followed by different letters are significantly different ($p < 0.05$).

Table 4-5. Gelatinization thermal properties of mung bean starches isolated using different steeping liquors (Chang et al., 2006)

Steeping liquor	Gelatinization temperature (°C)[A]			$T_C - T_O$ (°C)	ΔH
	T_O	T_P	T_C		(Jg^{-1},db)
LFS	65.3±0.1a[B]	70.9±0.1a	77.4±0.1a	12.0±0.2b	12.3±0.2b
NaOH	52.6±0.4b	64.7±0.4b	76.5±0.5b	23.9±0.6a	13.4±0.2a
Na_2SO_3	52.4±0.1b	64.0±0.2c	77.1±0.5ab	24.7±0.5a	13.4±0.1a
Distilled water	52.3±0.3b	64.5±0.1b	76.7±0.4ab	24.3±0.3a	13.3±0.2a

[A] T_O, T_P, T_C and T_C -T_O are the onset, peak, completion and temperature range of starch gelatinization, respectively. ΔH is the enthalpy change of starch gelatinization.

[B] Means within a column followed by different letters are significantly different ($p<0.05$).

Jane et al. (1999) investigated the relationship between chain length of amylopectin and gelatinization properties of starches with different X-ray patterns. They indicated that a higher average chain length of amylopectin or a lower proportion of short chains might contribute to higher gelatinization temperature of starch.

2.1.6. Molecular Structure (Molecular Weight and Chain Length Distribution)

Chang et al. (2006) investigated the molecular structure of mung bean starch isolated from different steeping liquor and indicated that the first fraction (F1) with a shorter retention volume corresponds to mung bean amylopectin, and the second fraction (F2) to the low molecular weight molecules consisting of mung bean amylose and low molecular weight amylopectin. The weight-average molecular weight (M_w) of F1 and F2 fractions among mung bean starches isolated by different steeping liquors were similar and ranged from 7.25 to 8.06×10^7 and 9.09 to 10.17×10^5 Da, respectively (Figure 4-11 and Table 4-6) (Chang et al., 2006). An increase on the proportion of low M_w fraction of corn starch when the concentration of lactic acid used in the steeping process increased from 0.2% to 1.5%, and concluded that depolymerization of starch during steeping was enhanced by higher concentration of lactic acid used. Therefore, the significant reduction in the amount of F1 fraction for the LFS-isolated starch should be attributed to the partial degradation of amylopectin during isolation. The chain-length distribution of mung bean starch (Figure 4-12) isolated using different steeping liquors was observed after debranching by isoamylase. The HPSEC profile was divided into four fractions. These fractions correspond to amylose (DF1), and longer B chains (DF2, B2 chains or longer), B1 chains (DF3) and A chains (DF4) of amylopectin (Hizukuri, 1986; Chang et al., 2006).

The weight-average degree of polymerization DPw of DF1, DF2, DF3 and DF4 for the isolated mung bean starches had ranges of 4516~5665, 61.1~63.8, 25.9~27.3 and 12.6~13.5, respectively (Table 4-7) (Chang et al., 2006). LFS-isolated starch had significantly higher percentage of fractions DF2 (longer B chains) and DF3 (B1 chains) than other starches. Consequently, the percentage of DF4 (A chains) for LFS-isolated starch was the lowest. Furthermore, the ratio of short-to-long chains (S/L ratio = [DF3 + DF4]/DF2) of amylopectin for LFS-isolated starch was 2.61, and was significantly lower than those (3.09~3.20) of starches isolated using different steeping liquors. The relatively lower values of the weight-percentage of DF4 fraction and S/L ratio for the LFS-isolated starch could be attributed to the degradation of amylopectin and amylose during steeping in lactic acid fermentation solution. The importance of molecular characteristics of starch on noodle quality had been reported by Mestres et al. (1988). They illustrated that the amylose and amylopectin macromolecules would reorganize within a new crystalline structure during processing. Furthermore, the extent of retrogradation would depend on the amylopectin structure as the starches shared similar amylose content. Starch with a higher proportion of long chains in its amylopectin fraction tends to have a higher extent of retrogradation (Wang et al., 2002). Therefore, this may be one of the reasons that the manufacturers of mung bean starch noodles tend to use LFS-isolated mung bean starch (Chang et al., 2006).

Table 4-6. Weight percentages and average molecular weights (Mw) of HPSEC fractions of mung bean starches isolated using different steeping liquors (Chang et al., 2006)

Steeping Liquor	Distribution (%, w/w)		M_W (Da)	
	F1	F2	F1 ($x10^7$)[A]	F2 ($x10^5$)[B]
LFS	66.3±0.7b[C]	33.7±0.7a	8.06±0.31a	9.09±0.53c
NaOH	70.0±0.1 a	30.0±0.1b	7.25±0.27b	10.17±0.31a
Na_2SO_3	69.7±0.5a	30.3 ±0.5b	7.97±0.10a	9.85±0.14ab
Distilled water	69.1±1.1a	30.9±1.1b	7.71±0.02a	9.48±0.29bc

[A] Molecular weight was determined by light scattering and refractive index detectors.
[B] Molecular weight was determined by refractive index detector based on pullulan standard curve.
[C] Means within a column followed by different letters are significantly different ($p<0.05$).

Table 4-7. Weight percentages and average polymerization degrees (DPw) of HPSEC fractions of isoamylase-debranched mung bean starches isolated using different steeping liquors (Chang et al., 2006)

Steeping Liquor	Distribution (%,w/w)				S/L ratio[A]	DP_w			
	DF1	DF2	DF3	DF4		DF1[B]	DF2[C]	DF3[C]	DF4[C]
LFS	32.5±0.1a[D]	18.7±0.7a	24.7±0.7a	24.1±1.4b	2.61±0.14b	4516±105b	63.1±0.6a	25.9±0.2c	12.6±0.1c
NaOH	32.3±1.2a	16.2±0.6b	23.2±0.6b	28.4±1.3a	3.20±0.10a	5444±115a	61.1±0.9b	26.3±0.4b	12.9±0.2b
Na_2SO_3	32.8±1.6a	15.9±0.8b	23.0±0.9b	28.4±0.5a	3.23±0.13a	5665±257a	63.2±0.6a	27.3±0.2a	13.4±0.1a
Distilled water	31.8±0.4a	16.7±0.3b	23.0±0.4b	28.5±0.3a	3.09±0.10a	5472±110a	63.8±0.2a	27.3±0.1a	13.5±0.0a

[A] S/L ratio=[(DF3%+DF4%)/(DF2%)].
[B] Molecular weight determined by light scattering and refractive index detectors.
[C] Molecular weight determined by refractive index detector based on pullulan standard curve.
[D] Means within a column followed by different letters are significantly different($p<0.05$).

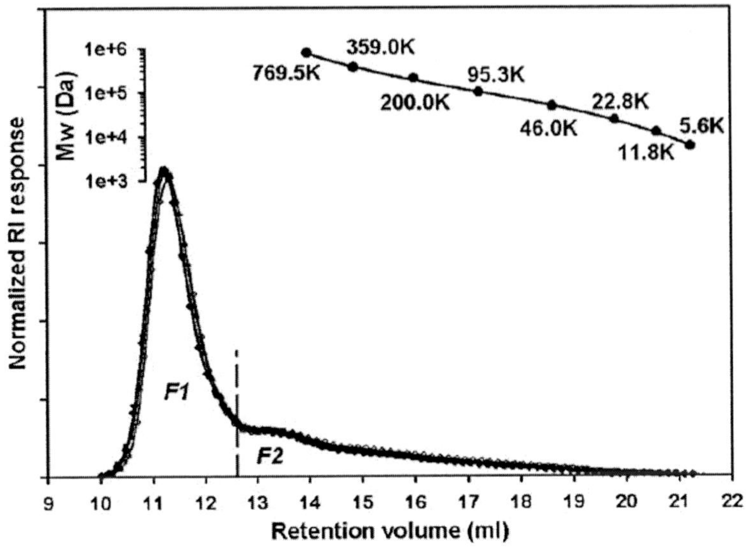

Figure 4-11. High-performance size-exclusion chromatograms of mung bean starches isolated using different steeping liquors: LFS (○), NaOH(◆), Na2SO3 (▲), and distilled water (▼), and molecular weight of pullulan standard (●). (Chang et al., 2006)

Chang et al. (2006) summarized the results of molecular weight distribution of starch, and indicated that mung bean starch was degraded during isolation with lactic acid fermentation solution. On the other hand, the LFS-isolated starch was found to have relative higher weight percentage of long chains (B2 or longer), lower weight percentage of A chains and lower S/L ratio than those of starches isolated using other liquors (Figure 4-12 and Table 4-7). Furthermore, a narrow and mono-modal gelatinization peak (Figure 4-10) with higher gelatinization temperature was also observed on the LFS isolated starch.

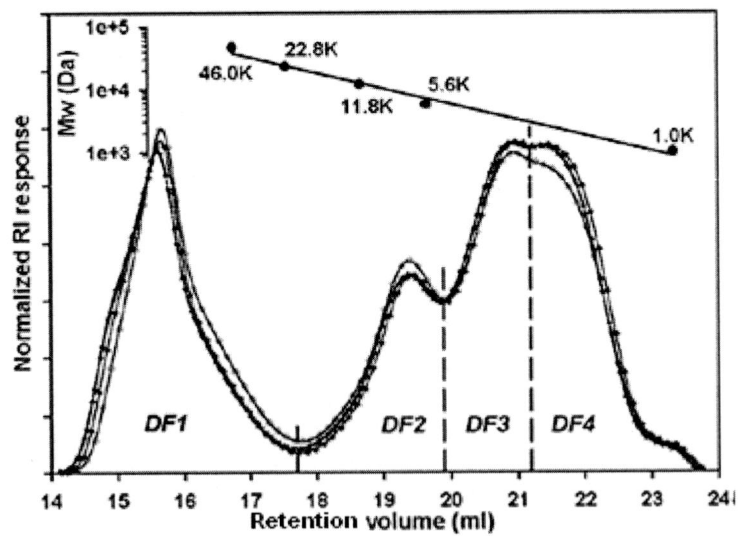

Figure 4-12. High-performance size-exclusion chromatograms of isoamylase-debranched mung bean starches isolated using different steeping liquors: LFS (s), NaOH (r), Na2SO3 (m), distilled water (.), and molecular weight of pullulan standard (•). (Chang et al., 2006)

2.1.7. Rheological Property of Mung Bean Starch

Starch noodles are produced from purified starch. The formation and quality of starch dough is a crucial step in the processing of starch noodles. The drop of starch dough and the formation of filament depend on the rheological properties of the dough itself, especially shear-thinning properties and gravity, which decreases viscosity, increases the fluidity of the starch dough and facilitates the dropping of filaments. In addition, process parameters such as the content of moisture and starch paste, stirring rate and temperature in starch dough, are also important. Tan et al. (2007) investigated the rheological behavior of mung bean starch dough (MBSD) under different conditions, which is essential for the production of starch noodles.

2.1.7.1. Thixotropic Flow Properties

The shear sensitivity of mung bean starch dough can be estimated from the hysteresis loops of the flow curves (Figures 4-13~14). After two serial sweepings of shear rate over the range from 0 to 500 s^{-1}(Uplink) were carried out, continued by a descending sweep from 500 to 0 s^{-1}(Downlink), MBSDs with different moisture contents, starch paste contents and temperatures, exhibited high thixotropy. The greater the area between the ascending and descending curves, the more sensitive is the starch dough to mechanical shearing. The flow curve of MBSD exhibited unclosed hysteresis loops with different area and yield stress, indicating the extent of restoration after breakdown in the inner structure of the starch dough. The structural breakdown process taking place in starch dough during shear was irreversible and the rebuilding of the inner structure of the sheared starch dough during shear was slow or negligible. This demonstrated that the starch dough was non-Newtonian, shear-thinning and thixotropic (Tan et al., 2007).

The MBSD exhibited unique rheological behavior which was dependent on factors such as starch paste content, moisture content, temperature, agitation rate and time. Stirring MBSD at 40 °C was suitable in view of its smallest hysteresis loop area and its lowest zero-shear viscosity, which gave rise to better fluidity. Increasing moisture content obviously led to a decrease in the starch dough viscosity, and the hysteresis loop area (Figure 4-14). The hysteresis loop area and zero-shear viscosity dropped markedly when moisture content in MBSD was 44 w/w%, which was suitable for stir and drop. A decrease in the content of starch paste led to an obvious decrease in the viscosity, and the area of hysteresis loop of the starch dough (Figure 4-15) (Tan et al., 2007). Because the hysteresis loop is partly interpretted in terms of time dependency at different shear stress levels (Härröd, 1989), Tan et al. (2007) measured the changes of viscosity of mung bean starch doughs with time at various constant shear rates (Figure 4-16). The higher the applied shear rate, the larger the rate and extent of viscosity reduction, and the more pronounced time-dependence level of the mung bean starch dough.

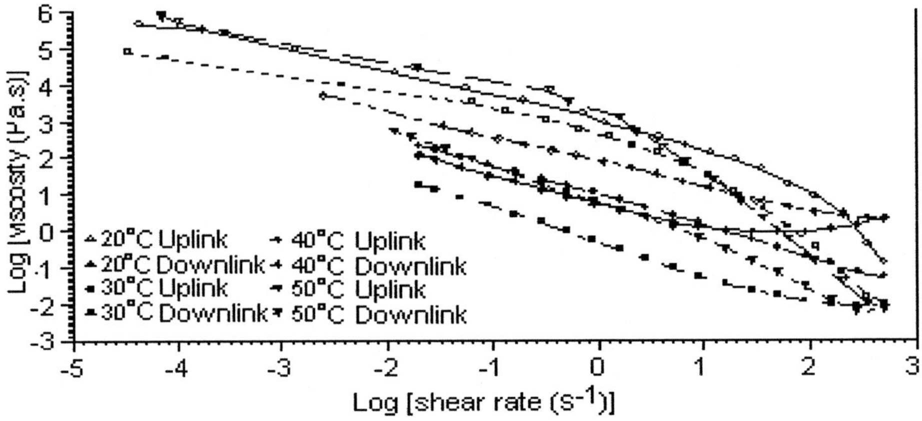

Figure 4-13. Variation of viscosity with shear rate of mung bean starch dough under different temperatures (Tan et al., 2007).

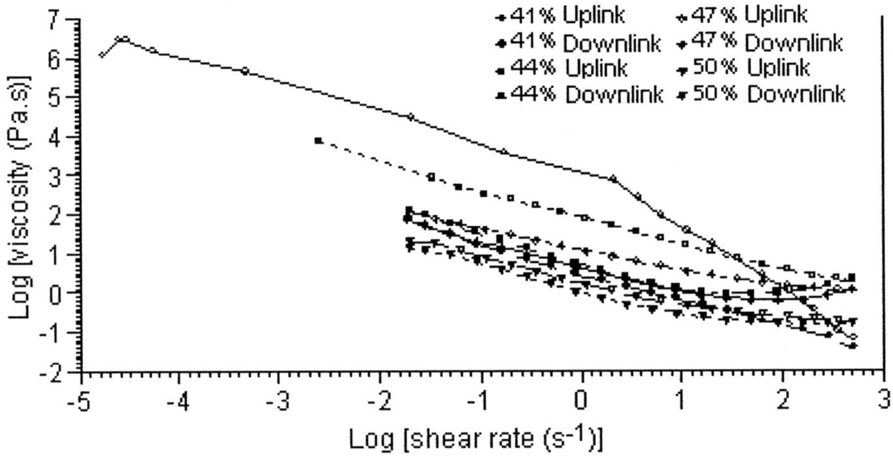

Figure 4-14. Variation of viscosity with shear rate of mung bean starch dough under different moisture contents (Tan et al., 2007).

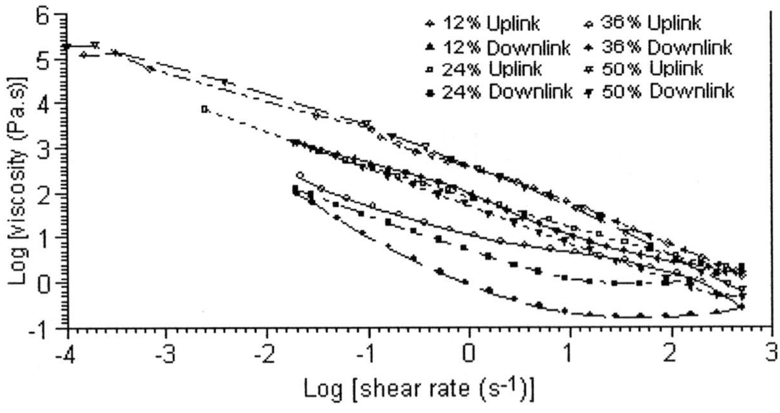

Figure 4-15. Variation of viscosity with shear rate of mung bean starch dough under different starch paste contents (Tan et al., 2007).

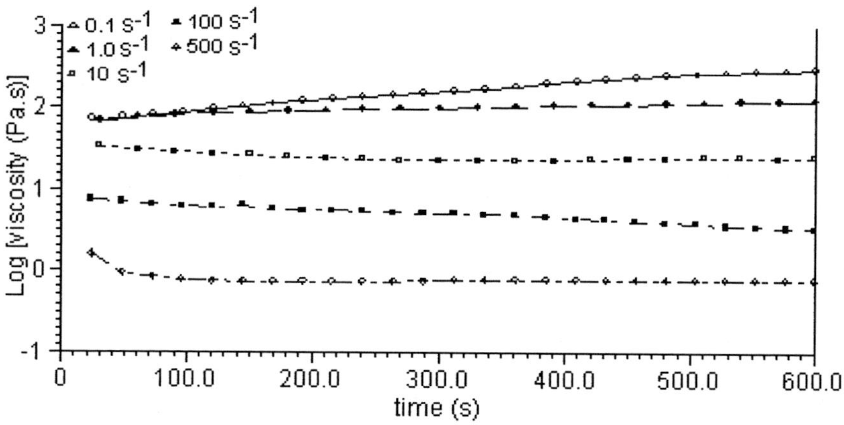

Figure 4-16. Variation of viscosity with time of mung bean starch dough under different constant shear rates (Tan et al., 2007).

2.1.7.2. Modeling of Flow Behavior for Mung Bean Starch Dough

The MBSD showed an initial Newtonian plateau (region where the viscosity remains approximately constant) and a relatively high yield stress. Therefore, the flow behavior of the MBSD was described by model-Cross model (Eq.4-1):

$$(\eta-\eta_\infty)/(\eta_0-\eta_\infty)=1/[1+(k\gamma)^m] \quad (4\text{-}1)$$

Where η_0 is the zero-shear or plateau viscosity (Pa·s), η_∞ is the infinite-shear viscosity (Pa·s), k is a constant with time-dimension (s) and m is a dimensionless parameter (Steffe, 1992). The values of viscosity at zero-shear rate can predict the yield intensity, which implied the energy required at the beginning of stirring for starch dough and offered some references for the design of stirring equipment, while the values of viscosity at infinite-shear rate can predict the maximum shear-thinning of starch dough, which implied the fluidity of starch dough. A high zero-shear viscosity produced a "damping" effect on shear rate at various locations, as reported by Prakash and Kokini (2000). It is essential to supply sufficient power to overcome these rheological effects and promote stirring efficiency. The starch dough requires a lower η_0 to minimize the energy when the stirring begins, and a higher η_∞ to contain starch dough glutinosity for the compactness of starch noodles under durative or infinite stirring (Tan et al., 2007).

2.1.7.3. Flow Behavior of Pure Mung Bean Starch Slurry without Starch Paste

The flow behavior of pure starch slurry (the mixture of starch and water), the main part of starch dough, exhibited a unique rheological behavior (Figure 4-17). After undergoing a process of shear-thickening, the viscosity of mung bean starch slurry dropped rapidly over 65 s^{-1}. When the shear rate decreased from 500 s^{-1} to 0 s^{-1}, the viscosity of starch slurry ascended slowly and restored to initial viscosity. The flow curve of mung bean starch slurry fitted the Herschel-Bulkley model (Eq.4-2) (Steffe, 1996):

$$\tau = \tau_y + K\gamma^n \quad (4\text{-}2)$$

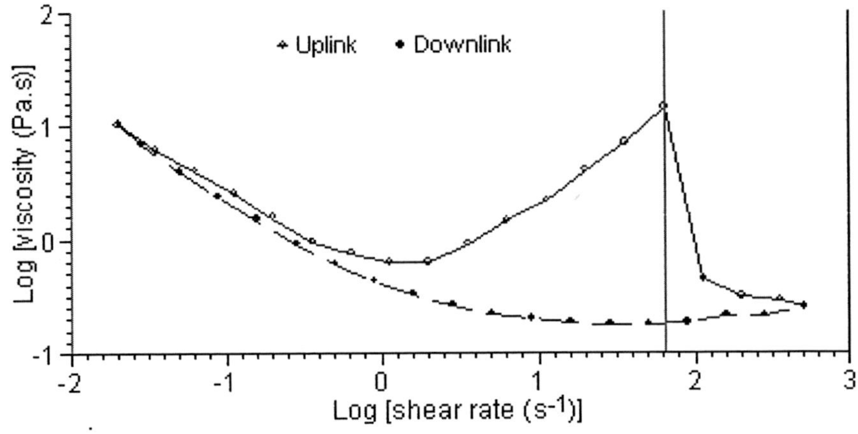

Figure 4-17. Variation of viscosity with shear rate of pure mung bean starch slurry without starch paste.

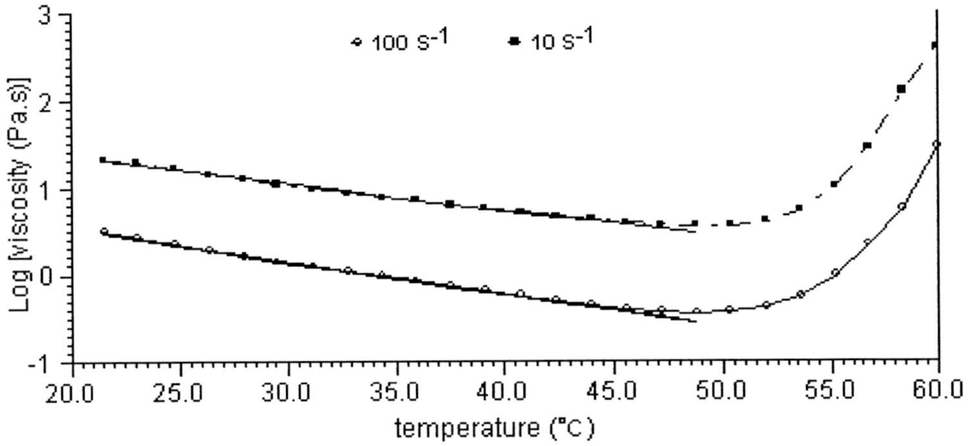

Figure 4-18. Variation of viscosity with continuously rising temperature of mung bean starch dough.

Where τ_y is the yield stress (Pa), K is the consistency index (Pa.s^n), n is the flow behavior index (dimensionless). The starch slurry, which was also a thixotropic fluid, displayed a close hysteresis loop and had only a little yield stress (τ_y=0.37Pa) during initial shearing. In the uplink of Figure 4-17, it can be seen that the shear-thickening behavior (n=1.98) of the starch slurry was preceded by shear-thinning behavior. The critical shear rate (γ_c) was 2.00 s^{-1} at which the flow behavior transformed from shear-thinning to shear-thickening (Christianson and Bagley, 1983).

2.1.7.4. Temperature-sensitivity of Mung Bean Starch Dough

The flow property for MBSD at the shear rate of 10 s^{-1} and 100 s^{-1} and the temperature of 20~60 °C exhibited the property of variation of viscosity with continuously rising temperature (Figure 4-18). For MBSD, the variation of viscosity dependent on temperature at 10 s^{-1} and 100 s^{-1} was evaluated by using the Arrhenius model:

$$\eta = A\ exp\ (\Delta E_\eta / RT) \qquad (4\text{-}3)$$

Where A is a frequency factor, exp is the natural logarithm base, ΔE_η is the activation energy (kJ/mol), R is the gas constant (8.314 kJ/kmol.K) and T is the absolute temperature (K). The activation energy increased during heating and led to the decrease in flow resistance and viscosity. The Arrhenius equation was suitable for MBSD within the range of 20~48 °C because 48~56 °C was a transition temperature range for forthcoming gelatinization of mung bean starch. The onset gelatinization temperature of mung bean starch (about 57 °C) (Tan et al., 2006) could help explain why the effect of working temperature on ΔE_η changed around 48~56 °C. The viscosity of starch dough rose rapidly and exhibited a resistance to flow after 57 °C. In general, the higher the activation energy, the greater is the effect of temperature on viscosity and vice versa (Sayar et al., 2001, Turhan and Gunasekaran 2002). Higher ΔE_η values for MBSD at 100 s^{-1} than 10 s^{-1} below 48 °C showed that some extra energy was needed for overcoming the resistance of the viscosity of starch paste and the dilatant flow behavior of raw starch granules for stirring at 100 s^{-1}. The starch dough at 10 s^{-1}, thus, was less sensitive to temperature than that at 100 s^{-1} and was more suitable for stirring in the process of starch noodle producing.

The MBSD exhibits thixotropic flow behavior, as characterized by a viscosity that decreases with increasing shear rate (shear-thinning or pseudoplastic) and increases with decreasing shear rate, but doesn't restore to the initial viscosity. These are typical properties of composite two-phase fluids with internal structures which undergo shear induced by changes with time, leading to changes in the macroscopic flow behavior (Nguyen et al., 1998). It may be predicted that the thixotropic structure in the MBSD breaks down irreversibly after two serial sweepings of shear rate (Tan et al., 2007).

Particle size distribution is an important factor affecting the flow behavior. Starch granules with different size, arranged orderly in a style where the small granules crammed into the interspace between large granules, formed a compact entity to resist flow. Starch granule shape was another important factor affecting the flow behavior. The pure starch slurry dough, consisting of countless irregular granules and water, was still difficult to flow when applying low shear rate and displayed dilatant flow behavior. As the shear rate increases continuously, the internal structure of the whole system of pure starch slurry will be disrupted, its viscosity thus drops rapidly and exhibits a section of disorderly curve (65 s^{-1}-500 s^{-1}) (Figure 4-15). Due to the rearrangement of starch granules in the pure starch slurry system and the infiltration and lubrication of water into the interspace between granules, the destroyed internal net structure in the pure starch slurry is rebuilt and leads to increased viscosity of the initial value during the process of decreasing shear rate (500-0 s^{-1}) (Tan et al., 2007).

2.2. Other Legume Starches

Legume starches have occupied an important place in noodle preparation in several countries of the world, so many researchers investigated the potential of other legume starches for noodle preparation. Lii and Chang (1981) prepared noodles from red bean *(Phaseolus radiatus var. Aurea)* starch and reported that noodles were of acceptable quality but not as good as mung bean starch noodles. The properties of kidney bean *(Phaseolus vulgaris)* starches, pigeonpea starch, and pea/lentil starch were examined by Yang et al.

(1980), Singh et al. (1989), Rask (2004) with special reference to noodle preparation, respectively.

2.2.1. Pea

The pea is a nutritionally important grain legume of the tropical and subtropical regions of the world. Pea starch is the second excellent material followed by mung bean starch for processing starch noodles. Ratnayake and Hoover (2002) and Rask (2003) provide an overview on the composition, structure and properties of pea starch.

There are two different seed phenotypes. They are genetically different so the morphologies and characteristics of the starches are different too (Table 4-8) (Ratnayake and Hoover, 2002).

The physical characteristics of pea amylose and amylopectin are listed in Table 4-9 and 4-10, respectively.

Table 4-8. Proximate composition and morphology of pea starch (Ratnayake and Hoover, 2002)

Phenotype	Yield pure starch (%)	Protein Content (%)	Ash Content (%)	Amylose (%)	Granular shape appearance	Granule size
Smooth pea	35-40	0.52-0.70	0.01-0.07	24-65	Large and small granules, oval or spherical	22.9-30.4μm 5-20μm
Wrinkled pea	18-22	0.34-0.46	0.01-0.08	60.5-88	Mixture of simple and compound granules 4-6 associated pieces in a ring formation	~10-40μm 17-30μm

Table 4-9. Physicochemical characteristics of pea amylose (Ratnayake and Hoover, 2002)

Starch source	Iodine binding capacity	Intrinsic viscosity (η[Ml/g])	Average degree of Polymerization (DP_n)	Number average Molecular Weight (M_n)	Weight average Molecular Weight (M_w)	B-amylolysis (%)	Branch points per molecules
Smooth pea	18.8-19.2	180-264	1300-1400	170000	N/A	81.6-86.9	3.2
Wrinkled pea	17.9-19.2	136-172	1000-1100	125000	1288000	79-85	2-3

Functional properties of pea starch have been described by Ratnayake and Hoover (2002). The swelling factor of smooth pea starch ranges from 4 to 27 in the temperature range of 50-95 °C. Wrinkled pea starch has a lower swelling power. Gelatinization parameters for smooth, wrinkled and mutant pea starches are listed in Table 4-11 (Ratnayake and Hoover, 2002).

Table 4-10. Physicochemical characteristics of pea amylopectin (Ratnayake and Hoover, 2002)

Starch source	Iodine affinity	Average branch chain length	Branch points per molecules	Molecular Weight (M_n)	Weight average degree of polymerization (DP)	B-Amylolysis (%)	Crystallinity (%)	B Polymorph (%)
Smooth pea	1.28	22-24.2	---	80.6 x 10^6	---	96-97	18.9-36.5	12.0-49.0
Wrinkled pea	5.26	34	8.2	19.4 x 10^6	6195	98	Not reported	Not reported

Table 4-11. Gelatinization parameters of wild and mutant pea starches (Ratnayake and Hoover, 2002)

Source	Transition Temperatures (°C)			Enthalpy ΔH [J/g]
	T_O	T_P	T_C	
Smooth pea	55-61.4	60-67.5	75-80	14.1-22.6
Wrinkled pea	117	133	138	2.9
Wild type	*	61.8	*	10.8
r	*	52.5-60.0	*	2.4
rb	*	66.1	*	12.6
rug3	*	70.0	*	7.5
rug4	*	65.4	*	9.8
rug5	*	49.0-57	*	5.1
lam	*	58.6	*	6.8

* not available; temperature of: T_O, onset; T_P mid-point T_C conclusion.

2.2.2. Red Bean

Red bean (*Phaseolus radiatus* var. *aurea*) is mainly used as one of the popular ingredients in oriental style desserts, bean jams and Japanese confectionery (Lii and Chang, 1981). However, Lii and Chang (1981) had used it to manufacture starch noodles and studied the characterization of red bean starch and its noodle quality. They made the following conclusions about it. Microscopically most red bean starch granules had irregular shapes with deep fissures, which may be due to the way in which the granules were packed within the protein matrix in the endosperm. Clearly centric birefringence was observed when the granules were examined under polarized light. The sizes of the granules ranged from 25–67 μm which were relatively larger than those of other legume starches (Lii and Chang, 1981).

The gelatinization temperature range of red bean starch was 63.0–66.5–70.0 °C. The swelling power of red bean starch is of the restricted type, like mung bean. The solubility pattern, as usual, paralleled the swelling power. The Brabender viscosity pattern of red bean gave no pasting peak during cooking. Neither did it show a breakdown of the hot paste. Such a pattern is similar to those of most legume starch pastes and could be classified into type C. The iodine affinity value was not high (4.83%), but red bean starch had high hot paste stability. This phenomenon seems peculiar. The result also indicated that the different steeping solutions used during starch isolation did not affect the viscosity pattern. However, they did influence the pasting viscosities. (Lii and Chang, 1981)

The degree of syneresis of red bean starch gel is higher than that of mung bean and less than that of kidney bean and pea starch gels. The degree of syneresis increased as the concentration of the starch gel decreased, which is similar to other legume starches. The gel strength of red bean starch was much weaker than those of mung bean, pea and kidney bean starches. This may be attributed to its low iodine affinity value and resulting lower amylose content. The X-ray diffractogram of the native starch granule of red bean showed an A-pattern (Lii and Chang, 1981).

2.2.3. Phaseolus Vulgaris (Common Bean)

In the literature, Phaseolus vulgaris refers equally to the terms common bean or dry bean, and to individual varieties such as kidney beans, navy beans (also known as white or pea beans), French beans, haricot beans, pinto beans, filed beans, China beans, frijol, marrow beans, snap beans, black beans or white beans, cranberry beans etc. (Rask, 2003).

There is a paucity of information in the literature regarding the physico-chemical characteristics and functional properties of *P. vulgaris* starch or its components, amylose and amylopectin (Table 4-12). The yield of pure starch varied from 18 to 40% among the five cultivars examined. Amylose content in these five biotypes of *P. vulgaris* showed a range of 30.2% (for pinto beans) to 37.3% (for black beans) (Hoover and Sosulski, 1985). Four Canadian varieties of milled and air-classified navy beans showed a range of 29.1–32.96% amylose (Hoover and Sosulski, 1991).

Starch granules for *P. vulgaris* ranged in size from 10–42 μm in width to 12–62 μm in length depending upon the cultivar.

Table 4-12. Proximate analysis of legume starches (Hoover and Sosulski, 1991)

Starch source	Yield of pure starch (%)	Protein (%)	Lipid (%)	Ash (%)	Amylase (%)	Iodine affinity	Water Binding Capacity (%)
Kidney bean	25	0.13-0.30	0.18	0.18	34.4-35.0	7.02-8.04	82.0
Northern bean	18-31	0.35-0.97	0.20-0.46	---	31.6	---	80.2
Navy bean	21-40	0.13-0.34	0.09-0.60	0.06-0.14	36	6.58-7.20	80.8
Black bean	32	0.55-1.12	0.15	0.11	35.1-37.3	6.82-7.20	84.1
Pinto bean	27-38	0.37-0.52	0.16-0.51	0.05-0.09	25.8-30.2	---	78.7

All legume starches studied by Hoover and Sosulski (1985) exhibited single stage restricted swelling and low solubility patterns that are indicative of strong bonding within the starch granule.

2.2.4. Pigeonpea

Pigeonpea (*Cajanus cajan* L.) is a nutritionally important grain legume of the tropical and subtropical regions of the world. Although India accounts for about 85% of the word's supply of pigeonpea, this legume is becoming popular in several countries of Africa and South East Asia. In these countries, the utilization of pigeonpea, including its alternative uses for humans and animals, is receiving increasing attention from food scientists and nutritionists. There were efforts on exploring some new foods for pigeonpea, such as starch-based food products and noodles (Singh et al., 1989). Singh et al. (1989) had studied physicochemical characteristics of the pigeonpea and its noodle quality. They found that the starch fraction contained 0.10–0.18% protein, 0.03–0.09% ash and 0.0–0.11% crude fiber, including high purity of the starch fractions. No large difference in the amylose content of pigeonpea (46.9%) and mung bean starch (47.0%) were observed. However, its amylose contents are considerably higher than those of black beans (35.1%) (Singh et al., 1989).

Most pigeonpea granules had irregular shapes, which varied from oval to round to bean-shaped. A large variability existed in the starch granule size of pigeonpea starch, ranging from 9.5 to 55.1 μm. The gelatinization temperature of pigeonpea starch (76 °C) was slightly higher than that of mung bean starch (72 °C). The range in gelatinization temperature is 65–71–76 °C. Gel strength of pigeonpea starch was lower than that of mung bean starch, whereas no large differences in gel consistency were observed. The degree of syneresis of pigeonpea starch gel was higher than that of the mung bean. The swelling power of pigeonpea starch at lower temperatures was also noticeably lower than that of mung bean starch. The extent of increase in viscosity on cooling to 50 °C reflected a retrograde tendency in the starch molecules. Pigeonpea starch showed a much lower set-back value than mung bean starch (Singh et al., 1989).

2.3. Sweet Potato Starch

The roots and foliage of the sweet potato (*Ipomoea Batatas LAM.*) are important commodities to small-scale farmers in Africa, Latin America and Asia. Over 90% of the production in developing countries is in Asia, especially China, where the crop has been estimated to provide up to 10% and 5% respectively of the intakes of calories and protein. In certain regions of Africa, South America, the Caribbean and the Pacific, the sweet potato is important as a staple calorie source (Tian et al., 1991). Total annual word production of sweet potato was 125 million tonnes, about 83% of which was produced in China. The annual plant area of sweet potato was 6.2 million hectares in China and account for 65% of the total world planting area (FAO; Collado and Corke, 1997).

Sweet potatoes are used as animal fed in China and the USA and for the manufacture of industrial starch in Japan (Tian et al., 1991). In a number of provinces in China, it is important for the production of noodles. Sweet potato starch noodles are extensively produced in China, where it is estimated that 28% of the processed sweet potato is made into starch noodles (Wang et al., 1995). This product is also widely consumed in Korea, Vietnam, and Taiwan province of China (Wang et al., 1995). Studies on noodles based on sweet potato starch are of interest to many developing and developed countries because it plays a vital role in food production, such as in substitution for expensive mung bean starch. A number of studies on the distinctive properties of sweet potato starch have been undertaken in the last three decades.

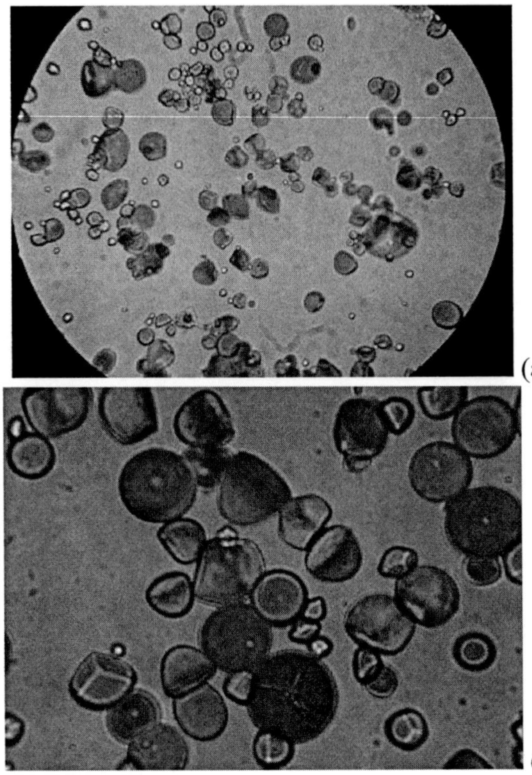

(a) (400×)

(b) (640×)

Figure 4-19. Polarization crosses of sweet potato starch granules (The author provided; Tan, 2007).

2.3.1. Isolation and Morphological Property of Sweet Potato Starch

The use of sweet potato starch is primarily determined by its physicochemical properties. The extraction of starch from sweet potato tubers is not simple. The presence of fibrous material and latex prevents easy settling of starch and this leads to extended residence time for the starch in the mother liquor. Since the mother liquor contains a lot of sugars, fermentation sets in, leading to deterioration of starch properties. Starch often possesses an off-colour if not processed properly due to the phenolics present in the tubers (Eliasson, 2004). Kallabinski and Balagopalan (1991) have used an enzymatic technique to extract starch from sweet potato tubers. The use of cellulose and pectinase resulted in an increased yield of starch without affecting the properties of the extracted starch. The purity of starch isolated from sweet potato roots grown in Japan, India, Indonesia, Philippines, Peru, and Ghana, varied from 88.1% to 99.8%. (Tian et al., 1991; Collado and Corke, 1997; Chen et al., 2003).

Sweet potato starch granules are small or large, smooth, oval spherical, oval round, round polygonal or polygonal and some of them had internal fissures. Sweet potato starches have polarization cross and obvious concentric hilum (Figure 4-19) (Tan, 2007).

Table 4-13. Chemical composition of isolated starches from sweet potatoes and mung bean (w/w %) (Tan, 2007)

Source	Starch	Moisture	Protein	Lipids	Phosphorus ug/g	Apparent A_m [a]	Absolute A_m	A_p [b]
Mung bean	88.45±0.33	8.99±0.13	0.68±0.08	0.74±0.07	-----	39.60±0.41	36.70	66.3
Xushu 18	89.94±0.46	9.75±0.21	0.36±0.11	0.67±0.10	1.20±0.06	30.20±0.35	28.90	71.1
Suyu 303	90.12±0.71	9.59±0.16	0.20±0.07	0.52±0.14	1.99±0.17	24.60±1.22	26.80	76.2
Ning 27-17	87.78±0.24	11.86±0.18	0.35±0.06	0.18±0.05	1.47±0.14	24.96±0.49	24.10	75.9
Sushu 8	89.11±0.52	10.43±0.14	0.28±0.06	0.49±0.05	2.98±0.21	28.83±0.60	27.00	76.0
Sushu 2	91.56±0.19	8.13±0.20	0.24±0.09	0.35±0.04	2.34±0.25	25.21±0.12	24.90	75.1
Sushu 9	91.93±0.20	8.63±0.18	0.42±0.12	0.50±0.07	6.23±0.09	22.48±0.39	22.20	77.8
AB94 001-8	90.09±0.48	9.32±0.15	0.37±0.12	0.23±0.09	5.02±0.24	26.46±0.29	26.30	76.7

A_m: Amylose.
A_p: Amylopectin.

The size of the starch granules may be estimated by the rate of sedimentation, by the use of an instrument such as a Coulter or by microscopic analysis. The granular size of sweet potato starch ranged from 2 to 42 μm (Tan et al., 2007). Sweet potato starch granules are of a similar size to those of cassava and maize but are smaller than those of potato which also have a large range of granular sizes. There is a negative correlation among sweet potato cultivars between particle size and susceptibility to α–amylase and acid degradation (Tian et al., 1991).

2.3.2. Proximate Analysis of Sweet Potato Starch

The protein content of the 3 sweet potato starches (Chen et al., 2003) and the 7 sweet potato starches (Tan, 2007) in China were 0.17~0.23% and 0.20~0.42% respectively. It was higher than that found for Irish potato starch but lower than that found for mung bean starch. The lipid content of these sweet potato starches was lower than that of mung bean starch. High lipid contents may result in low clarity of the starch paste (as with cereal starches) and repressing starch granule swelling (Kasemsuwan et al., 1998). Sweet potato starch is also similar to cassava starch in its lipid and phosphorus content and hence its properties are quite similar to cassava starch (Eliasson, 2004). Like potato starch, the amylose of sweet potato starch contains less phosphate than the amylopectin. High levels of phosphate ester groups give amylopectin of potato starch a slight negative charge, resulting in some coulombic repulsion that may contribute to the rapid swelling of potato starch granules in warm water and to several properties of potato starch pastes like high viscosity, high clarity, and low rate of retrogradation (Bemiller and Whistler, 1996).

2.3.3. Physical Property of Sweet Potato Starch

2.3.3.1. Swelling and Solubility

Swelling and solubility tests on starch provide evidence for the associative bonding within the granule. The extent of swelling can be plotted against pasting temperature to monitor the progressive relaxation of the bonding forces within the granule. This permits comparison of relative bond strengths in starch and the temperature (i.e., energy level) necessary to cause relaxation. Data on the swelling power has been compared by Tian et al. (1991) and the values vary considerably not only among varieties, but also at different temperatures. The mean swelling volume of the different genotypes of sweet potato starch was 33.0 mL/g, in a fairly narrow range from 30.9 to 35.2 mL/g, while mean solubility was 12.7% (ranging from 10.7% to 14.4%) (Collado and Corke, 1997). The swelling power at 90 °C of mung bean starch was low (10%) compared to that of sweet potato which ranged from 26 to 33% (Tan, 2007). This indicates that the associative bonding forces within the granules are rather weak as compared to mung bean starch, which is strong even at high temperatures. The comparatively lower swelling volume of sweet potato starch has been attributed to a higher degree of intermolecular association compared to cassava or potato starch. Collado et al. (1999) have examined the swelling volume of starch of a number of Philippine accessions and found the range to be between 24.5 to 32.7 mL/g with a mean value of 29.9 mL/g showing weaker associative forces compared to legume starches. There was no significant correlation between amylose content and swelling volumes.

Swelling volume was correlated with solubility of starch. The solubility of starch extracted from seven sweet potato collections from Peru indicated that solubility increased with temperature and reached nearly 10%, while for commercial starch, it was 28% (Garcia and Walter, 1998). The authors found that the selection index did not have a noticeable effect, but location had significant influence at temperatures about 60 °C. Collado et al. (1999) found the solubility to be in the range 12 to 24% (average 16.9%). It was presumed that the bonding forces might be tenuous but comparatively extensive, immobilizing the starch within the granules even at high levels of swelling.

2.3.3.2. Water-binding Capacity

The water-binding capacity of sweet potato starch ranged from 66.3 to 211.6 as shown in the summing-up from Tian et al. (1991). In general, tuberous starches have higher water-binding capacities than those of cereal origin, and the majority of workers have demonstrated that sweet potato starch has a higher water-binding capacity than potato (93%) and cassava starches (72–92%).

2.3.3.3. Syneresis

Syneresis, in general, is related to “freeze-thaw” stability, and the latter can be used as an indicator for the tendency of starch to retrograde (Eliasson and Kim, 1992). Chen et al. (2003) found that the syneresis values (without a freeze-thaw treatment) of sweet potato starch were lower than that of mung bean starch but higher than that of potato starch. The retrogradation tendency measured by the syneresis of freeze-thaw stability and by the syneresis without a freeze-thaw treatment (stored at 2 °C) did not agree with each other. The amount of water excluded in the freeze-thaw phase would be the result of increased intermolecular and intermolecular hydrogen bonding due to the interaction between starch chains (amylose-amylose, amylose-amylopectin, and amylopectin-amylopectin) during frozen storage. The retrogradation tendency, as measured by setback ratio of paste viscosity at the higher starch concentration (6% in Brabender amylogram and 8% in RVA profile), agree well with the results measured by the syneresis without the freeze-thaw treatment (stored at 2 °C) (Chen et al., 2003).

2.3.3.4. Crystalline Structure

Sweet potato starch has a variable X-ray pattern of “A” pattern, “C” or intermediate between “A” and “C” (Eliasson, 2004). Takeda et al. (1986) observed “A” pattern for two varieties while it was “C_A” for another variety. The absolute crystallinity for this starch was 38%. Type A starches tend to have higher levels of crystallinity (33~45%) and higher gelatinization temperature (Tian et al., 1991).

2.3.4. Pasting and Gelatinization Properties

Some genotypes of sweet potato starch showed a broad peak almost like a plateau, which reflected in the P_{time}, and the stability ratio (Tan, 2007). While some sweet potato starches have a distinct and sharp peak. The average P_{time} was 1.4 min, and the average stability ratio was 0.43. For the sweet potato genotypes evaluated, P_{time} was highly correlated with stability ratio (Collado and Corke, 1997). The average peak viscosity (PV) was 385 RVU, ranging from 331 to 428. PV was significantly negatively correlated with amylose content. The hot-paste viscosity (HPV) was significantly correlated with the cool-paste viscosity (CPV). The average HPV was 163 RVU ranging from 127 to 203 RVU, while the average CPV was 251 RVU ranging from 208 to284 RVU. The average P_{temp} was 80.4 °C ranging from 78.3 °C to 84.1 °C (Collado and Corke, 1997). The value of RVA parameters for these sweet potato starches was far lower than those of mung bean starch (Tan, 2007).

Most researchers have examined the DSC characteristic of 70 sweet potato genotypes from China and the Phillipines, and obtained considerable variation in all the parameters (Collado and Corke, 1997; Collado et al., 1999; Tan, 2007). The mean T_{onset} was 64.6 °C and ranging 61.3–70 °C, mean T_{peak} 73.9 °C(range 70.2–77 °C) and mean T_{end} 84.6°C, range being 80.7–88.5 °C and the mean gelatinization range was 20.1 °C with 16.1 °C to 23 °C. Sweet potato has been reported to gelatinize between 58 °C and 90 °C, with a gelatinization enthalpy ranging from 10.0 to 16.3 J/g. The P_{temp} correlated with the T_{end}, but values from the RVA were lower than those from DSC. From Tan’s (2007) research results, the gelatinization enthalpy of 7 sweet potato starches in China have significant variety ranging from 0.71 to 11.9 J/g (Figure 3). The starch from fresh tubers and freeze-dried sweet potato tubers gave nearly equal values (67–73 °C), but the small granules gelatinized between 75 °C and 88 °C (Eliasson, 2004).

The pasting temperature of sweet potato starch varied between 66.0 and 86.3°C by viscography while microscopic determination gave values of 57–70 °C to 70–90 °C. Sweet potato starch behaves similarly to cassava starch in its viscosity characteristics, viz., peak viscosity, viscosity breakdown and setback viscosity (Eliasson, 2004). Chen et al. (2003) also reported that the gelatinization temperature range of the 3 sweet potato starches was obviously higher than those of potato starch and mung bean starch.

2.3.5. Molecular Structure

Noda et al. (1996) used HPAEC-PAD on sweet potato starch and found the amylopectin to have peaks at DP=12 and DP=8. The concentrations of the peaks at DP=6 and DP=7 were 7.1–7.5% and 6.7–7.0% respectively. Takeda et al. (1986) found a trimodal pattern for the sweet potato amylopectin while Hizukuri (1969) reported a bimodal distribution. They conclude that sweet potato has a higher proportion of "A" chains and short "B" chains compared to potato. Seog et al. (1987) reported alkali number values between 7.66 and 12.13 for six Korean sweet potato varieties compared to 5.33 for cassava starch.

Tan et al. (2006) studied the structure of sweet potato starch and compared it with mung bean starch, as a standard starch for the production of starch noodles. They concluded that the amylopectin in sweet potato (A_p-SP) possessed a molecular weight of 2.23×10^7 Da, corresponding to approximately 137600 (DP) which was characteristic of hydroglucose residues. In comparison, the amylopectin in mung bean (A_p-MB) possessed a molecular weight of 1.82×10^7 Da, which corresponded to a chain length mass of approximately 112300 (DP) characteristic of hydroglucose residues, and implied smaller amounts of branches for the A_p-MB fraction than the A_p-SP fraction. The intermediate materials in sweet potato starch (SPS) were present in higher yields (approximately 11.0%) than that in mung bean starch (MBS) (approximately 4.0%).

The amylose content of starch isolated from sweet potato roots grown in China, Japan, India, Indonesia, Philippines, Peru, and Ghana, ranged from 8.5% to 37.4% (Tan et al., 2006; Tian et al., 1991; Collado and Corke, 1997; Chen et al., 2003). In general, sweet potato can have amylose content slightly higher than that of cassava but less than that of wheat, maize or potato (Tian et al., 1991).The absolute amylose content in MBS was 33.7%, greater than that of SPS (28.9%). Both amylose molecules possessed 9.0 and 1.8 chains with various chain lengths 226 and 2250 for amylose in sweet potato (A_m-SP) and amylose in mung bean (A_m-MB), respectively, indicating that the Am-MB contained low molar fractions of branched molecules whereas the Am-SP contained a high molar fraction (Tan et al., 2006). Takeda and Hizukuri (1987) also reported that sweet potato amylose was composed of 9.8 chains. Sweet potato amylose appears to have more branches per amylose molecule than that from legume, cassava, potato, wheat or maize. This is one of the reasons for the lower retrogradation tendency of sweet potato amylose. 70% of sweet potato amylose molecules were branched compared with 42% in cassava and 27% in wheat (Tian et al., 1991).

A_p-MB has a longer peak chain length of long-branch chains than A_p-SP (DP 40) compared with DP35 at peak. The resolution of the linear oligosaccharide peak fractions revealed 5 populations for A_p-SP of chain length distributions from the amylopectin molecules. A_p-SP contained more short chains than long chains. The chemical analysis indicated that the long chains of A_p-MB (DP40) were longer than A_p-SP (DP35), but the short chains of A_p-SP (DP8) were shorter than those of A_p-MB (DP15) (Tan et al., 2006).

2.3.6. Rheological Properties

The rheological properties of sweet potato starch extracted using an enzymatic process did not vary among the different concentrations of enzyme up to 0.1% (Moorthy and Balagopalan, 1999). Guraya et al. (1998) reported the apparent viscosity of a large number of sweet potato varieties to vary considerably from 71–442 cPs and storage led to reduction in viscosity. The rheological properties of sweet potato starch have been examined using a

Bohlin rheometer (Garcia and Walter, 1998). Storage modulus, G′, Loss modulus, G″ and tan d summed over different starch samples were determined. During heating, the G′ and G″ increased while phase angle decreased indicating change from sol to gel. The initial increase has been attributed to progressive swelling of starch granules leading to close packing. When the starch granules became very soft, deformable and compressible, decreases in G′ and G″ were observed. The elastic nature prevailed over the viscous nature of the paste. In terms of the summing-up on a number of starch rheology researches from Tian et al. (1991), sweet potato amylose has a limiting viscosity higher than that of wheat but lower than that of cassava or Irish potato amylose. Similarly, sweet potato amylopectin has a lower limiting viscosity number than Irish potato amylopectin, suggesting smaller or more spherical molecules. Varietal differences in viscosity have been reported as significant.

2.4. Potato Starch

The potato was introduced to China probably several times via various routes during the seventeenth century. Nowadays the potato plays an important role in the food industry, especially its starch which is used to produce starch noodles in China. The noodles maintain a clear and shiny appearance after cooking, have a smooth and slippery texture, and high absorption of soups and sauces (Singh et al., 2002). Potato starch is used for its characteristics, which differ significantly from those of starch from other plant sources. Identification of native starch sources is required for desired functionality and unique properties (Singh and Singh, 2001).

2.4.1. Morphological Property of Potato Starch

Starch from different potato varieties differed significantly in granule size and shape. Starch granules ranged from large to small and oval to irregular or cuboidal with diameter ranges between 15–20 μm and 20–45 μm, respectively (Figure 4-20). The surface of the granules appeared to be smooth when viewed at 400 × magnification. They showed the presence of irregular or cubiodal granules in large number and very much less or negligible numbers of small and oval granules or large numbers of small and large oval granules. The variation in size and shape of starch granules may be due to the biological origin. The morphology of starch granules depends on the biochemistry of the chloroplast or amyloplast, as well as physiology of the plant (Singh and Singh, 2001).

2.4.2. Physico-chemical Characteristics of Potato Starch

The amylose content of potato starches ranged from 25.1% to 31.6% (Kaur et al., 2002, Singh et al., 2002). Mealy potatoes have a higher amylose content than waxy potatoes (Kaur et al., 2002).

The difference in swelling powers and solubility of different starches may be attributed to the difference in viscosity patterns and weak internal organization resulting from negatively charged phosphate groups within the potato starch granules (Table 4-14) (Kim et al., 1996).

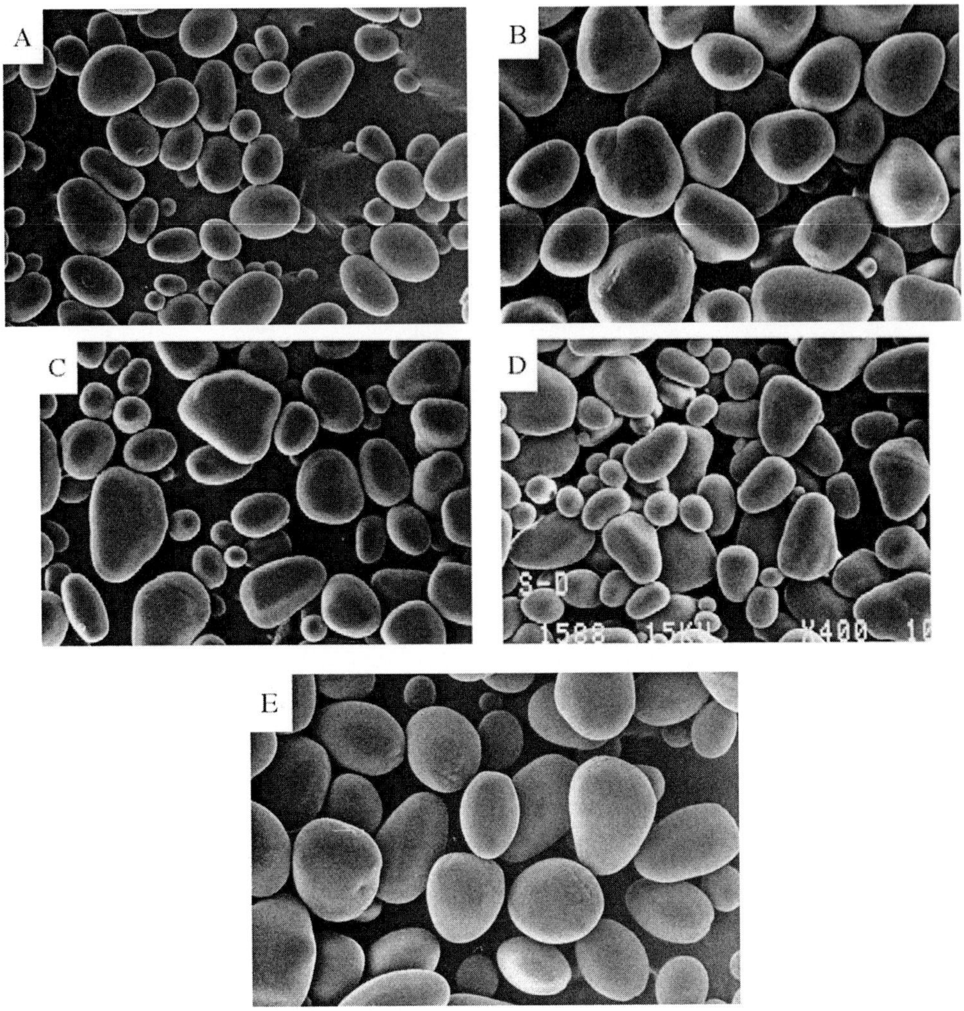

Figure 4-20. Scanning electron micrographs (SEM) of starches separated from different potato cultivars (A) Kufri Chandermukhi, (B) Kufri Badshah, (C) Kufri Jyoti, (D) Kufri Sindhuri, (E) S1. (Singh and Singh, 2001).

The difference in morphological structures of granules may also be responsible for the difference in swelling power and solubility of the three starches (Singh and Singh, 2001). The turbidity values of gelatinized starch suspensions from the three potato cultivars differed significantly (Tables 4-15). This may be due to the presence of fewer granule remnants in the starch paste, which in turn depends on the starch granule morphology.

The covalently bound phosphate groups in potato starch granules also contribute to the differences in the light transmittance values. The light transmittance values of starch suspensions from all the potato cultivars decreased while turbidity values increased progressively during storage. The granule swelling, granule remnants, leached amylose and amylopectin, amylose and amylopectin chain lengths have been reported to be responsible for turbidity development in starches during storage. Starches from the mealy potato cultivars having larger sized granules showed higher transmittance and lower turbidity values. WBC of the three starches also differed significantly (Table 4-14). It may be attributed to the variation in granular structure. Loose association of amylose and amylopectin molecules in the native starch granules has also been reported to be responsible for high water bonding capacity (Kaur et al., 2002).

Table 4-14. Swelling power, solubility, water binding capacity and amylose content of starches separated from different potato cultivars [a](Kaur et al., 2002)

Cultivar	Solubility (%)	Swelling Power (g/g)	Water Binding Capacity (%)	Amylose content (%)
Pukhraj	0.093a	56.22a	99.8c	25.2a
Kufri Jyoti	0.127b	64.7c	93.4a	31.2c
Kufri Badshah	0.099a	59.74b	97.41b	29.8b

[a] Values with similar letters in column do not differ significantly($p<0.05$).

Table 4-15. Effect of storage duration on the turbidity of starches separated from different potato cultivars [a] (Kaur et al., 2002)

Cultivar	Turbidity (Absorbance at 640 nm)				
	0h	24h	48h	72h	96h
Pukhraj	1.25a	1.47b	1.68c	1.77d	1.81e
Kufri Jyoti	1.247a	1.33b	1.44c	1.5d	1.57e
Kufri Badshah	1.238a	1.3bc	1.34cd	1.38d	1.43e

[a] Values with similar letters in column do not differ significantly($p < 0.05$).

2.4.3. Thermal Properties of Potato Starch

The transition temperatures (T_o; T_p; and T_c), range (T_c–T_o), enthalpies of gelatinization (ΔH_{gel}) and peak height indices (PHI) of starches from different potato cultivars differ significantly (Table 4-16). The ΔH_{gel}, T_o value of various potato starches ranged from 12.55–13.85 J/g, 59.72–60.69 °C, respectively. T_p and T_c of starches from different cultivars ranged between 63.26–64.58 °C and 67.28–68.35 °C, respectively. Kaur et al. (2002) reported similar ranges of transition temperatures and enthalpies of gelatinization for starches from 3 potato cultivars. Double helical and crystalline structures are disrupted in starches during gelatinization. This order-disorder phase transition showed melting of crystals which was illustrated by DSC endotherms, in the range 50–70 °C, for various native starches. The ΔH_{gel} reflected the loss of double helical rather than crystalline order. High transition temperatures have been reported to result from a high degree of crystallinity which provided structural stability and made the granule more resistant to gelatinization. The starch from potato cultivars having smaller starch granules showed lower ΔH_{gel} and vice versa. Granule shapes, percentage of large and small granules, and presence of phosphate esters have been reported to affect the gelatinization enthalpy values of starches (Singh and Singh, 2001).

Table 4-16. Thermal properties of starch separated from different potato cultivars [a] (Singh and Singh, 2001)

Cultivar	T_O (°C)	T_P (°C)	T_C (°C)	ΔH_{gel} J/g	PHI	R	ΔH_{ret} J/g	%R
Kufri chandermuki	60.27bc	63.39a	67.28a	12.55a	4.022c	7.01a	6.42a	51.50a
Kufri badshah	59.72a	63.45a	68.35c	13.85c	3.713ab	8.63c	8.61d	62.16c
Kufri jyoti	59.86ab	63.26a	67.66ab	13.68c	4.023c	7.80b	7.53c	55.04c
Kufri sindhuri	60.70c	64.58b	70.34b	13.38b	3.439a	9.65d	7.84c	58.59d
S1	59.78a	63.41a	68.00bc	13.36b	3.680a	8.21bc	7.12b	53.3b

[a] T_O= onset temperature, T_P=peak temperature, R=gelatinization range(T_C-T_O); ΔH_{gel}=enthalpy of gelatinization(dwb, based on starch weight), PHI=peak height index $\Delta H_{gel}/(T_P - T_O)$, ΔH_{ret}=enthalpy of retrogradation, % R=percentage of retrogradation(ratio of enthalpy of gelatinization to enthalpy of retrogradation).values with similar superscripts in column do not differ significantly($p<0.05$).

2.4.4. Rheological Properties of Potato Starch

The three cultivars studied by Kaur et al. (2002), showed TG′ of 60.1–62.7 °C during the heating cycle, which proved the difference between gelatinization temperatures of these starches (Table 4-17). The G′ and G″ of three potato starches increased progressively to a maximum, and then dropped during the heating cycle. The difference in the G′, G″ and tan σ during the heating cycle may be attributed to the difference in the starch granular structure which in turn depends on their biological origin. The extent of breakdown in G′ was measured as the degree of disintegration of starch granules (Singh and Singh, 2001). The greater breakdown in potato starch may be attributed to the presence of more large-sized starch granules which are fragile in nature. During cooling of the heated starch pastes from 75 to 25°C, G′ and G″ values increased and tan σ value decreased. That potato starch have higher G′, G″ and lower tan σ, means the formation of the most rigid gel structure. A decrease in tan σ values during cooling of starches has been reported to be evidence of gel formation (Redy and Seib, 2000). Kufri Jyoti potatoes with highest mealiness scores resulted in starch paste showing highest consistency coefficient and lowest flow behaviour index (Table 4-17). The mealiness of the cooked potatoes correlated with the relative starch viscosities (Kaur et al., 2002).

Table 4-17. Rheological properties of starches measured using dynamic rheometer (during heating) and Brookfield viscometer[a] (Kaur et al., 2002)

Cultivar	*TG'* (°C)	Peak *G'* (Pa)	Peak *G''* (Pa)	Breakdown in *G'*(Pa)	Peak tan δ	*K* (Pas$^{\eta}$)	*N*	R^2
Pukhraj	62.7c	8519a	2186a	3521a	0.2566c	25.0a	0.33a	0.997
Kufri Jyoti	61.5bc	12804b	2471b	6894b	0.193b	38.1c	0.29a	0.988
Kufri Badshah	60.1ab	16100c	2617c	9889c	0.1626a	26.0b	0.38b	0.995

[a] Values with similar letters in column do not differ significantly($p<0.05$).

2.5. Corn Starch

China has the second largest output of corn in the world. The demand for corn is growing in China with the setting up of food processing units involved in the processing of corn (ISY, 2006/2007). Corn provides a high-quality starch used widely in the food industry in many applications requiring particular viscosities and textures (Eliasson, 2004). Using corn in starch noodle making will be a good trial, but the traditional production experience and the previous study showed that the corn starch noodle is not as good as the mung bean starch noodle. Yuan et al. (2008) introduced spontaneous lactic acid fermentation to corn starch to improve the texture of the corn starch noodle. Starches from different corn types differ widely with respect to the morphological, rheological, functional and thermal properties (Sandhu et al., 2004).

2.5.1. Physicochemical Characteristics of Corn Starches

Starch is the major carbohydrate of corn, making up 72–73% of the kernel. Normal maize starch consists of 75% branched amylopectin; the remaining 25% is linear amylose. Amylose content of starches separated from different corn types ranged between 15.3% and 25.1% (Table 4-18) (Sandhu et al., 2004). An amylose content of 22.1% in corn starches has been reported earlier by Singh and Singh (2003). Cluskey et al. (1980) observed that amylose content of dent corn starch granules, fractionated according to size ranged from 24% for the largest to 22% for the smallest granules.

The ability of the starches from different corn types to swell in an excess of water and their solubility is presented. The swelling power and solubility of starches from different corn types ranged from 14.9 to 17.9 g/g and 12.5 to 20.3%, respectively. Highest swelling power was observed for dent corn bold grain starch. The swelling power of starch has been reported to depend on the water holding capacity of starch molecules by hydrogen bonding (Table 4-18, Sandhu et al., 2004). Among various pop corn grain fractions, medium grain fraction had highest swelling power and solubility. Dent corn bold grain fraction had higher swelling power as compared to its counterpart dent corn long grain fraction. Baby corn starch has a low amylose content and swelling power, so lower amylose content-higher swelling power applies only for starch granules obtained from same corn type (Sandhu et al., 2004). The starch granules with higher amylose content, on the other hand, being better reinforced and thus more rigid, probably swell less freely.

Water bonding capacity (WBC) of the starches from different corn types ranged between 96% and 107%, lowest for starch from dent corn bold grain fraction and highest for baby corn and pop corn medium starch was observed (Table 4-18) (Sandhu et al., 2004). The differences in WBC of starches from different corn types may be attributed to the variation in their granule structure. Pop corn medium grain fraction had higher WBC than its counterpart's small and large grain fractions. Dent corn long grain starch had higher WBC than dent corn bold grain starch.

The turbidity values of the starch paste from all corn fractions increased progressively during storage (Figure 4-21 and Table 4-17) (Sandhu et al., 2004). The increase in turbidity during storages has been attributed to the interaction between leached amylose and amylopectin chains that led to development of function zones, which reflect or scatter a significant amount of light (Perera and Hoover, 1999). Turbidity development in starch pastes during storage have been reported to be affected by factors such as granule swelling, granule remnants, leached amylose and amylopectin, amylose and amylopectin chain lengths (Jacobson, Obanni, and BeMiller, 1997). Pop corn starch pastes showed highest turbidity values whereas lowest values were observed for dent corn starch pastes after 120 h of storage at 4 °C. The dent corn bold grain starch paste showed lower turbidity than its counterpart long grain starch paste which may be due to the presence of fewer granule remnants in the starch paste, which in turn depends on the granule morphology (Sandhu et al., 2004).

2.5.2. Morphological Properties of Corn Starches

The form of starch granules separated from different corn types range from small to large and oval to polyhedral (Figure 4-22) (Sandhu et al., 2004). Singh et al (2003) reported angular shape for corn starch granules. The figure clearly indicates that diameter of majority of starch

Table 4-18. Swelling power, solubility, water binding capacity, amylose content, mean diameter and turbidity of starches separated from different corn types and their fractions (Sandhu et al., 2004)

Corn type	Fraction	Swelling power (g/g)	Solubility (%)	Water binding capacity (%)	Amylase content (%)	Mean diameter (μm)	Turbidity(nm)				
							0h	24h	48h	72h	120h
Dent corn	Bold	17.9	18.9	96	20.7	13.35	0.89	1.12	1.33	1.44	1.56
Dent corn	Long	17.7	20.3	104	23.4	13.13	0.93	1.22	1.38	1.50	1.58
Pop corn	Small	16.7	18.3	102	25.1	12.77	0.93	1.44	1.56	1.60	1.66
Pop corn	Medium	17.3	18.8	107	22.4	13.64	0.91	1.44	1.52	1.60	1.64
Pop corn	Large	16.4	18.2	105	24.4	13.42	0.95	1.50	1.56	1.62	1.64
Baby corn	-	14.9	12.5	107	15.3	6.33	1.32	1.40	1.48	1.56	1.66

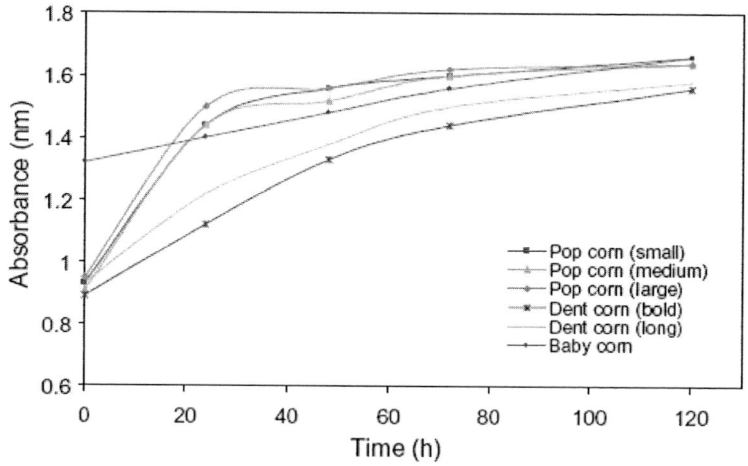

Figure 4-21. Effect of storage duration on the turbidity of starch pastes from different corn types (Sandhu et al., 2004).

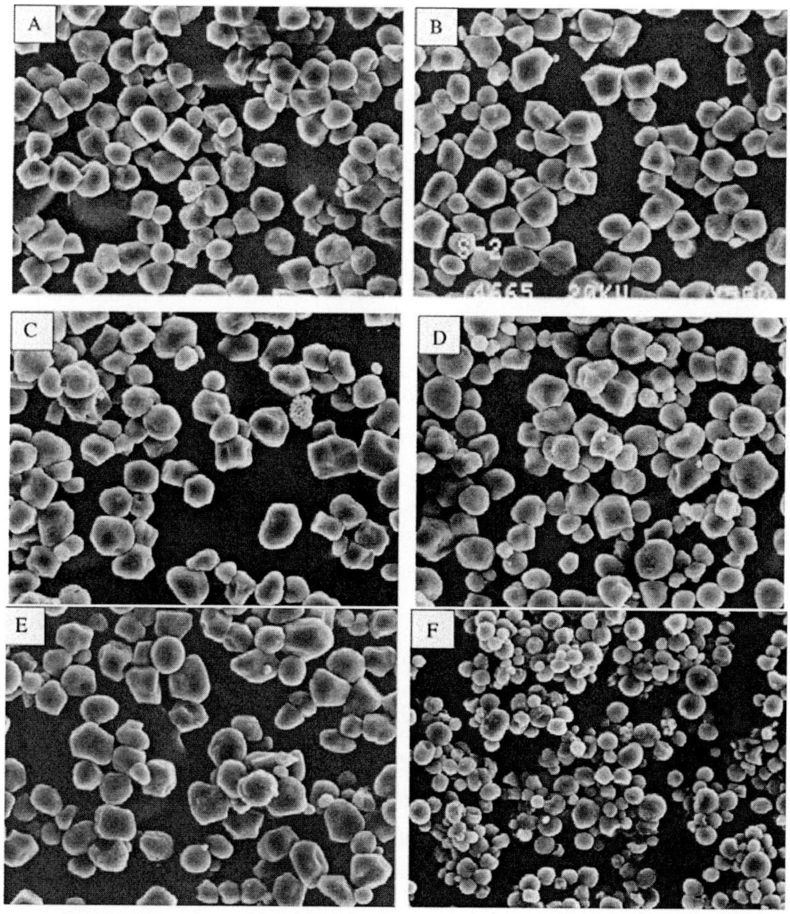

Figure 4-22. Scanning electron micrographs (SEM) of starches separated from different corn types: (A) pop corn (small), (B) pop corn (medium), (C) pop corn (large), (D) dent corn (bold), (E) dent corn (long), (F) baby corn (dent type) (Sandhu et al., 2004).

granules ranged between 6 and 30 μm with some granules having diameter in the range of 0.4-4 μm (Sandhu et al., 2004). Singh et al (2003) reported average size of individual corn starch granules in ranges from 1 to 7 μm for small and 15 to 20 μm for large granules. Baby corn starch showed the presence of smallest size granules with mean diameter of 6.33 μm whereas pop corn medium grain fraction starch had largest granules with mean diameter of 13.64 μm (Table 4-18). Dent corn bold grain starch had a higher mean diameter than its counterpart long grain fraction starch. Baby corn starch had small oval shape granules whereas starches from other corn types showed the presence of polyhedral shape granules. When viewed under scanning electron microscope, the surface of the granules showed the presence of surface pores. Fannon and BeMiller (1992) also observed the presence of pores on the surface of corn, sorghum and millet starch granules.

Table 4-19. Thermal properties of starches separated from different corn types and their fractions (Sandhu et al., 2004)

Corn type	Fraction	T_O(°C)	T_P (°C)	T_C (°C)	$\triangle H_{gel}$(J/g)	PHI	R
Dent corn	Bold	69.3	73.1	77.7	10.9	2.9	7.5
Dent corn	Long	69.2	73.1	78.0	10.1	2.6	7.7
Pop corn	Small	67.9	71.9	77.2	9.7	2.4	8.0
Pop corn	Medium	68.1	71.9	76.5	9.5	2.5	7.6
Pop corn	Large	68.1	71.9	76.9	10.2	2.7	7.6
Baby corn	---	66.3	71.5	77.8	8.9	1.7	10.3

T_O=onset temperature, T_P=peak temperature, T_C=conclusion temperature, R=gelatinization range2(T_P - T_O), $\triangle H_{gel}$=enthalpy of gelatinization(dwb, based on starch weight), PHI=peak height index$\triangle H_{gel}$ /(T_P -T_O)

2.5.3. Thermal Properties of Corn Starches

$\triangle H_{gel}$ of corn starches ranged from 8.9 to 10.9 J/g (Table 4-19)(Sandhu et al., 2004). The lowest and highest $\triangle H_{gel}$ values among different corn types were in the starches isolated from baby corn and dent corn bold grain fraction. The gelatinization enthalpy value of starch was affected by factors such as granule shape, percentage of large and small granules, and the presence of phosphate esters. The lower $\triangle H_{gel}$ of baby corn starch may be attributed to its small granule size and lowest amylose content. T_o, T_p and T_c of starches from different corn types ranged between 66.3-69.3 °C, 71.5-73.1 °C and 76.5-78.0 °C, respectively. No significant differences were observed in T_o and T_p values among different fractions of dent corn and pop corn (Sandhu et al., 2004). Perera et al (2001) reported value of T_o for normal corn starches to be 64.4 °C. Highest T_p and T_c of 73.1 and 78.0 °C, respectively was observed for starch separated from dent corn long grain fraction. Li et al (1994) reported values of T_o, T_p and T_c and $\triangle H_{gel}$ among several maize populations in the range of 64.3-69.6 °C, 70.1-73.9 °C, 76.8-79.6 °C and 2-2.9 cal/g, respectively. Baby corn starch showed maximum R value of 10.3 while the PHI value (1.7) was narrow for the same. Highest PHI of 2.9 and lowest R value of 7.5 was observed for dent corn bold grain fraction. The differences in the R values among the starches from different corn types may be due to the presence of crystalline regions of different strength in the granule. Li et al (1994) reported gelatinization ranges of 35 tropical and subtropical maize populations to vary between 9.8 and 13 °C. The gelatinization ranges for starches from five open pollinated corn populations has been reported by White et al. (1990) to vary between 8.7 °C and 16.4 °C. Starches from both fractions of dent corn and pop corn large grain fraction with higher T_o, T_p, ΔH_{gel}, PHI and narrower R may have a higher degree of molecular order than starches from other fractions. Similar observations for corn starches have been reported earlier by Krueger et al (1987). The variation in T_o, ΔH_{gel} and R in starches from different corn types might be due to differences in amounts of longer chains in amylopectin(Sandhu et al., 2004).

2.5.4. Rheological Properties of Corn Starches

Sandhu et al (2004) studied the rheological properties of starches separated from different corn types during heating. They illustrated changes in storage modulus (G′), loss modulus (G ″) and loss factor (tan σ), respectively of the starches as a function of temperature during heating of suspensions in a dynamic rheometer (Table 4-20). The temperature at which G′ was maximum (TG') ranged from 73 to 73.7 °C, highest for dent corn bold grain fraction and baby corn starch gels and lowest for starch gels from pop corn small and large grain fraction was observed. The temperatures observed for peak G′ and G″ of corn starch gels ranged between To and Tc obtained with DSC. Peak G′ and G″ values of different corn starch gels ranged between 2172–5354 and 383–920 Pa, respectively (Sandhu et al., 2004).

Table 4-20. Rheological properties of starch gels from different corn types and their fractions during heating (Sandhu et al., 2004)

Corn type	Fraction	*TG′* (°C)	Peak *G′* (Pa)	Peak *G″* (Pa)	Breakdown in G′(Pa)	Peak tan δ
Dent corn	Bold	73.7	2919	427	1177	0.146
Dent corn	Long	73.4	2463	383	1393	0.155
Pop corn	Small	73.0	2172	393	1102	0.181
Pop corn	Medium	73.4	3620	438	2020	0.122
Pop corn	Large	73.0	5354	920	3184	0.172
Baby corn	---	73.7	4884	813	2944	0.166

Starch gel from dent corn bold grain fraction had higher TG', peak G′ and G″ than starch gel from dent corn long grain fraction. The extent of breakdown in G′ is a degree of disintegration of starch granules (Singh et al., 2002). Pop corn large grain fraction starch gel showed maximum breakdown in G′, followed by baby corn starch gel whereas it was lowest for pop corn small grain starch gel. The differences in breakdown values among corn starch gels may be attributed to the differences in morphological characteristics of starch granules and peak G′ values. Peak tan σ values of starch gels from all corn types were <1 (Sandhu et al., 2004). Peak tan σ value was 0.181 for pop corn small grain fraction and 0.122 for pop corn medium grain fraction starch gels. Among dent corn fractions, starch gel from long grain fraction had higher values of breakdown and tan d than starch gel from bold grain fraction (Sandhu et al., 2004). The differences in G′, G″ and tan σ during the heating cycle may be attributed to the difference in the starch granule structure which in turn depends on their biological origin (Svegmark and Hermansson, 1993). Therefore the rheological properties of starch depended mainly on the interaction among close-packed granules and their rigidity during the heating process. Lii et al (1996) reported that rheological behavior of gelatinized starch suspension was primarily due to intergranular interaction, such as entanglement between surface molecules of adjacent granules and the properties of the granules themselves.

3. Processing Technology for Starch Noodles

The best mung bean threads may keep their original shape and remain intact for about two hours after cooking and being kept in soup. This is because of its unique starch gelling properties, which also provide very good *al dente* properties (Hui, 2007). The characteristics of starch noodles, unlike wheat-based noodles, depends heavily upon the functional properties of the starch as it undergoes one or two heat treatments during processing. The heat treatment may involve boiling or steaming that gelatinizes the starch and the subsequent retrogradation sets the structure of the starch noodles. The processing technology is unique and divided into three parts, namely, dropping, extruding and cutting (Figure 4-23).

Soaked cleaned mung bean in water (4–5 h in summer, 10 h in winter)
↓
Finely grind soaked mung beans with added water
↓
Dilute the slurry with 3×water
↓
Ferment the diluted slurry for 8–9 days (change water as needed)
↓
Filter out mung bean starch in cloth bag by gravity
↓
Divide mung bean starch into two portions
↓
Add cold water to first half mung bean starch to form slurry
↓
Add boiling water to make a thin paste
↓
Add the second half of mung bean starch to form thick and elastic paste
↓
Press think paste through perforated funnel
↓
Drop extruded threads into boiling water to form transparent threads
↓
Recover transparent threads and form bundles
↓
Dry bundles of mung bean threads
↓
Pack dry mung bean threads into plastic bags

Figure 4-23. Steps in the production of traditional mung bean threads (Hui, 2007).

3.1. Traditional Processing Technology

The dropping method is the most traditional one in China. About 5% of starch is cooked in water using a double boiler to prepare starch paste and used as dough binder. The cooked gelatinized starch (starch paste) is then mixed with 95% of starch and water to give 50% moisture content in the dough and then mixed and stirred at the rate of 100 r/min about 10min using a blender to distribute water evenly and to obtain a smooth ball (starch dough) that does not stick to the hands. The dough was extruded through the holes (about 0.5~1.5 cm diameter) of the stainless steel cylinder by gravity, directly into hot water (98–100 °C), and heated for 30–60 s before transferring into cold water (when noodles are floated on the surface of water then transfer them into cold water). After rinsing in cold water, the strands were drained, subsequently, separated and hung to partially dry, kept at 4 °C for 2 h and −10 °C for overnight, dried at 40 °C in convection dryer, and then packed in polyethylene bags and stored at room temperature (Tan et al., 2006).

Figure 4-23, 4-24 describes the procedures used to make traditional mung bean threads. Recently, broad strips made from mainly mung bean are also available (Hui, 2007).

Starch isolation →

Forming the starch dough

→Dropping

Hanging→

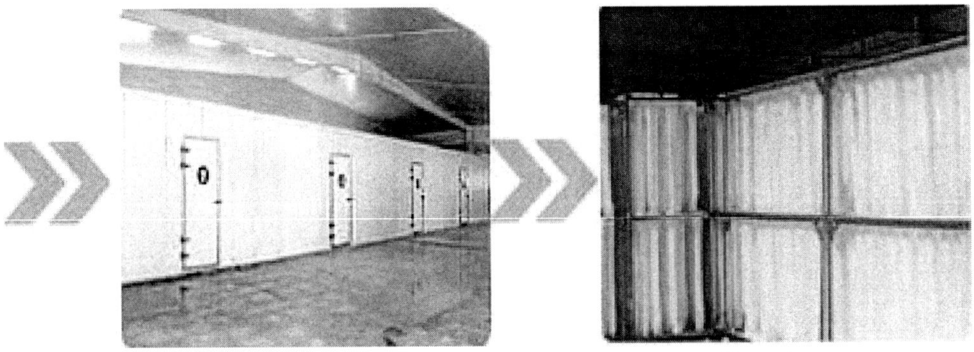

Freezing→ Thawing→

Figure 4-24. (Continued).

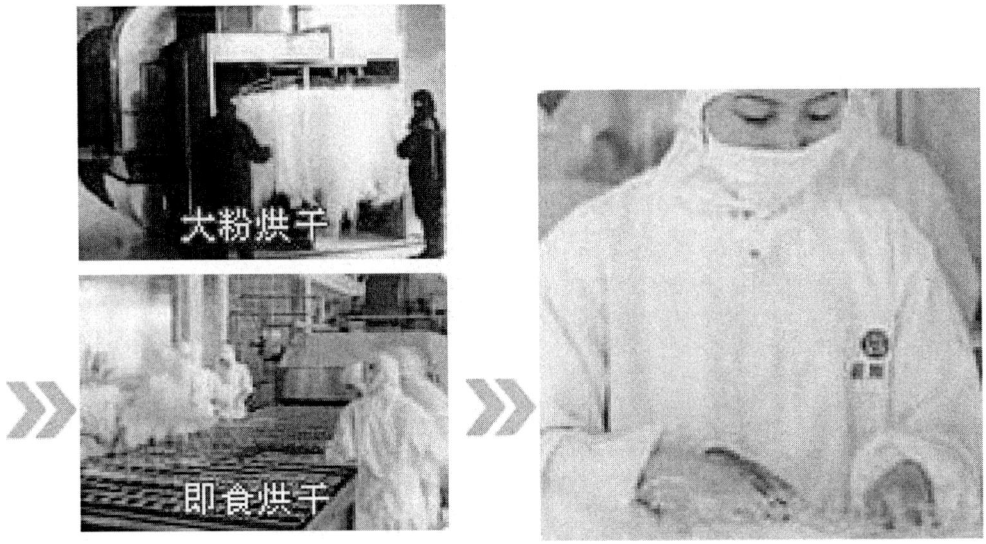

Drying → Selecting→

Packing--→ Production (starch noodles)

Figure 4-24. Production Process Flow Chart of starch noodles in modern manufactory in China (http://www.vermicelli-longkou.com).

3.1.1. Forming of Starch Dough

The forming and quality of starch dough is a crucial step in the processing of starch noodles. The drop of starch dough and the formation of filament depend on the rheological properties of the dough itself, especially shear-thinning properties and gravity, which decreases viscosity, increases the fluidity of starch dough and facilitates the dropping of filaments. In addition, process parameters such as the content of moisture and starch paste, stirring rate and temperature in starch dough, are also important. Tan et al (2007) investigated the rheological behavior of mung bean starch dough (MBSD) under different conditions.

The MBSD exhibited unique rheological behavior which was dependent on a wide range of factors such as starch paste content, moisture content, temperature, agitation rate and time (Figs.4-13~15). The zero-shear viscosity (the viscosity at zero-shear rate, η_0) of MBSD decreased while the hysteresis loop area reduced with an increase in temperature from 20 °C to 40 °C. However, the MBSD exhibited the highest zero-shear viscosity and the largest hysteresis loop area at 50 °C due to the forthcoming gelatinization of starch which led to a too high viscosity to flow, while those corresponding values at 20 °C and 30°C were also higher than those at 40 °C. It could be explained that the short-term retrogradation of amylose in starch paste at 20~30 °C induce the difficulty to flow for MBSD. It thus indicated that stirring MBSD at 40 °C was suitable in view of its smallest hysteresis loop area and its lowest zero-shear viscosity, which gave rise to better fluidity (Tan et al 2007).

Increasing moisture content obviously led to a decrease in the starch dough viscosity, and the hysteresis loop area. The viscosity of starch dough with lower moisture content (41 w/w %) at zero-shear rate reached 1.72E6 Pa·s, then dropped sharply on 7-8 magnitude over the range of shear rate of 0-500 s^{-1}, and the hysteresis loop area reached 4.20E5 s^{-1}·Pa·s. Under these conditions with lower moisture content (≤41 w/w%), for blender, higher energy was needed to stir starch dough in starch noodle production. However, the hysteresis loop area and zero-shear viscosity markedly dropped when moisture content in MBSD was 44 w/w%, which was the suitable value for stir and drop. Some broken streams during drop due to high moisture contents (47 w/w% and 50 w/w%) in MBSD, although their hysteresis loop areas and zero-shear viscosities were lower than those of with moisture content (44 w/w%) (Tan et al., 2007).

The moisture in the final starch dough, which had a statistically significant main effect on all physical properties of the noodles, was considered the most important factor that affects all response variables measured, followed by holding temperature and cooking time. Note that the amount of moisture in the starch dough also indicates the amount of total starch which is (100-%moisture) (Galvez et al., 1994). Noodles with low moisture content (50%) had low cooking loss, high L-value (hence, opaque), and high maximum cutting stress and work to cut. Correspondingly, those with higher moisture had higher cooking loss, lower L-value or higher transparency, and lower maximum cutting stress and work to cut (Galvez et al., 1994).

A decrease in the content of starch paste obviously led to a decrease in the viscosity, and the area of hysteresis loop of starch dough. The MBSD with the lowest content of starch paste (12 w/w%) showed the lowest viscosity, the smallest area of hysteresis loop and the largest fluidity, while the viscosity of starch dough with 36 w/w% and 50 w/w% of starch paste content dropped sharply from 5~6 magnitude and exhibited the high zero-shear viscosity values (1.55E5 and 1.93E5 Pa·s, respectively), the large areas of hysteresis loops (2.0E4 and 3.3E4 s^{-1}·Pa·s, respectively) and the low fluidities. It might be attributed to the adhesiveness of starch paste, which endued starch dough with the higher yield stress. If there was no or only small amount (eg.12 w/w%) of starch paste used in dough, the starch dough would display too large fluidity to form starch noodles in view of the lack of glutinosity. Thus MBSD with 24 w/w% starch paste not only exhibited a small hysteresis loop area and a low zero-shear viscosity but also formed streams (Tan et al., 2007).

Mung bean starch dough has the characteristics of highly time-dependence (Figure 4-17). The viscosity of starch dough decreased rapidly with time within the first 60 s and then approached to a constant value at fixed shear rates of 500 s^{-1}. This implied that the starch dough had too low time-dependence. At fixed shear rates of 100 s^{-1} and 10 s^{-1}, the viscosity of starch dough decreased rapidly with time within the first 120 s and 300 s, respectively, and then approached to a constant value corresponding to an equilibrium state. This moderate time-dependence fitted the mixing of starch dough in the starch noodles processing. While at fixed shear rate of 1.0 s^{-1} and 0.1 s^{-1}, the viscosity did not decrease but slightly increased within 10 min of shear and then did not approach a constant value. This implied that the mixing of starch dough was time-consuming in processing. The rate and extent of viscosity reduction appeared to depend on the applied shear rate. The higher the applied shear rate, the larger the rate and extent of viscosity reduction, and the more pronounced time-dependence level of the mung bean starch dough (Tan et al., 2007).

The flow behaviour of MBSD within the range of 20~48 °C can be described by Arrhenius equation because 48~56 °C was a transition temperature range for forthcoming gelatinization of mung bean starch (Figure 4-18). The onset gelatinization temperature of mung bean starch (about 57°C) (Tan et al., 2006) could help explain why the effect of working temperature on ΔE_η changed around 48~56 °C. The viscosity of starch dough rose rapidly and exhibited a resistance to flow after 57 °C, where overrun the working temperature of stirring starch dough and was not analyzed further in the present research. The different starch dough obtained at 10 s^{-1} and 100 s^{-1} were all temperature-sensitive due to their high activation energies (2.8E4 and 3.4E4 kJ/mol, respectively). Higher ΔE_η values for MBSD at 100 s^{-1} than 10 s^{-1} below 48 °C showed that it was needed some extra energy for overcoming the resistance of the viscosity of starch paste and the dilatant flow behavior of raw starch granules for stirring at 100 s^{-1}. The starch dough at 10 s^{-1}, thus, was less sensitive to temperature than that at 100 s^{-1} and was more suitable for stirring in the process of starch noodle producing (Tan et al., 2007).

The MBSD with moisture content of 44 w/w%, starch paste content of 24 w/w%, shear rate of 10 s^{-1} and temperature of 40 °C exhibited a better flow performance to stir and hang during starch noodle production. The starch dough at 10 s^{-1} was less sensitive at lower temperature than that at 100 s^{-1} and was more suitable for stirring (Tan et al., 2007).

3.1.2. Cooking and Cooling of Starch Noodles

3.1.2.1. Cooking

After obtaining a smooth ball (starch dough) that does not stick to the hands, the starch dough is extruded using a dropper into boiling water for 30 s. This course, virtually, is the gelatinization of starch. Noodles are dropped into boiling water and removed after they are sufficiently cooked as they floated up to the surface of the water. This is due to the change in specific gravity of the noodle strand as it is cooked or gelatinized. Uncooked starch granules have a specific gravity of about 1.5, so uncooked noodles settle directly to the bottom of the cooking container, but as they gelatinize, the granules swell as they absorb more water and float (Tam et al., 2004).

Cooking temperature is immobile because starch strands drop into boiling water (100 °C or close to 100 °C). Cooking time, therefore, is a variable parameter in the cooking step. Galvez et al. (1995) studied the formulation and process optimization of mung bean noodles using response surface methodology. They found that as cooking time increased the region that satisfied the operating specifications drastically decreased in size. This was primarily due to decreased acceptance cores for the texture of cooked noodles as cooking time was increased. When the cooking time was 20 s, the region of overlap represented a moisture content in the final slurry or dough between 48 % to 53 % and a holding temperature between 4-12 °C. When the cooking time increased to 30 s, the region of overlap represented a very narrow range of moisture content in the final slurry or dough of 48 % to 49 % and holding temperature of 11-12 °C. When the cooking time was further increased to 40 s, there was no region of overlap. No combination of moisture content or holding temperature would satisfy all the required operating specifications (i.e. acceptance scores greater than both commercial samples) (Galvez et al., 1995).

But, if the solid content in the starch noodle is too high, water content in the noodles may become insufficient for starch to fully gelatinize (Lee et al., 2005). The noodles which contained 38-45% solids provided a uniform and translucent appearance. Under polarized microscopy, no starch granules with birefringence were observed, indicating full gelatinization. Therefore, the high solubility at high solid content was not from insufficient gelatinization but simply from the excess presence of starch in the noodle matrix.

Water uptake during cooking was closely related to the texture and cooking qualities of starch noodles (Lee et al., 2005). Insufficient water uptake (swelling) usually results in noodles with hard and coarse texture, but excess water uptake often results in noodles too soft and sticky (Jin et al., 1994).

Takahashi et al. (1987) determined the degree of gelatinization of mung bean starch noodles during the process of Harusame noodle manufacture and preservation. They found that the degree of gelatinization was 56% immediately after extrusion at 80 °C, 83% after heating in boiling water for 3 min and 74% after direct drying. The degree of gelatinization of mung bean starch noodles at each stage was 5 to 16% lower than that of noodles made of a potato and sweet potato starch mixture (1:1), so they easily underwent retrogradation as compared with other starch noodles.

3.1.2.2. Cooling

The cooked starch strands are transferred to cold water, and drained. Strands are separated and hung to partially dry, kept at 4 °C for 2 h and -10 °C for overnight. A series of processing steps, theoretically, are the retrogradation of starch when cooling. Starch retrogradation occurred during aging and effectively stabilized the starch chains in the gel matrix. Retrogradation is responsible for stability of the starch noodles and the capacity to withstand boiling temperature. During retrogradation, cooled gelatinized starch goes back to an ordered system. Process such as low-temperature conditioning was applied after the gelatinization of the noodle strands to enhance retrogradation in the production of starch noodles (Tam et al., 2004). These may involve a simple washing in water as in rice noodles, or freezing and thawing treatments as in mung bean starch noodles. Earlier research revealed that amylose crystallization in retrograded B-form kept the structure intact in rice noodles and mung bean starch noodles, which are able to withstand boiling temperatures (Mestres et al., 1988).

In the manufacture of starch noodles, retrogradation is achieved by holding at temperature (-18~5 °C) for a certain period of time (12~24 h) (Galvez et al., 1994). Lee et al (2005) studied the effect of processing variables on texture of sweet potato starch noodles. They found that the cooking loss of the noodles decreased as aging time increased but increased as the solid content increased. The firmness (or hardness) of starch gel increased linearly with aging time or solid content of noodles. Starch retrogradation rate was highly dependent on the starch content in a gel. In a starch gel, maximum rate of retrogradation was observed at a solid content of 50-55% (Longton and Legrys, 1981). Thus, as the starch solid content in the noodles increased up to 45%, the starch retrogradation rate might continuously increase.

Lower cooking loss was demonstrated in mung bean noodles kept at a higher holding temperature (Galvez et al., 1994). Among process variables studied by Galvez et al (1994), holding temperature (cooling temperature) had a significant main effect on cooking loss and transparency.

3.1.3. Drying of Starch Noodles

After starch strands were retrograded by cooling, it should be dried at 40 °C in a convection dryer, and cooled to room temperature, then finally packed. Lee et al. (2005) studied the effect of drying temperature on the quality of starch noodles. They found the drying temperature had no significant effects on the cooking loss of noodles. The surface firmness of pasta increased as drying temperature increased (Pavan, 1979). However, Aktan and Khan (1992) reported no significant difference in noodle firmness between drying at 40°C and 70°C. Lee et al. (2005) also found that the effect of drying temperature on noodle texture was far less significant than those of solid content and aging time.

There are a number of researchers who focused on the processing variables of starch noodles. In preparing starch noodles from mung bean and red bean, Lii and Chang (1981) used 5% gelatinized starch, 54% moisture in the final dough, cooking time 10-20 s, holding temperature -10 °C and holding time of 24 h. However, the optimum conditions obtained in Galvez's (1995) study require lower moisture in the final dough or slurry and high holding temperature. In another study, Singh and coworkers (1989) prepared starch noodles from mung bean and pigeonpea with much higher moisture content (1:7 starch: water) and the holding temperature (5 °C). Galvez et al. (1995) indicated that moisture content and cooking time were the most important factors that affected consumer acceptance of mung bean noodles. Products with better quality than commercial samples were obtained when moisture content of final dough was 48-53%, holding temperature 4-12 °C and cooking time 20 s or when moisture content of dough was 48–49%, holding temperature 11-12 °C, and cooking time 30s when using 5% total starch as gelatinized starch and holding time 36h. Mung bean noodles processed at optimized conditions had the predicted sensory and physical properties the models established by Galvez et al. (1995). Lee et al. (2005) concluded that the starch noodles prepared from slurry of 45% solids, aged for 21 h, and then dried either at 25 °C or 65 °C were most comparable to the commercial starch noodles in textural properties and cooking loss. In conclusion, starch noodles from different materials starches should be manufactured from different processing variables.

Kuzukiri, a similar type of starch noodle produced by cutting in a traditional method in Japan (Kim et al., 1999; Lee et al., 2005), has been manufactured without freezing, in Japan. Starch slurry is cooked on a steel belt that moves into a steam chamber. The gelatinized starch is then quickly chilled and moved from the belt in a sort of elastic sheet. The starch sheet is subsequently aged in a refrigerator and then cut into thin noodle strands which are then dried in an air oven. This non-freezing process is simple and cost-effective and produces straighter strands than the conventional methods. However, the noodles from this process are often inferior with regard to texture and quality after cooking when compared with conventional noodle products produced by freezing.

3.2. Modern Processing Technology

Extrusion cooking has become a popular processing method for starch-based foods and for producing pregelatinized starches. It has also been used for the production of pre-cooked cereal-based blends and pasta products (Li and Vasanthan, 2003). The extruding method involves this process for producing starch noodles, which comprises adding 45–55 parts by weight of hot water to 100 parts by weight of starch obtained from at least one member-selected from the group consisting of various starches and a product thereof followed by being mixed to prepare large particles of dough, extruding the dough under degassing at degrees of vacuum of not less than 650 Torr to produce a dough sheet, gelatinizing the dough sheet with steam, retrograding the gelatinized dough sheet by cooling below about 8 °C without freezing the dough sheet; cutting the dough sheet into noodles. Starch noodles can be produced efficiently in simple procedures without separately preparing starch paste and without using special rollers. Further, starch noodles thus produced are highly transparent and less melted by boiling (US Patent 5916616).

3.2.1. Forming of Starch Dough

The materials starch used is at least one member from potato, sweet potato, tapioca, corn, wheat and a product thereof. Starch dough is first prepared by adding hot water to starch and kneading it. Starch dough is prepared by adding 45 to 55 parts by weight of hot water to 100 parts by weight of said starch under stirring in a mixer. Adding hot water in an amount of less than 45 parts by weight results in small, hard and brittle particles of dough that cannot be formed into a dough sheet by extrusion through an extruder. Adding hot water more than 55% of weight results in a soft and sticky dough sheet and causes inconvenience such as adherence to a roller in the rolling step. Hot water is preferably at a temperature of not less than 90 °C, otherwise it leads to small, hard and brittle particles of dough that is formed by extrusion into a readily broken dough sheet which will cause inconvenience in the subsequent rolling step (US Patent 5916616).

3.2.2. Extruding

Starch dough prepared is then subjected to extrusion into a dough sheet in an extruder. The extrusion should be conducted under degassing at degrees of vacuum of not less than 650 Torr or otherwise the dough will not form a firm dough sheet by extrusion owing to the voids present. Further, there occurs the nonuniform distribution of the water in the dough sheet, resulting in lack of uniform transparency in starch noodles. By degassing at degrees of vacuum of not less than 650 Torr, the dough sheet can be made uniform and set firm. The apparatus used in extrusion may be any one which can be operated under degassing to extrude the dough, an example being a vacuum extruder. Starch dough is preferably passed through the degassing zone for a period of time of not less than 15 seconds, more preferably 25 to 45 seconds, so it can be sufficiently degassed and set firm to give an excellent dough sheet. The degrees of vacuum at the time of degassing should be 650 Torr or more to permit the particles of dough to form a too soft dough sheet upon extrusion, which will be troublesome in rolling and cause poor transparency for starch noodles. (US Patent 5916616).

The extrusion die used in the extruder may be any of the conventional rectangular type die, but a preferable example is a cone shaped die (i.e. trumpet-shaped die) which is provided with a cutter in the outlet. In case the rectangular die is used, the dough becomes harder to reach the end of the die as the width of the die is made larger with respect to the diameter of the screw in an extruder, and thus a wider dough sheet is difficult to produce. In case the above cone-shaped die is used, however, the dough is uniformly distributed in the die. Another advantage of the cone-shaped die is that because the dough extruded into a cylindrical shape through the die is cut in one position in its perimeter with the cutter, the resulting dough sheet possesses a width being equal in length to the circumference of the outlet in the die and being wider than the diameter of the screw in the extruder. Hence, such cone shaped die provided with a cutter in the outlet can be used in extrusion to improve the efficiency of the subsequent rolling step (US Patent 5916616).

The outlet of the cone-shaped die should usually possess a 1- to 2-fold inner diameter that of the barrel diameter. However, the outlet of the die will usually not be required to possess an inner diameter of not less than 200 mm because the maximum width is about 600 mm with respect to the dough sheet that can be handled by the rollers. The cone-shaped die can be used to extrude the dough into a dough sheet of usually 10–20 mm in thickness. This cone-shaped die can be used for the production of a dough sheet about 3 times as wide as that by a rectangular die. The wider the dough sheet becomes upon extrusion, the thinner it becomes, so the number of rollers in the subsequent rolling step can be reduced to enable efficient production. The dough sheet obtained by extrusion is rolled through a series of rollers until it reaches the desired thickness. The dough sheet can be rolled at a rolling ratio of as high as 80% or more through a first set of rollers because it has previously been set uniform and firm under degassing. The process can thus reduce the number of rollers in the rolling step; for example, the dough sheet can be made 2 mm or less in thickness by rolling through a few sets of rollers (US Patent 5916616).

3.2.3. Gelatinization and Retrogradation of Dough Sheet

The dough sheet thus rolled is then placed in an immersion chamber to add water to it, followed by complete gelatinization with steam in a steamer. Subsequently, it is cooled and retrograded. The dough sheet is cooled, preferably by refrigeration, although any suitable cooling means is acceptable if the dough sheet is cooled without being frozen, which would occur at approximately 0 °C. The specific temperature to which the dough sheet is cooled can vary depending upon the particular dough composition. Generally, the dough sheet is cooled until it reaches a temperature just above 0 °C to about 10 °C, preferably to about 1 °C to 8 °C. It is also preferred to use a refrigerator capable of maintaining a substantially constant temperature while cooling, although conventional refrigerators that typically have some amount of temperature fluctuation are also suitable as long as they do not freeze the dough sheet (US Patent 5916616).

3.2.4. Cutting and Drying of Starch Noodles

Retrograded dough sheet is cut linearly into noodles with a cutting roller. The noodles are dried and cut into dried starch noodles in suitable length. The noodles are dried to a water content of not more than 14.5% by weight, preferably 10-14.5% by weight, to permit occurrence of mold, while water content of less than 10% by weight causes breakage in the starch noodles during transport, which results in a decrease in the value of the product. After the immersion step, the dough sheet that was made uniform and firm under degassing can be transferred on a conveyer to be subjected successively to the above steps in series. When the dough sheet of same thickness is processed in the apparatus of same throughput capacity, the wider the dough sheet becomes, the slower the transfer speed of the dough sheet can be made. Because a wide dough sheet can be obtained, the transfer speed can be slowed down and thus permits a reduction in the length of apparatus relative to the processing time necessary for the respective steps including steaming, cooling, etc. This is advantageous for a reduction in the manufacturing cost of apparatus (US Patent 5916616).

There were a number of researchers whom interested in the extruding method of starch noodle. Takashi et al. (1985) stated starch noodles manufactured by extruding as follows: Starch + water → Extrusion (Nozzle diameter 0.9 mm, at 80 °C) →Drying(into starch nooldes) →Heating in boiling water for 3 min → Cooling in cold water → Draining (→Freezing →Thawing) →Drying →Harusame (starch noodles). Li and Vasanthan (2003) also used an extrusion cooker to prepare starch noodles with hypochlorite oxidation of field pea starch. Starch was mixed well with water to 48% moisture and extruded in a co-rotating twin-screw extruder with a 1.0 mm die opening and a screw speed of 40 rpm at 70 °C. Noodles were collected after torque and die pressure reached steady state, and stored at 4 °C for 24 h prior to drying at 40 °C overnight. Li and Vasanthan (2003) thought the noodles extruded at higher temperatures (80 °C) were chalky in appearance due to the presence of small air bubbles in the finished products. Maintaining the mixing chamber and screw under vacuum to remove air trapped in the dough/ slurry may minimize this problem. After extrusion cooking, starch noodles were held at 4 °C to accelerate retrogradation of starch, which would contribute to development of mouth-feel, texture and flavor. Unlike in conventional noodle making, extruded noodle making does not require the use of pre-gelatinized starch, cooking in boiling water, and cooling (in cold water) after the extrusion step. Therefore, extrusion cooking may greatly simplify the traditional noodle making procedure.

4. Structure and Nutrition of Starch Noodles

4.1. Structure Property of Starch Noodles

The mung bean starch noodle (MBSN) is favored for its desired appearance and excellent texture. However, other starch noodles produced from sweet potato starch, potato starch, corn starch and so on, are moderately elastic or dull, opaque, or have high cooking loss and swelling in cooking. Why do these non-mung bean starch noodles have poor cooking quality compared to transparent, glossy and elastic mung bean starch noodles? An understanding of the structure of the starch noodle is a prerequisite to undertaking additional efforts to improve the quality of non-mung bean starch noodles. Traditionally, these differences in the quality of starch noodles have been attributed to the content of amylose (Cheng and Shuh, 1981), the ratio of amylose and amylopectin (Kim et al., 1996), fat and protein in starch (Kim et al., 1996), and starch granule size (Chen et al., 2003). However, chemical structures of both starches, such as amylose molecular size, chain length, and branched property of amylose and amylopectin also differ. Mestres et al. (1988) and Xu and Seib (1993) investigated the structure of MBSN by hydrolyzing MBSN with acid and enzymes, and then described MBSN as a ramified three-dimensional network held together by short segments of strongly retrograded amylose that melts at temperatures above the boiling point of water. Tan et al. (2006) investigated elaborately the structure of starch noodles made from mung bean and sweet potato by utilizing the methods used by Mestres et al. (1988) and Xu and Seib (1993), who analysiced of the properties of sweet potato starch (SPS) and mung bean starch (MBS).

4.1.1. Gel-permeation Chromatography of Starch Noodles

The gel-permeation chromatography of the acid-resistant molecules in MBSN showed two peaks with $\overline{DP}$ 68 and 49, whereas those in sweet potato starch (SPSN) showed five peaks with $\overline{DP}$ 68, 55, 49, 41 and 22, respectively; This implied that retrograded amylopectin in SPSN was degraded partly to shorter chain segments during acid treatment, whereas retrograded amylopectin in MBSN was difficult to degrade to shorter chains, and retained a large number of long chains. The α-amylase resistant residues in MBSN showed four peaks with $\overline{DP}$ 57, 50, 43 and 35, respectively; whereas those in SPSN showed six peaks at $\overline{DP}$ 57, 50, 43, 31, 14 and 6. It implied that the population of long chains in SPSN decreased and the fraction with short chains increased during α-amylase treatment. The oligosaccharides with very short chains may be represented by segments of α-amylase-degraded long chains. In MBSN, the long chains were still dominant, which may be due to differences in the arrangement of the long chains in the amylopectin clusters of the mung bean starch noodle compared to that of SPSN. The β-amylase and pullulanase resistant residues in MBSN showed four peaks with $\overline{DP}$ 70, 57, 40 and 30, whereas those in SPSN showed seven peaks at $\overline{DP}$ 70, 61, 45, 36, 25, 16 and 8; This implies that SPSN were hydrolyzed more rapidly than MBSN because of more A chains (external chain) in Ap-SP than in Ap-MB and the greater ratio of long chains to short chains in MBS than in SPS. The residues from acid and enzymes in MBSN contained mainly high molecular weight fractions which appeared at the void volume, and some low molecular weight fractions such as limit dextrins, indicating the difficulty to hydrolyze MBSN. Those high molecular weight fractions may be the short amylose chains, generated by the degradation of amylose, which can form double helical again to resist hydrolysis. This phenomenon was analogous with enzyme-resistant retrograded starch, and based on restricted enzyme access to potential substrates arranged in double helical aggregates (Gidley et al., 1995). Gidley et al (1995) found that X-ray diffraction and C CP/MAS NMR spectroscopy indicated levels of crystalline and double helical order to be 25-30 % and 60-70 %, respectively, in enzyme-resistant retrograded starches. The residues from acid and enzyme treatment of SPSN contained some high molecular weight fractions and large amounts of low molecular weight fractions such as limit dextrin, including maltotriose and maltose (Inouchi et al., 1987; Eliasson, 2004), indicating the facility to hydrolyze SPSN.

Both starch noodles hydrolyzed by acid contained fewer small molecular weight materials than those hydrolyzed by enzymes, indicating the possibility to attack starch noodles by enzymes (Tan et al., 2006).

4.1.2. Microscopic Observation of Starch Noodles

The surfaces of both starches were crimpy to a different extent due to shrinkage during drying. The smoother surface of MBSN than that of SPSN (Figure 4-25a, c) might be due to a stronger gel strength and elasticity of MBSN, which can withstand shrinkage better during drying. The inside of MBSN (Figure 4-25d) contained long, thick and orderly filaments that may be cellulose-like crystalline areas because a higher amylose content and longer chain length of amylopectin in MBS lead to ease of retrogradation. The leakage of water during cooling generated a compact structure inside MBSN, while there were many pore spaces on the inside of SPSN (Figure 4-25b) because a higher amylopectin content and shorter chain length of amylopectin lead to less retrogradation and loose inside structure; and the leakage of water after freezing and drying generated many pores on the inside of SPSN (Tan et al., 2006).

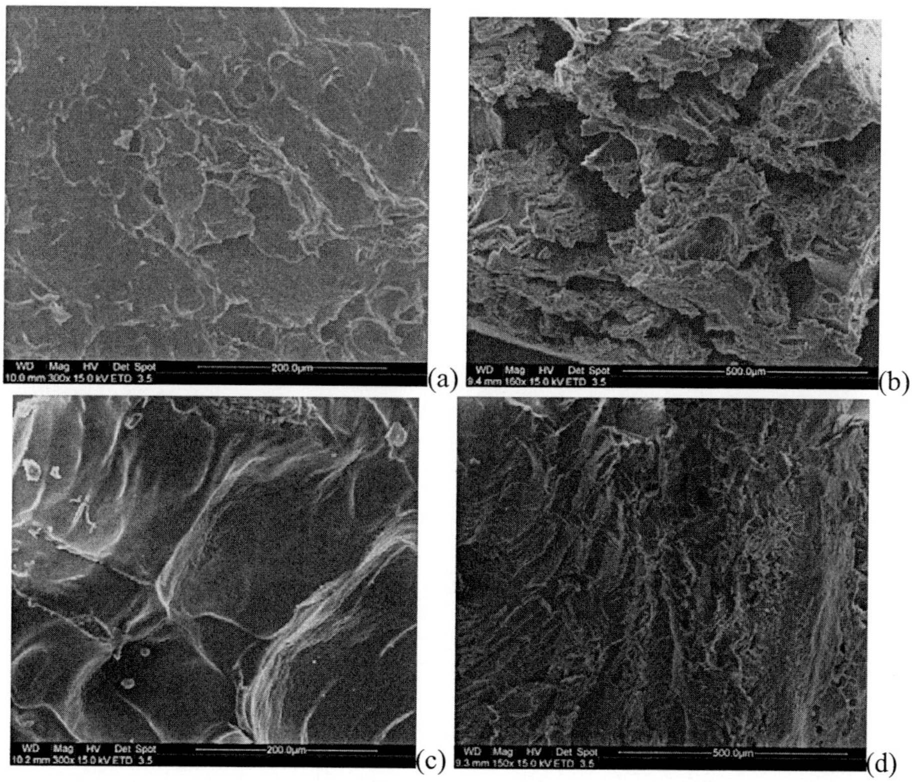

Figure 4-25. Scanning electron micrographs(150~300×) of both uncooked starch noodles. (a)the surface of SPSN; (b) the cross section of SPSN; (c) the surface of MBSN; (d) the cross section of MBSN(Tan et al., 2006).

4.1.3. Thermal Properties of Starch Noodles

The DSC thermogram of original SPSN at 70 % moisture level between 10–180 °C showed a single and faint endotherm at 47.7–54.7–61.2 °C(T_o-T_p-T_c) with ΔH 0.97 J/g, and was much smaller than SPS at 64.6–72.1–80.7 (T_o-T_p-T_c) with ΔH 1.5 J/g. The SPSN, which is composed mainly of retrograded amylopectin, gelatinized easier than its original starch. An endotherm at ~50 °C, characteristic of crystalline retrograded amylopectin (Ring et al., 1987) was also observed in our thermogram of uncooked SPSN. Many researchers had also reported that the endothermic transition for retrograded starch began at a temperature about 20 °C lower than that for gelatinization of starch granules in waxy maize starch with high amylopectin content (Yuan et al., 1993; White et al., 1989; Shi and Seib, 1992). During storage at 4 °C, gelatinized starch molecules reassociate in the SPSN, but in less ordered and hence less stable forms than in the native starch granular state. The resistant residues after HCl-hydrolysis showed the largest endotherm (99.0–106.7–112.6 °C with ΔH 24.5 J/g) among these resistant residues with acid and enzymes. This is indicative of the fact that the retrograded sweet potato starch was more resistant to acid than to α-amylase, β-amylase and pullulanase especially in the initial stages up to 5 days (Tan et al., 2006).

The DSC thermogram of uncooked mung bean starch noodles at 70 % moisture level between 10–180 °C also showed a single and broad peak at 68.3–72.5–83.5 °C(T_o-T_p-T_c) with ΔH 5.4 J/g, which was higher than those of original mung bean starch (at 57.6–64.8–75.9 °C with ΔH 2.6 J/g) (Tan et al., 2006). This shows that it is difficult to gelatinize MBSN, which is composed of mainly retrograded amylose. These findings are in agreement with previous findings that the mung bean starch noodles have crystals which melt at 67–72–78 °C (Xu and Seib, 1993). The resistant residues after HCl-hydrolysis showed the highest endotherm at 104.3–111.2–115.5 °C with ΔH 44.8 J/g. This also indicated that the retrograded mung bean starch was more resistant to acid than to enzymes, which was consistent with those of SPSN. A possible explanation might be that more long B-chains from amylopectin were released by acid hydrolysis than by enzyme. Those chains could behave like short amylose chains, capable of forming lipid complex and double helices (Chung et al., 2003), both of which required a higher enthalpy to melt. Mestres et al (1988) reported ΔH 7.9 J/g at T_p 119 °C for the acid-resistance residue from uncooked mung bean starch noodles while Xu and Seib (1993) reported ΔH 18 J/g at T_p 128 °C for the same sample but cooked.

The α-amylase–resistant residues showed only one faint peak at 96.7-99.4-104.2 with ΔH 0.07 J/g for SPSN, and at 98.5-108.3-110.4 with ΔH 0.72 J/g for MBSN. The findings of Xu and Seib (1993) show that α-amylase–resistant residues of MBSN do not show a peak in the temperature range tested (7–147 °C). The residues resistant to the combination of β-amylase and pullulanase from cooked MBSN gave a higher peak temperature (T_p 105.9 °C) and a higher enthalpy of gelatinization(ΔH 6.3 J/g) than those of SPSN (T_p 103.0 °C and ΔH 2.0 J/g), resulting from high amylose and low amylopectin content in MBS than those in SPS. After the surface density of the amylopectin has been reduced by beta-amylase, the task of pullulanase in penetrating the interior must become progressively easier, because the relative density of the branch point in space decreases. With cooking, the melting of amylopectin crystallites in starch noodles accelerated a successive attack by β-amylase and pullulanase, while the difficulty of melting amylose crystallites in starch noodles when cooking prohibited β-amylase and pullulanase from attacking the crystalline zones (Tan et al., 2006).

4.1.4. X-ray Analysis of Starch Noodles

For original SPSN three peaks were observed at 2θ values of 16.3, 22.0 and 27.2 Å, corresponding to d-spacing (inter planar distances) of 5.4, 4.1 and 2.6 Å, respectively (Figure 4-26). For resistant residues hydrolyzed using a mixture of *β*-amylase and pullulanase, one peak disappeared and two peaks remained at 2θ values of 17.1 and 21.9 Å, corresponding to d-spacing of 5.2 and 4.1 Å, respectively. It can be inferred that crystallites within enzyme-resistant residues from SPSN were smaller and /or less perfectly packed than in original SPSN because of their weaker retrograded amylopectin state of crystallinity (Tan et al., 2006).

The X-ray diffraction pattern of the MBSN gave strong peaks at 2θ=17.0, 23.0 and 22.1 Å, corresponding to d-spacing of 5.2, 4.0 and 3.9 Å, respectively, which can be attributed to different crystalline structures which are typical patterns of B-type peak (Mestres et al., 1988) and should be distinguished from that of SPSN. Upon cooking and then hydrolysis with *β*-amylase and pullulanase, the X-ray diffraction pattern changed and was indicated by three smaller peaks at 2θ of 16.8, 19.4 and 22.0 Å, corresponding to d-spacing of 5.3, 4.6 and 4.0 Å, respectively. This can be attributed to the case that retrograded amyloses are still partly hydrolyzed by enzymes. Such a description is in line with model studies on amylose gels and enzyme-resistant material from amylose gels which show weak X-ray diffraction (Cairns et al., 1990). Cairns et al (1990) suggested that network disruption by enzyme hydrolysis did not allow increased crystalline packing to occur. Similar observations were made on both acid- and *α*-amylase-treated starch noodles. The acid/ enzyme-resistant residues exhibited weaker diffraction peaks than original starch noodles, which showed the presence of poor B-patterns, especially MBSN in our research. This was in agreement with findings of Sievert et al. (1991). The appearance of broad diffraction lines strongly suggested that smaller and /or less perfect crystallites were present in acid/enzyme-resistant residues than in MBSN, where the sharp, well-resolved pattern reflected a higher degree of crystallite perfection. However, generally, crystallinity is a property of the amylopectin fraction. X-ray diffraction pattern showed weaker crystallinity of SPSN than that of MBSN, resulting from insufficient amylose crystallinity and more short chain crystallinity in Ap-SP, and more amylose crystallinity and long chain crystallinity in Ap-MB, respectively (Tan et al., 2006).

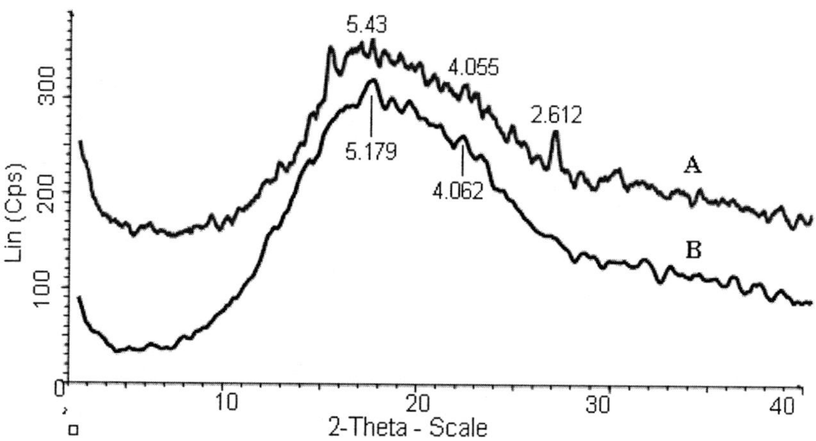

(a)

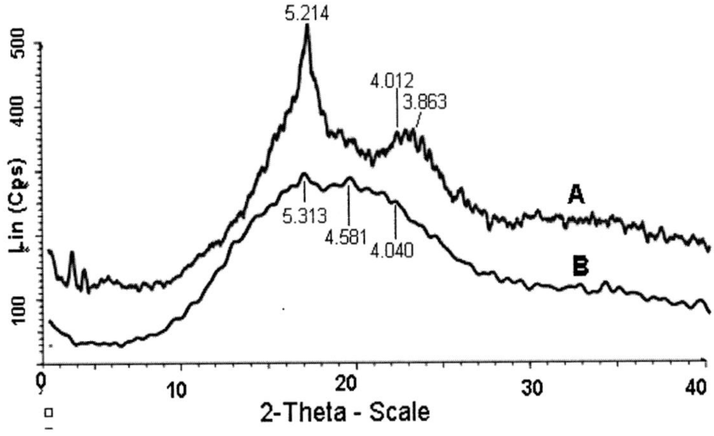

(b)

Figure 4-26. X-ray diffraction patterns of both starch noodles and their resistant-residues hydrolyzed with a mixture of β-amylase and pullulanase at 35°C for 60 hrs. (a) SPSN; (b) MBSN; A: original starch noodles; B: hydrolyzed residues from starch noodles (Tan et al., 2006).

4.1.5. Structure of Starch Noodles

The resistant residues from both starch noodles after HCl and enzyme hydrolysis all show a broad endotherm peak near 100 °C (96–115 °C). Apparently this is difficult to reconcile with the results from Mestres et al (1988), who reported $\triangle H$ 7.9 J/g at T_p 119 °C for the acid-resistance residue from uncooked mung bean starch noodles, and Xu and Seib (1993), who reported $\triangle H$ 18 J/g at T_p 128 °C for the same sample but cooked. We deduce that it may be due to the presence of the complexes of amylose-lipid and lipid-(long chains in amylopectin). It agrees with the findings by Morrison et al. (1993), Jacobson et al. (1998) and Chung et al. (2003), in which the acid/enzyme-resistant residues had a greater tendency to form amylose-lipid complex. A similar result was also found by Godet et al. (1995), who reported the melting temperature of the different amylose-lipid complexes was in the range 78–115 °C. The acid/enzyme hydrolysis might produce amylose chains of reduced chain lengths, which have increase mobility and thus complex more readily with lipids.

Morrison et al. (1993) reported that the residual amount of amylose-lipids complexes (Single V_6-amylose helices) increased by acid hydrolysis. It was because the amylose-lipids complex was resistant to the acid/enzyme hydrolysis. In accordance with their result, the resistant-acid/enzyme residues contain the single helices of amylose-lipids complexes. This phenomenon is also supported by Sievert et al. (1991) and Chung et al. (2003), who reported the reflection of amylose-lipid complexes appeared at 0.449 nm (about 22 Å) and about 20 Å, respectively. If we accept that the 2θ value of about 22 Å reflection arose from amylose-lipid complexes, then the reduced intensities of this peak could be interpreted as amylose-lipid crystallites being melted out near 100 °C. During the DSC scanning of 10–180 °C, we observed a transition of melting enthalpy at about 96.4–112.6 °C for the acid/ enzyme-resistant residues in SPSN and about 98.5–115.5 °C for those residues in MBSN. This is in agreement with results obtained from Sievert et al (1991), who reported that about 105 °C corresponds to dissociation of amylose-lipid complexes.

Mestres et al. (1988) and Xu and Seib (1993), based on their findings of acid and enzyme hydrolysis of uncooked and cooked MBSN at 35 °C, proposed that junction zones anchor the three-dimensional structure. The cause of SPSN loose structure compared to MBSN, allows further speculation based on the three-phase theory (micelle, paracrystalline fringe and filler mass) proposed by Xu and Seib (1993).

Tan et al. (2006) conjecture that SPSN has a loose structure due to its crystalline inferiority to MBSN. In MBSN, the micelle contains retrograded segments of amylose molecules and is resistant to acid and enzymes (Xu and Seib, 1993). The most highly organized zone containing crystallites is caused by moderate chain length in A_m-MB, in order in close juxtaposition due to fewer amylose branches comprised 1.8 branch chains per molecule, facilitating chains juxtapose closely. However, much shorter chains in A_m-SP and more amylose branches comprising 9.0 chains per molecule were adverse to ordered and juxtaposed chains. Thus SPSN did not have a more compact micelle than MBSN. The hydrolysis-resistant crystalline zone is considered to be the structural center, a composite of intensity features from ordered (double helical), which is produced by amyloses and long chains in amylopectin, and a small amount of non-ordered (amorphous single chain) materials, which consists of amylose-lipid and lipid-(long chains in amylopectin). Attached to the micelle is the paracrystalline finger composed of less organized material. Xu and Seib (1993) argued that the molecules in this zone are all linear; that zone does not swell sufficiently. But our findings provide additional information that the second zone is composed of branched amylopectin, which can form network-like framework due to its cohesiveness. Both amylopectins possess five fractions and different length branched chains, but large amounts of short branched chains in SPS form network-like framework, and decrease the ability for crystallization in SPSN, while large amounts of long branched chains in MBS can crystallize so that this zone is still organized in MBSN. The third, and most prominent zone in the starch noodle is the filler mass or amorphous zone. The filler mass is composed of cracked gelatinized starch granules and their fragments, which exhibit good viscosity and cling tightly to the other two zones. Besides occupying a large volume in a starch noodle, the filler mass would be hydrolyzed by acid and enzymes in SPSN and MBSN. The structure of starch noodles is thus composed of three phases (Tan et al., 2006): hydrolysis-resistant crystalline zone (double helical and amorphous single chain), network-like framework (amylopectin) and filler mass (cracked gelatinized starch granules and their fragments). Because of a low content of branched amylose and much more amylopectin in SPS, SPSN have lower crystallinity and higher adhesiveness; whereas there is a high content amylose with little branching and moderate amylopectin in MBS, thus, MBSN has higher crystallinity, good cohesiveness and excellent quality.

4.2. Nutrition and Function of Starch Noodles

Starch is the major component of the starch noodle, and it could improve its nutritional value after being gelatinized and retrograded, principally by improving in vivo starch digestibility. Many factors can affect native starch digestibility. The rate of starch digestion in legumes is lower both in vitro and in vivo, than that of cereals. In vivo, starch is hydrolyzed by salivary and pancreatic a-amylase. However, a proportion of starch in starchy foods generally escapes complete digestion. This fraction is called 'resistant starch' (Hoover and Zhou, 2003). Rice noodles were demonstrated to have lower glycemic blood index of diabetic patients (Panlasigui et al., 1990). Starch noodles are retrograded and are, therefore, a source of resistant starch (RS). There is considerable interest in the nutritional implications of RS in foods, since a relatively slow rate of starch hydrolysis in the gastrointestinal tract of humans may have some of the physiological effects of dietary fiber (Englyst et al., 1992).

4.2.1. Digestibility of Starch

Sandhua and Lim (2008) investigated the digestibility of common legumes in India (black gram, chickpea, mung bean, lentil, field pea and pigeon pea) and related to their structural (amylose content and crystallinity) properties. They found that all legume starches exhibited a characteristic C-type diffraction pattern with relative crystallinity ranging between 27.2% and 33.5%. Slowly digestible starch (SDS) content followed the order: mung bean > chickpea > field pea > lentil > black gram > pigeon pea, whereas, the resistant starch (RS) content followed the following order: pigeon pea > lentil > black gram > field pea > chickpea > mung bean. The hydrolysis indices (HI) of the legume starches ranged from 8.2 to 20.0, and the estimated glycemic indices (GI) based on the HI were between 44.2% and 50.7%. Several significant correlations were observed among different starch properties as revealed both by Pearson correlation (PC) and principal component analysis (PCA). Together, the first two PCs represent 86.6% of total variability. Digestibility of starch was negatively correlated with starch granule diameter and M_w of amylopectin and amylose. A negative correlation between relative crystallinity and amylose content was observed. M_w of amylopectin was positively correlated to relative crystallinity and negatively correlated to amylose content (Sandhua and Lim, 2008).

Apolonio et al. (2004) studied the regarding starch digestibility of five common bean varieties after cooked. They found that cooking time of different cultivars ranged between 2.55 and 5.92 h. Available starch (AS) values decreased with the storage time and the bean sample that had the lowest AS content (control sample, without storage) showed the shortest cooking time. A similar pattern was found for resistant starch (RS); the varieties that had the longest cooking time presented the widest range in RS values, measured as the difference between the control sample and the value obtained in the sample stored during 96 h. The retrograded RS (RRS) depended on the variety and even more on the molecular structure of each starch. The in vitro α-amylolysis rate decreased with the storage time; the samples with the smallest hydrolysis percentage had the highest RS content. These results suggested that some bean varieties could be recommended depending on the specific dietetic use of beans (Apolonio et al., 2004).

Comparatively, the sweet potato starch was better digestible with glucoamylase than some of the legume and cereal starches (Madhusudhan et al., 1996). The poor digestibility of the latter, particularly the legume starches, has been ascribed to their high amylose content which is considerably branched and is of a relatively high molecular weight, as well as due to the presence of very highly branched amylopectin and the intermediate fraction (Madhusudhan and Tharanathan, 1996). On the other hand, the high digestibility of cereal (and some tuber) starches could be due to their low amylose values (therefore more of amylopectin) and comparatively less branching and low molecular weight of the constituent fractions. Zhang and Oates (1999) studied the relationship between α-amylase degradation and physico-chemical properties of sweetpotato starches. They found that susceptibility to pancreatic α-amylase varied between starches produced by the different clones. Structural characteristics at various levels, such as ratio of major fractions, size of amylose, gelatinization temperature and granule morphology, were also different between clones. Correlating structural attributes with susceptibility led to the suggestion that granule structure, including amylopectin/amylose ratio and molecular associations were important critical factors in the hydrolysis of sweet potato starch granules. High amylopectin content of sweet potato starch was associated with a high gelatinization temperature and correspondingly less susceptibility to α-amylase attack. The hydrolysis pattern was correlated with degree of hydrolysis. Extensive surface erosion was shown to indicate a high degree of hydrolysis, whereas less surface erosion indicated less degradation.

4.2.2. Hydrolysis Property of Gelatinized and Retrograded Starch

Gelatinization converts starch into a physical form that is desirable in many food systems such as the starch noodle. Starch gels are, however, thermodynamically unstable and undergo changes affecting their technological suitability. Upon cooling, starch molecules reassociate in a complex recrystallization process known as retrogradation, which is often associated with water separation from the gel. These changes may result in textural and visual gel deterioration. Retrogradation is also important from a nutritional point of view, since most of the resistant starch occurring in processed foods consists of retrograded α-glucans. Tovar et al. (2002) investigated the possible relationships between resistant starch formation and other phenomena associated with retrogradation, such as syneresis, by hydrating and gelatinizing starches from three cereals (maize, sorghum and rice), two legumes (jack bean and lentil) and arracacha roots (Arracacia xanthorrhiza). Drained gels were stored for 24 h at 4 °C before the analyses. The results indicated that neither apparent amylose contents nor water exclusion values showed clear correlation with RS-III content in the overnight stored gels. Legume starches reached 6–7% (dmb) RS-III levels, while the lowest values (2–3.6%) were recorded for maize, rice and arracacha samples. Jack bean starch gels showed the greatest syneresis indices, followed by the cereals, arracacha and lentil preparations. Data support the perceived idea of different mechanisms governing syneresis and RS-III formation in gelatinized starches (Tovar et al., 2002). The results summarized by Faulks and Bailey (1990) showed that the extent of hydrolysis of gelatinized legume starches ranged from 70.5% for wrinkled pea to 90.4% for red lentil (Table 4-21).

Table 4-21. Sum of the hydrolysis products of gelatinized and retrograded starches after treatment with porcine pancreatic a-amylase for 4 h

Starch source	Gel age [a] (h)	Hydrolysis [b] (%)
Smooth pea	0[c]	89.8
Wrinkled pea	0[c]	70.5
	24[a]	58.9
	48[a]	53.4
Red kidney bean	0[c]	84.1
	24[a]	70.5
Mung bean	0[c]	80.0
Red bean	0[c]	90.4
Broad bean	0[c]	80.0

Sum of the oligosaccharides up to maltopentose. Adapted from Faulks and Bailey (1990).

[a] Gelatinized starch prepared above, was aged for 24 and 48 h at 1 °C, prior to treatment with a-amylase;

[b] Results expressed as a percentage of the total, starch;

[c] The starches were heated in a boiling water bath for 2 h, quickly cooled to 37 °C and then treated immediately with a-amylase (Hoover and Zhou, 2003).

However, the extent of hydrolysis of retrograded starch gels was lower than that of their freshly gelatinized counterparts (Table 4-21). The authors have postulated that in gelled starches, there is a hierarchy of structures of differing susceptibility to amylolysis, and that retrogradation leads to an increase in degree of ordering, resulting in a decrease in the extent of hydrolysis.

4.2.3. Hydrolysis Property of the Starch Noodle

Tan et al. (2006) investigated the hydrolysis property of starch noodles from mung bean and sweet potato. The two-stage hydrolysis pattern was quite obvious in cooked MBSN and SPSN. A fast hydrolysis rate during the first 6 days followed by a slower rate between 7 to 20 days for both starch noodles hydrolyzed with 1M HCl at 35 °C was observed (Figure 4-27). When both starch noodles were hydrolyzed with α-amylase at 35°C they displayed a pattern which was a considerably fast hydrolysis rate during the first 3 days followed by a slower rate between 4 to 20 days. Another fast hydrolysis rate during the first 12 h followed by a slower rate between 13 to 60 h for both starch noodles hydrolyzed with a mixture of β-amylase and pullulanase at 35°C was also observed. Comparatively, the SPSN had a higher digestibility with 1 M HCl, α-amylase, β-amylase and pullulanase than those of the MBSN. The lower digestibility of the latter can be attributed to its high amylose content (~40 %), which is of a relatively high molecular weight, as well as due to comparatively less branching. On the other hand, the high digestibility of SPSN could be due to its low amylose content and the presence of very highly branched amylopectin and low molecular weight of the constituent fractions.

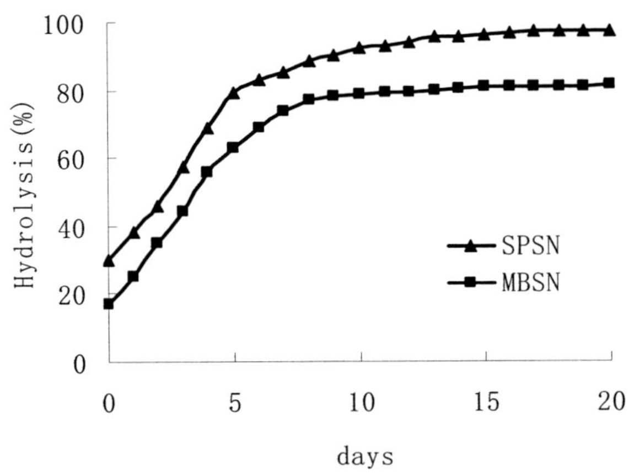

(a)

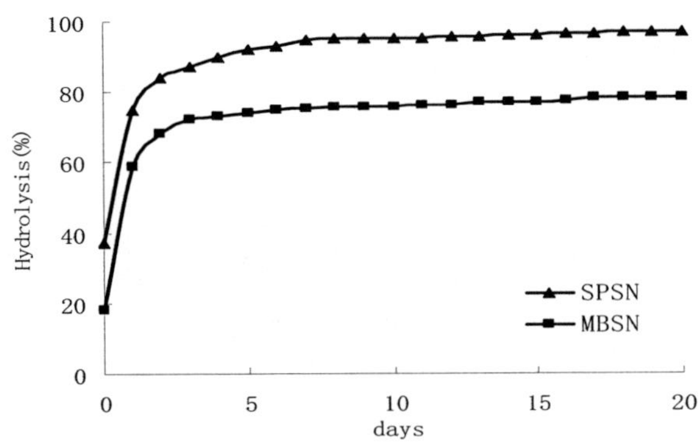

(b)

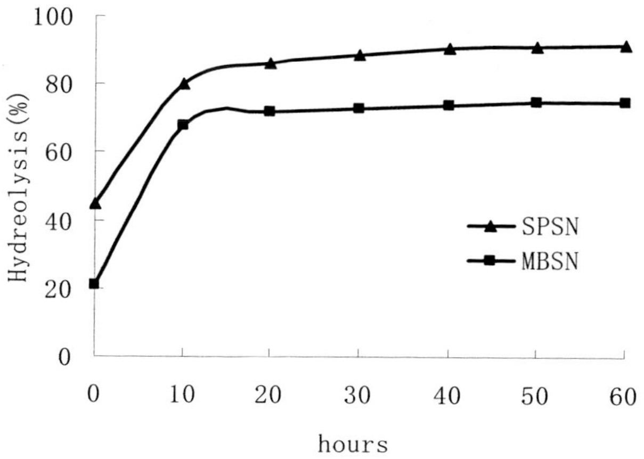

(c)

Figure 4-27. Hydrolysis of both cooked starch noodles using 1M HCl at 35 °C for 20 days (a); α-amylase at 35 °C for 20 days (b); a mixture of β-amylase and pullulanase at 35 °C for 60 h(c).SPSN: sweetpotato starch noodle; MBSN: mung bean starch noodle (Tan et al., 2006).

The faster hydrolysis pattern corresponds to the hydrolysis of the more amorphous parts of all starch noodles. During the second stage, the crystalline starch is slowly degraded (Tan et al., 2006). This is analogous to the phenomenon observed with cellulose and a number of semicrystalline synthetic polymers. Hydrolytic action in these materials occurs most rapidly in the disordered regions, whereas the crystalline areas are more resistant (Banks and Greenwood, 1975). The slower hydrolysis rate of the crystalline parts of the starch noodles may be due to two reasons. First, the dense packing of starch chains within the crystallites of starch noodles does not readily allow the penetration of HCl and enzymes into these regions. Second, acid hydrolysis of a glucosidic bond may require a change in conformation for the glucose unit, from chair to half-chair (Tan et al., 2006). Obviously, if the hydrolyzed bond exists within a crystallite, this change in conformation would require a high energy of activation. All glucosidic oxygens are buried in the interior of the double helix in starch crystallites and are, therefore, far less accessible to acid or enzyme attack (Biliaderis et al., 1981).

5. Quality Evaluation of Starch Noodles

Noodle qualities are defined by visual attributes of the uncooked and cooked noodles. The cooking and eating qualities such as absence of discoloration, high glossiness, and high transparency are important considerations of consumers when purchasing dry starch noodles. Fine straight strands, whiteness, translucency, and absence of broken strands contribute to better-priced noodles. In cooked starch noodles, mouthfeel and texture were the most important characteristics. The noodles should remain firm, chewy and not sticky on standing after cooking. Starch noodles should also have a short cooking time with little loss of solid in the cooking water (Galvez and Resurrection, 1992). It was recently demonstrated that starch noodle quality has three distinct aspects: sensory property (appearance of dry starch noodles), cooking property (eating quality) and texture property of cooked starch noodles (Collado and

Corke, 1997; Muhammad et al., 1999; Collado et al., 2001; Baek et al., 2001; Chen et al., 2002; Tam et al., 2004; Kaur and Singh, 2005; Lee et al., 2005; Tan, 2007).

5.1. Sensory Property

Sensory property is defined as the acceptance of the sensory attributes of a product by consumers who are the regular users of the product category (Galvez and Resurrection, 1992). There are many methods of sensory evaluation of starch noodles. Galvez et al. (1995) used the following method to evaluate the sensory property of starch noodles. Screened to be regular users of mung bean starch noodles, 76 consumers of oriental origin participated in the tests. The tests, designed so that each sample was evaluated by at least 24 consumers, were conducted in two parts: (1) evaluation of dry samples and (2) evaluation of cooking samples. Dry samples were evaluated by consumers for acceptability of appearance. Cooked samples were evaluated for acceptability of appearance and texture/mouthfeel. Nine-point hedonic scales were used where 1=dislike extremely, 5=neither like nor dislike, and 9=like extremely. Dry noodle samples were cut onto strands approximately 6 cm long and presented in coded plastic petri plates arranged on table tops. Participants evaluated 10 samples each. They were allowed to open the petri plates for closer examinations of the samples. Cooked samples were cut into 2–3 cm lengths, and presented in 20-g amounts in coded 1-oz covered plastic cups. Two sets of 5 samples each from the 26 treatment combinations were presented to participants who were asked to place a spoonful of the sample in their mouths when evaluating for acceptability of texture/mouthfeel. Participants rinsed their mouths with water between samples and took a compulsory 10-min break between each set of samples.

After freshly cooked noodles were prepared by boiling them in water for 10 min and then cooling them in tap water (about 20 °C), Kasemsuwan et al. (1998) arranged 10 trained panel members to evaluate the firmness, chewiness, clarity, flavor, and general acceptability of starch noodles, using an unstructured 6 inch line-scale. Panelists tasted the noodles under red lights (to mask possible color differences). Noodles were evaluated in sets of five samples per plate and each set was replicated twice; scores of each characteristic were averaged.

Table 4-22. Sensory attributes evaluated in sensory evaluation (Kim et al., 1996)

Sensory attributes	Definitions
Transparency	Extent of visibility through the cooked starch noodle strands of objects lying behind them.
Slipperiness	Extent to which the product slides across the tongue.
Firmness	Amount of force required to bit through the starch noodle strands.
Chewiness	Length of time required to masticate one strand of sample at a constant rate of force application to reduce it to a consistency suitable for swallowing.
Tooth packing	Amount of starch noodle left on teeth after masticating one strand of noodle.

Muhammad et al. (1999) arranged 20 trained panelists to evaluate the elasticity, stickiness and taste of cooked noodles. Noodles were cooked in 200 mL of boiling distilled water for 1 min, drained for 30 s, cooled for 2 min, and were served to the panelists in 3–4g portions. Elasticity of the cooked noodles was judged by stretching them until they broke, and stickiness was evaluated by tasting whether the noodles adhered to the tongue or not. Samples were scored on a five-point scale as follows: elasticity (1= extremely non-elastic; 5=extremely elastic), stickiness (1= extremely sticky; 5=extremely not sticky) and taste (1= not acceptable; 5= highly acceptable).

Among these sensory attributes, transparency was demonstrated to be a very important appearance characteristic of dry or uncooked mung bean noodles which affect their marketability (Galvez, 1992). Transparent noodles are perceived as high-quality products by consumers. Very low values for maximum cutting stress and work are not desirable. These two physical attributes have a significant positive correlation with sensory mouthfeel attributes of hardness or firmness. Sensory mouthfeel attributes of hardness or firmness have a significant positive correlation with maximum cutting stress and work. A specific range of hardness is required in mung bean noodles (Galvez et al., 1994).

5.2. Cooking Property

In the cooking stage, small parts of the starch noodles will be separated from the noodle itself and suspended in the water. The noodle becomes weaker and less slippery while the cooking water becomes cloudy and thick. This is usually quantitatively described by the term "cooking loss" (Chen et al., 2002). During cooking or keeping in water the starch noodles will also absorb water constantly and the starch noodle will become swollen. This is normally quantified by "swelling index" or "cooked weight".

The cooking loss and cooked weight of starch noodles were measured by the following method. Noodles (5 g) were cut into 3-5 cm lengths and cooked in 200 mL of boiling distilled water for 1 min more than the optimum cooking time. The beaker was covered with aluminum foil to minimize evaporation losses. The optimum cooking time was determined by crushing cooked noodles between a pair of glass plates until the white hard core in the noodles strand disappeared. This indicated that starch in the center of noodle strands was cooked. The cooked noodles were then filtered through a nylon screen, rinsed with distilled water, and drained for 5 min. Cooking loss (CL) was determined by evaporating the combined cooking water and rinse water to dryness at 110 °C and expressed as the percentage of solid loss during cooking. Cooked weight (CW) was calculated as the weight of cooked noodles as a percentage of dry noodle weight prior to cooking. (Li and Vasanthan, 2003)

There was another method of cooking test from Mestres et al (1988). Spring water (150 mL) was heated under reflux in a 250 mL beaker. When the water was boiling, 5g cut noodles (2 cm long) were added. Optimum cooking time was determined with the crushing test. Cooking was continued 1 min more than the optimum cooking time. The sample was then drained for 5 min and rapidly weight (W1, g).Cooked product was predried in an IR oven and dried in an oven at 130°C to constant weight (W2, g). Cooking water was centrifuged (7500×g) for 10 min. Then dry matter contents of the sediment and supernatant (W3, g and W4, g, respectively) were determined as previously reported. Total cooking losses, which include solid losses and soluble losses during cooking were calculated with the following equations (DM=dry matter ratio of crude samples):

$$\text{Total cooking loss (TCL, \%)}=(5\times DM-W2)\times 100/(5\times DM) \quad 4\text{-}4$$

$$\text{Solid loss (SL1,\%)}=W3\times 100/5\times DM \quad 4\text{-}5$$

$$\text{Soluble loss (SL2,\%)}=W4\times 100/5\times DM \quad 4\text{-}6$$

Swelling index after cooking was calculated by the equation:

$$\text{Swelling index (SI, \%)}=(W1-W2)\times 100/W2 \quad 4\text{-}7$$

Cooking loss is a measure of cooking quality of noodles. This may be considered a measure of resistance of the noodles to disintegration upon prolonged boiling. It is desirable to have as low cooking loss as possible. The Chinese Agriculture Trade Standards for starch noodles set ≤10% solid loss during cooking as accepted (NY 5188-2002). The Thai Standards for transparent noodles, however, state that solid loss during cooking should be ≤ 9% (Sisawad and Chatket, 1989). Galvez et al (1994) considered a cooking loss of 10% or less as acceptable. In general, the cooking loss of mung bean starch noodle is the lowest among various pure starch noodles. The swelling indexes of starch noodles from sweet potato, potato, or corn, were higher than that of mung bean starch noodle, which showed a more favorable behavior. Mung bean starch noodle absorbed water slowly in the first 0.5 h but more rapidly during the period of 0.5-1h. Cooking loss and swelling index are affected by recrystallization of the starch which also influences starch gel properties. The high firmness of the starch gel can predict low swelling index of the starch noodle (Chen et al., 2002).

5.3. Texture Property

There are many attributes to reflect the texture property of starch noodles, such as cohesiveness, adhesiveness, extension, cutting behavior (hardness, firmness), and strength. The cohesiveness, extension and cutting behavior are important attributes which can directly reflect the characteristic of starch noodles.

5.3.1. Cohesiveness of Starch Noodles

The cohesiveness of starch noodles was determined by attaching 2 noodle strands to each other and pulling them apart using a texture analyzer. The test speed was 1.00 mm/s and the 5-kg force transducer was used (Chen et al., 2002).

One of the important factors in starch noodle production, also influencing the quality of the final product, is stickiness. Fresh mung bean starch noodles are known to have a low degree of stickiness and are easy to separate from each other during the drying process. Noodles made from other starches, including sweet potato starch, potato starch and cassava starch, is easy to stick strongly to each other, thus causing more difficult to separate during drying. Therefore, the cohesiveness of starch noodles at various stages of the preparation process may not only provide information on the separation ability of different kinds of starch noodles, but also exhibit the effects of treatments in the noodle-making process, such as freezing (Chen et al., 2002). Cohesiveness is an indicator of the extent of disruption of the noodle structure during first compression and is the ratio of the peak areas of first and second compressions of the fore-time plot in the Texture Profile Analysis (TPA) (Singh et al., 2002). Strictly speaking, measuring the stickiness is not the cohesiveness between starch noodles, but the adhesiveness between the instrument probe and the starch noodles (Chen et al., 2002). Chen et al (2002) found that the cohesiveness of the sweet potato starch noodles decrease significantly by freezing treatment. This confirmed that freezing is an important step in starch noodle manufacture. A better separation of the noodles at this stage is not only due to the ice-crystal formation between the starch noodle strands but also due to cohesiveness reduction of the starch noodle strands themselves. The cohesiveness of the cooked starch noodle not only affects the cooking property but also affects the mouthfeel of the starch noodle, such as slipperiness (Chen et al., 2002).

5.3.2. Extension of Starch Noodles

The extension of dried and cooked starch noodles (a single strand) was measured by using 25-kg and 5-kg force transducers, respectively, using the texture analyzer The extension modulus (E) and the relative extension (r_e) were calculated from the following equations: $E=(F/\triangle L)(L/A)$ and $r_e = \triangle L/L$. Here F is the extension force, and A is the cross-sectional area of the starch noodle. $\triangle L$ is the increased length, while L is the original length of starch noodle. The test speed was 1.00 mm/s (Chen et al., 2002).

The extension modulus (E) represents the stretch firmness of starch noodles, while the relative extension (r_e) of the noodle strand is a measure for the stretchability of the starch noodle. The stretch firmness of the dried mung bean starch noodle, in general, is higher than other starch noodles. No clear correlation was found for the stretch firmness and stretchability between dried and cooked stages of starch noodles (Chen et al., 2002).

5.3.3. Cutting Behavior of Starch Noodles

Cutting behavior was measured using a 0.3-mm-dia wire cutting probe to cut a single noodle strand, stabilized on the platform at 2 sides. Force transducers of 25 kg and 5 kg were used for dried and cooked starch noodle measurements, respectively (Chen et al., 2002).

The cutting behavior is usually measured by using a cutting probe to cut the noodle strands placed on a metal platform. Chen et al (2002) found that it was difficult for the cutting probe to cut the noodle strands completely without inevitably touching the platform. Thus, the platform also gave a force to the cutting probe, which made it rather difficult to measure values for the real cutting force of noodle strands. The cutting force (F_c) and the increased length ratio (r_c) of dried noodles is a measure of the cutting firmness and the flexibility of the dried noodle strands. For the cooked noodles the cutting force (F_c) exhibits the firmness of the noodle strands which mimics the bite behavior during consumption. The firmness and flexibility of mung bean starch noodle is higher than that of sweet potato, potato and corn starch noodles.

5.4. Correlation between the Physical Properties of Starch and the Sensory, Cooking and Texture Property of Starch Noodles

The characteristics of both the dried and cooked starch noodles are affected by the properties of the original starch. However, no significant correlation of either the preference or the attributes (color, transparency, and glossiness) between the dried and cooked starch noodles was found according to Chen et al. (2002).

The color, transparency, and glossiness are attributes that play important roles in the appearance of both dried and cooked starch noodles. However, no statistically significant correlation was found between the color, transparency, and glossiness of starch noodles evaluated by the sensory panel, and their starch color and paste clarity. The transparency of the starch noodle is not affected by the degree of starch retrogradation.

Since no correlation was found between the noodle quality and the physicochemical properties, the starch gel properties appear to be more suitable for predicting final noodle quality. High firmness and elasticity of the starch gel also can predict high stretch and bite firmness of the cooked starch noodles. Cooking loss was significantly correlated with cohesiveness, while swelling index was significantly correlated with stretch firm and bite firmness of the cooked starch noodles (Chen et al., 2002).

Comparing sensory evaluation results with texture analysis results, only a significant correlation between flexibility and preference of sensory evaluation and the cutting force of texture instrumental measurement of the dried starch noodles, and significant correlation between sensory chewiness and instrumental cohesiveness of the cooked starch noodles, were found. The attempt to use instrumental results to objectively quantify sensory attributes for foods is not easy. For the time being, both methods of sensory (subjective) evaluation and instrumental (objective) measurement are necessary and important to measure food appreciation (Chen et al., 2002).

5.5. Correlation between the Quality and Processing Variables of Starch Noodles

The texture properties of the starch noodles were affected by the processing variables. Chewiness, gumminess, and hardness, as determined by the texture analyzer, were positively related to solid content and aging time. However, the drying temperature (25–60°C) exerted no significant effects on the textural properties of the cooked noodles. The elasticity, measured by a sensory analysis, positively correlated with solid content and was the highest after 12 h of aging. It was assumed that moisture loss occurred on the noodle surface when aged for an extensive period and that this caused the decrease in elasticity. (Lee et al., 2005)

The stickiness of the surface of the noodles, as measured by sensory analysis, correlated negatively with aging time and drying temperature. While the elasticity increased consistently as the solid content increased, stickiness of cooked noodles was the lowest with the solid content of 41%. Excess solid in the noodles included greater starch leaching and thus the noodle surface became stickier. Surface stickiness exhibited a positive correlation with solubility. None of the processing variables (aging time, solid content, and drying temperature) exerted significant effects on the water uptake of the starch noodles (Lee et al., 2005).

5.6. Prediction of the Quality of Starch Noodles

Viscoamylograph pasting profiles of starches are used in the evaluation of suitability for starch noodles. It was suggested that the ideal starch base is one with a type C viscoamylograph pasting profile characterized by absence of peak viscosity and one which remains constant or even increases during continued heating and shearing, indicative of good hot paste stability and high cold paste viscosities, such as in those generally observed in legume starches. Collado and Corke (1997) claimed that RVA viscoamylography proved to be a sensitive method for monitoring quality of starch for sweet potato starch noodle production. Type C starches show restricted swelling and behave like chemically cross-linked starches (Schoch and Maywald, 1968). This pasting pattern can be observed in legume starches such as lima bean, lentils, garbanzos, yellow peas, and navy bean, chick peas, filed bean, azudki bean, pigeonpea, pinto, navy bean, and mung bean (Collado and Corke, 1997). A type C pasting profile of starch was also observed in some genotypes of potato (Red Pontiac and Mainechip) with stability ratio of 0.95–1.00 (Wiesenborn et al., 1994). The starch noodle produced from these was comparable to the quality of noodle produced from mung bean starch (Kim and Wiesenborn, 1996). Legume starch noodles such as mung bean noodles are known for desirable qualities of greater clarity, glossiness, and high tensile strength as compared with the other tuber and cereal starch substrates (Tam et al., 2004).

Amylose has been indicated as the component of starch that enables it to maintain the integrity of starch noodles. In order to illuminate the contribution of amylose in the production of starch noodles, Tam et al (2004) used maize starches extracted from selected maize cultivars with 0.2–60.8% amylose contents to produce *bihon*-type noodles. The results indicated that the normal maize starches with amylose content of ≈28% were successfully used for *bihon*-type noodle production, but way maize starch 0.2–3.8% amylose content failed to produce *bihon*-type noodles. High-amylose maize starches (>40% amylose) cannot also be used to advantage because they do not sufficiently gelatinize at the boiling temperature of water at 100°C under normal atmospheric pressure. Without gelatinization, the amylose molecules are not released to participate in the retrogradation process that sets the noodle structure. Because amylose content was very highly but negatively correlated to peak viscosity and peak time of the RVA pasting profile, as well as the swelling volume of maize starch, these parameters may be used to indicate whether amylose content of maize starch are at a suitable level for *bihon*-type noodles (Tam et al., 2004).

However, Chen et al (2003,b) investigated the chemical compositions, physical properties and suitability for starch noodle making of different granule size fractions from potato and sweet potato starches. They found that the ash content, amylose content, phosphorus content, gel firmness, and freeze-thaw stability of small-size granule factions (< 20μm) were significantly different from those of the large-size granule fractions. The processibility and the qualities evaluated by objective and subjective methods of both dried and cooked starch noodles made from small-size granule fractions were significantly better than those made from their initial starch preparations and much better than those made from the large-size granule fractions. Their findings show that a simple fractionation method on starch granule size is sufficient to use potato starch (fraction) for starch noodle preparation, whereas sweet potato starches can perform better with decreasing granule size. Granule size dimension plays a very important role in starch noodle making and noodle quality. High amylose content and

C type of viscoamlygram-pasting profile of starches are not necessary for making good-quality starch noodles, although several earlier publications stressed that these are necessities of ideal starches for starch noodle preparation. Starch gel firmness showed a significant correlation with starch noodle quality. Noodles made from small-size granule fractions (< 20μm) had better processibility (fluidity of starch dough for noodle making) and better quality, which may be attribute to their large specific surface area of granules. (Chen et al., 2003)

Surprisingly it has now been found that selected small granular starch is very well suited for the preparation of translucent foods (Semeijn et al., 2004). For example it has been found that with small granular potato starch, in spite of not having a "C" type gelatinization, a glass noodle can be prepared having a satisfactory clarity and translucency in the dried state, which has earlier been found to be impossible using non-modified potato starch. Accordingly, the invention (US patent 4,871, 572) relates to a translucent food prepared from granular starch having preferably 90% of the granules are smaller than 20 μm. As is illustrated in the appended examples, small granular starch imparts superior dough rheological properties, clarity and elasticity to translucent foods. Translucent food products based on small granular starch according to the invention (Semeijn et al., 2004) have furthermore excellent organoleptic characteristics. In addition, the use of small granular starch leads to a low cooking loss during preparation of the translucent food. Besides potato starch also other starches such as potato starch, sweet potato starch, banana starch, kanna starch, kidney bean starch, red bean starch, tapioca starch, maize starch, wheat starch and various bean starches can be used. With preference potato starch, sweet potato starch, banana starch, kanna starch, kidney bean starch, or red bean starch is used. It is further possible to use starches with varying amylose content (0–90%), as long as they are treated in such a way as to fulfill the criteria about granule weight average and size (Semeijn et al., 2004).

5.7. The Quality Standards for Starch Noodles in China

There is the Chinese Agriculture Trade Standard for Starch Noodles, namely "The food without social effects of pollution----starch noodles" (NY 5188-2002), which is suitable for starch noodles from mung bean, pea, broad bean and other legumes (Table 4-22, Table 4-23).

There is a Chinese National Standard for Starch Noodles, namely "Product of designations of origin or geographical indication---- Longkou vermicelli" (GB 19048-2003), which is suitable for starch noodles from mung bean and pea (Table 4-24 and Table 4-25).

There is *anothe*r Chinese National Standard for Starch Noodles, namely "Product of designations of origin or geographical indication—Lu long vermicelli" (GB 19852- 2005), which is suitable for starch noodles from sweet potato (Table 4-26 and Table 4-27).

Table 4-23. Sensory attributes of starch noodles (NY 5188-2002)

Terms	Request
Color	White and shiny, or owing themselves color
Odor and flavor	Owing corresponding smell and flavor with mung bean, pea, broad bean, and other legumes starch. Without peculiar smell.
Configuration	Uniformity, don't stick to each other, no broken strands, tender, and stretchy, semi-transparency
Impurity	No eyeable impurity from outside

Table 4-24. Physicochemical attributes of starch noodles (NY 5188-2002)

No.	Terms	Request
1	Moisture (%)	≤15
2	Starch (%)	≥75
3	Soluble substances after dried (%)	≤10

Table 4-25. Sensory attributes of Longkou vermicelli (GB 19048-2003)

Terms	Request
Color	White, shiny, semi-transparency
Shape	Uniformity, don't stick to each other
Handle	Flexible, stretchy.
Mouthfeel	Tender, gliding, and stretchy after cooking
Impurity	No impurity

Table 4-26. Physicochemical attributes of Longkou vermicelli (GB 19048-2003)

Terms	Request
Starch (%) ≥	75.0
Moisture (%) ≤	15.0
Diameter of strand (mm) ≤	0.7
Rate of rupture (%) ≤	10.0
SO_2(mg/kg) ≤	30.0
Ash (%)	0.5

Table 4-27. Sensory attributes of Lu Long vermicelli (GB 19852- 2005)

Terms	Request
Color	Nature, shiny, semi-transparency
Shape	Uniformity, don't stick to each other
Mouthfeel	Tender, gliding, stretchy and no peculiar smell after cooking
Impurity	No visible impurity

Table 4-28. Physicochemical attributes of Lu Long vermicelli (GB 19852- 2005)

Terms	Request
Starch (%) ≥	75.0
Moisture (%)	13.0~17.0
Rate of rupture (%) ≤	10.0
Ash (%) ≤	0.9

6. Quality Improvement of Non-Mung Bean Starch Noodles

Glass noodles are translucent both before and after cooking, are resilient after cooking, and have a bland taste. Mung bean starch provides unique properties for this application and is the ideal material for noodle manufacture. In recent years, the demand for starch noodles is gradually increasing in China and abroad, and the limited output of mung bean cannot meet this demand. Furthermore, mung bean starch is much more expensive than other starches. So looking for other materials to totally or partly substitute for the mung bean will be valuable. The utilization of different substrates for starch noodles that have been prepared includes other legumes, tuber and tuber-legumes starch blends. Some well-known starches for this application is pea starch, broad bean starch, sweet potato starch, potato starch, corn starch, and so on, but the qualities of noodles based on these starches are generally inferior to that of noodles based on mung bean starch. Proposals to use leguminose starches have been published, but the availability of this type of starch is often even more limited. Another frequently described possibility is a partial or complete replacement of mung bean or sweet potato starch by chemical or genetically modified starches, in particular by starches derived from tapioca and potato (Semeijn et al., 2004).

6.1. Looking for Other Materials to Substitute Totally or Partly for Mung Bean Starch (Starch Noodles from Various Sources)

6.1.1. Noodles from Red Bean Starch

Lii and Chang (1981) investigated the quality of red bean starch noodles and compared it with that of mung bean starch noodles. The results indicated that solid loss was higher for red bean starch noodles than for mung bean starch noodles and noodles prepared from mixed (1:1) red bean and mung bean starches. However, the 5.77% solid loss was still far below the acceptable 10% level set by the Chinese Agriculture Trade Standard for Starch Noodles, namely "The food without social effects of pollution—starch noodles" (NY 5188-2002). The tensile strengths of the noodles decreased in the order: mung bean, mixed 1:1 red bean and mung bean, and red bean. This may be explained by the lower content of linear fractions in red bean starch which may cause less retrogradation of the starch in the noodle. Organoleptic evaluation indicated that noodles made from mung bean, and mixed bean starches had similar scores based on texture. Red bean starch noodles were slightly softer in texture, but gave fairly good quality, although not as good as mung bean starch noodles.

6.1.2. Noodles from Pigeonpea Starch

Singh et al. (1989) investigated the quality of pigeonpea starch noodles and compared it with that of mung bean starch noodles. Sensory properties such as color, texture, clarity, and general acceptability, were evaluated. Starch extracted from whole seed and dhal samples of both legumes showed noticeable differences in their noodle qualities. The whole-seed starch isolated from pigeonpea produced noodles with poor to fair quality, with an average score of 1.9 on general acceptability, whereas the noodles of whole-seed starches of mung bean were rated as fair to good with an average score of 2.8. The scores on noodle clarity and color from whole seed starch of pigeonpea were lower than those of the mung bean. The scores on noodle clarity and color from whole seed starch of pigeonpea were lower than those of the mung bean. Dhal starch of pigeonpea produced noodles with better quality than that of mung bean, as revealed by various sensory properties and noodle color. This was due to the brighter color of pigeonpea dhal starch as no pigments were extracted along with the starch. On the other hand, some starch bound pigments might have been extracted in the case of mung bean dhal starch. No marked differences were observed in the quality of hard noodle of mung bean and pigeonpea dhal starches. These results indicated that in the case of whole seed starch, noodle quality was better for mung bean than of pigeonpea whereas the reverse was true, except for texture, for dhal starch. Quality of hard noodle made from dhal starch of pigeonpea or mung bean was comparable.

Starch from pigeonpea dhal was as good for noodle preparation as that from mung bean dhal perhaps even better. It was apparent that pigeonpea could be used as a potential starch source for making transparent noodles.

6.1.3. Noodles from Edible Canna Starch

The overwhelming portion of edible canna starch production in Vietnam is processed into transparent starch noodles ("cellophane noodles"), a luxury food of south-east Asia and traditionally made of costly mungbean starch. Good cellophane noodles are about 1 mm thick; they display high tensile strength and good transparency. Dry matter loss during prolonged cooking is less than 10% (Hermann, 1996). Starch noodles of non-canna origin are usually produced through extrusion cooking, which requires the extruded noodles to pass through a cooling water bath. By contrast, canna noodles are manufactured by a different, and previously undescribed, process involving the steam-sheeting of a starch/water dough. The resulting gel sheets are stretched and semi-dried on bamboo frames. The gel sheets are then folded and cut into straight noodles. They are finally dried to a moisture content of about 18% to 21% (Hermann, 1996).

Canna noodles in Vietnam have excellent eating quality, much superior to extrusion noodles made experimentally from sweet potato and cassava starches which are widely available in Southeast Asia. Special but as yet poorly understood functional properties of canna starch make it a substitute which has totally replaced expensive mung bean starch as the raw material for cellophane noodles in Vietnam. The high amylose content (25% to 30%) of canna starch as compared with other root starches has been proposed to explain the high peak viscosity observed during gelatinization, which permits the sheets to be easily handled. Canna starch also displays high gel retrogradation (recrystallization) and transparency which is critical to noodle quality (Hermann, 1996). Canna processing in Vietnam provides employment to many thousands of people in rural communities with as little as 500 m^2 of arable land per capita. Canna use in Vietnam shows how product development can provide new perspectives for crop utilization and stimulate demand for otherwise obsolete crops (Hermann, 1996).

6.1.4. Noodles from Pea/lentil Starch

Rask (2004) evaluated the Canadian pea/ lentil starch extraction and noodle preparation. Starch noodles using the isolated legume starches from 4 varieties of peas and lentils were successfully manufactured in their laboratory scale. The optimum processing parameters using our machines were: moisture content of dough less than 50%, 40 seconds cooking and 2 hours cooling at 6 Celsius. Sensory evaluation using a focus group with Asian background should be carried out to get comparative evaluations of starch noodles made with mung bean starch and legume starches. Starch noodles made from legume starches need to have sensory properties that are superior or the same when compared to the starch noodles made from mung bean starch in order to successfully use as an ingredient replacement. Starch isolated from peas and lentils would be very competitive with the mung bean starch on a cost basis. If isolated starches are suitable for the manufacture of starch noodles, a new market for Canadian legumes could be realized (Rask, 2004).

6.1.5. Noodles from Mixed Potato and Mung Bean Starch

Potato starch plays a very important role in the production of another type of oriental noodle—glass noodles. Many noodle manufacturers made the starch themselves, but then they were left with the problem of fiber and protein disposal. They found that using potato starch in place of part of the mung bean starch lessened this disposal problem. At first, they used 50% potato starch; then they found they could go as high as 80% potato starch while still keeping the quality of the glass noodles. Mung bean starch and water are made into a slurry and cooked. Then water is added to cool the slurry down below the gelatinization temperature of potato starch. Potato starch is added and the mixture is kneaded to form a dough. Noodles are extruded, then cooked to gelatinize the starch. Then they are held at -12 °C for 12~24 h while still moist. Freezing is thought to accelerate retrogradation of the starch, which contributes to development of mouthfeel, texture, and flavor. Bundles of noodles are hung up to air dry. Glass noodles are boiled and eaten in soups or dishes with abundant sauce. At present it is still necessary to use 20% mung bean starch to achieve texture and transparency. Research programs are underway to develop glass noodles made from 100% potato starch (Labell, 1990).

6.1.6. Noodles from Potato Starch (Comparing with Edible Bean Starch Noodles)

Kim et al. (1996) prepared starch noodles from two types of bean (navy and pinto) and three sources of potato starch (ND651-9, Mainechip, and commercial potato starch). Physicochemical properties of those starches and cooking quality parameters and sensory characteristics of the noodles were investigated. Potato starches contained significantly less amylose and more phosphorus when compared to bean starches. Amylograph pasting properties showed lower pasting temperature and peak viscosity for potato starches than for bean starches, but more shear stability for bean starches. Swelling and solubility of potato starches was significantly higher than for bean starches. Noodles made from bean starches exhibited cooking quality similar to that of commercial starch noodles with respect to cooking loss and cooked weight. Texture profile analysis (TPA) results showed starch noodles made from dean starch had high hardness values, but lower cohesiveness values when compared to those from potato starches. Sensory panelists scored noodles made from potato starches higher in transparency than those made from bean starches. Both transparency and overall acceptability by sensory evaluation were significantly correlated with cohesiveness by TPA. With respect to texture characteristics of starch noodles, starch noodles made from potato starches were more suitable than navy and pinto bean starch noodles.

This conclusion illustrate that amylose content is not the only factor which decide the quality of starch noodle. Noodles prepared from high-amylose starch are known to be too firm, resulting from a rigid and tight structure that inhibits water absorption. Therefore, an optimum amylose-to-amylopectin ratio is desirable for good noodle quality. Protein act as an essential structural component in pasta products, causing noodle strands to integrate and maintain their form during cooking. Lipids form an amylose-lipids complex, result in minimized cooking losses. Phosphorus content also contributes to the high cohesiveness of potato starch noodles. In additional, other starch properties are more important than amylose content.

6.1.7. Noodles from Corn Starch (Comparing with Potato Starch Noodles)

Singh et al. (2002) analyzed the quality of the corn starch noodle and compared it with the potato starch noodle. They found that the cooked weight of noodles made from corn starch was lower than for that from potato starch. Noodles made from starches with higher swelling power exhibited higher cooked weight and *vice versa*. Corn starch noodles had lower cooking loss than that of potato starch noodles. The lower cooking loss might be due to the presence of lipids, the high gelatinization temperature and the more stable granular structure of corn starch. The noodle made from corn starch had lower value of hardness and cohesiveness than that from potato starch. The insufficient release of amylose due to strong internal bonds may have caused the lower cohesiveness of noodle made from corn starch. The contribution of lower solubility and swelling power in decreasing the cohesiveness of corn starch noodles cannot be ruled out. In conclusion, corn starch is not more suitable for the production of noodle than for potato starch.

6.1.8. Noodles from Sweet Potato Starch

Generally pure sweet potato starch is considered inferior, relative to other starches like mung bean, for the production of noodles, and this is normally overcome at least partially by additives and other treatments. The formulation of sweet potato starch often includes the use of potash alum or the addition of elephant yam flour to improve the quality of noodle produced from it in the past in China. Collado and Corke (1997) investigated the qualities of starch noodles made from 14 sweet potato genotypes in the Philippines. They found that there were significant differences in the texture and cooking quality of the starch noodles produced from the different genotypes. There is an important finding in that the quality of both dried and cooked starch noodle of *Sushu 8* variety is the best among all sweet potato starches in China studied by Chen et al (2002). They found that starch with high firmness and elasticity of its gel will result in good quality starch noodle. Starch noodle quality can be predicted by starch gel properties. The qualities of dried and cooked starch noodles made from the Chinese sweet potato varieties determined by both texture analyzer and sensory evaluation showed some difference. It can be said that dried starch noodle made from *Sushu 8* sweet potato had a final quality well comparable to the noodle made from mung bean starch. This was surely not the case for another starch noodles made from other sweet potato varieties. For the cooked noodles, the quality of *Sushu 8* was even better than that of cooked mung bean starch noodle. Therefore, the statement in the literature cited that sweet-potato starch is not very suitable for starch noodle making is generally incorrect. Obviously this depends in variety. Whereas the quality of cooked starch noodle of *Sushu 8* variety is just far better than that of other sweet potato starches in China studied by Tan (2007), but still inferior to the quality of mung bean starch noodle. The best performing sweet potato variety in China for preparing roast sweet potato food and may not have yet been tested for starch noodle preparation.

6.2. Adding Chemically Modified Starches

Although attempts have been made to substitute mung bean starch with starches from various resources, we found that starch noodles made from these starches were not as good as mung bean starch noodles. A number of researchers considered that better quality noodles can be obtained by substitution of mung bean starch with chemical modified starch. For instance, WO-00/55605 describes the partial replacement of mung bean starch with a genetically modified potato starch with elevating amylose content. US patent 4871572 described the application of crosslinked potato starch in glass noodles. Process for producing glass noodles and demoldable gels using genetically modified starch, preferably from potatoes, was also reported (US Patent Issued on July 8, 2003; http://www.patentstorm.us/ patents/ 5916616.html).

6.2.1. Phosphorylate Starch

The phosphorylated tapioca starch used undergoes less breakdown during cooking and also has lower swelling power and solubility compared to native tapioca starch. Muhammad et al. (1999) reported the results of substituting potato starch with native or phosphorylated tapioca starch in the production of starch noodle. Substituting potato starch with up to 17% native tapioca starch or tapioca starch phosphate, or up to 35% MTS283 (a commercial tapioca starch), improved the strength of uncooked noodles, reduced the stickiness and cooking loss and resulted in the noodles being able to retain their shape in comparison to noodles containing potato starch only. However, substitution with native tapioca starch reduced transparency and the noodles tended to swell more when cooked. Cooked noodles containing either type of phosphorylated tapioca starch were less sticky, more elastic and retained more of their shape than noodles produced using native tapioca starch. Of the two phosphorylated starches substitution with MTS283 is preferable, due to it resulting in noodles with quality comparable to mung bean noodles in terms of cooking loss, swelling index and stickiness. However, the application of native or phosphorylated tapioca starch instead of potato starch will reduce the flowability of the dough, and the extrusion technique used must be able to cope with this problem (Muhammad et al., 1999).

6.2.2. Hypochlorite Oxidate Starch

Oxidized starches are widely used in the food, paper, and textile industries. Oxidation of starch with alkaline hypochlorite is one of the most common methods used. Oxidation causes depolymerization of starch, which results in lower gel viscosity and minimizes retrogradation of amylose by introducing carbonyl and carboxyl groups (Li and Vasanthan, 2003).

Li and Vasanthan (2003) investigated the effect of hypochlorite oxidation on the Brabendar pasting properties of field pea starch and the suitability of native and oxidized starch for noodle making by extrusion cooking. As the degree of oxidation increased from 0.02 to 0.20%, the cooking loss increased substantially and the noodle diameter, cooked weight, firmness, tensile strength and breaking distance decreased. Upon substitution of native field pea starch with native potato starch (10-40%), noodle diameter, cooking loss and breaking distance increased and cooked weight, firmness and tensile strength decreased. However, when oxidized field pea starch was substituted with potato starch at the 40% level, cooked noodles prepared from oxidized field pea starch with higher degree of oxidation had higher diameter, cooked weight and firmness and lower cooking loss, tensile strength and breaking distance. At similar degree of oxidation (0.2%), increasing level of potato starch substitution from 20 to 40% increased the noodle diameter, cooked weight and firmness and decreased the tensile strength and breaking distance. A marginal change was observed in cooking loss (Li and Vasanthan, 2003).

Li and Vasanthan (2003) concluded that field pea starch was oxidized with sodium hypochlorite at a level of active chlorine ranging from 0.89 to 3.28% (starch db). The degree of oxidation was determined and expressed in terms of percentage of carboxyl and carbonyl groups, which ranged from 0.02 to 0.38% and 0.06 to 0.19%, respectively. Hypochlorite oxidation of field pea starch influenced its Brabendar pasting properties. Starch recovery and peak viscosity, hot paste viscosity, cool paste viscosity, and setback of oxidized starches decreased with increasing degree of oxidation. The cooking quality attributes of noodles prepared from native field pea starches were acceptable but were negatively influenced by hypochlorite oxidation. Substitution of potato starch (40%, db) for field pea starch yielded more glossy noodles with better cooking quality (Li and Vasanthan, 2003).

6.2.3. Cross-linked Starch

Noodles prepared from unmodified tapioca starch were too soft and not acceptable as replacements for mung bean starch noodles. Tapioca starch is also a good candidate to manufacture clear noodles because of its low cost and the clarity of its starch paste. Ways to simulate the making of clear noodles from mung bran starch were investigated by studying the molecular structures of mung bean and tapioca starches (Kasemsuwan et al., 1998). The results of the molecular structure study and physical properties were used to develop acceptable products using mixtures of cross-linked tapioca and high-amylose maize starches. Tapioca starch was cross-linked by sodium trimetaphosphate (STMP) with various reaction times, pH values, and temperatures. The correlation between those parameters and the pasting viscosity were studied using a visco/amylograph. Starches, cross-linked with 0.1% STMP, pH 11.0, 3.5 h reaction time at 25, 35, and 45 °C (reaction temperature), were used for making noodles. High-amylose maize starch (70% amylose) was mixed at varying ratios (9, 13, 17, 28, 37, and 44%) with the cross-linked tapioca starches. Analysis of the noodles included: tensile strength, water absorption, and soluble loss. Noodles made from a mixture of cross-linked tapioca starch and 17% high-amylose starch were comparable to the clear noodles made from mung bean starch (Kasemsuwan et al., 1998). In conclusion, tapioca starch cross-linked by using STMP with varying reaction temperature, reaction time, and reaction pH, viscosities, and pasting properties indicated that the different cross-linked treatments affected the functional properties of starch. Native and cross-linked tapioca starch alone produced noodles that were unacceptable. The noodles prepared from mixtures of cross-linked tapioca starch and high-amylose starch indicated good quality at both the dry and cooked stages. The sensory evaluation indicated that panelists preferred the noodles made from the mixtures of tapioca and high-amylose starch rather than mung bean noodles (Kasemsuwan et al., 1998).

The acceptance of genetically and chemically modified food ingredients is, however, low. Although recipes tend to be cheaper using these starches, the price is still rather high. Another disadvantage in acceptance for the public is the label as "food starch modified" on the packaging of the food stuff. The reason why genetically or chemically modified starches are applied is that it is generally accepted that for this type of application the starch needs to be gelatinised according to a "C"-type of gelatinization curve as disclosed in Chen et al (2002, a.b) and Semeijn et al (2004). Thus, physically treatment on starch emerges timely on this background.

6.3. Adding Physically Modified Starches

The term “hydrothermal treatment” was used by Stute (1992) to describe physical modification of starch resulting from various combinations of moisture and temperature conditions that affect starch properties without visible changes in granule appearance. Physical modification of starch slurries in excess water at temperatures below gelatinization were referred to as annealing. Heat-moisture treatment (HMT), on the other hand, refers to the exposure of the starch to higher temperatures normally above the gelatinization temperature (80 to 120°C) at very restricted moisture content (<35%). Results on heat moisture treatment may also have been influenced by partial gelatinization (Eerlingen et al., 1996). There is considerable to be more natural and safe as compared to chemical modification. Stute (1992) investigated the impact of HMT on viscoamylograph of potato starch. Either a higher onset of temperature for viscosity development, a lower peak viscosity, or a higher or lower end viscosity was observed, depending on treatment conditions. The same observations were made for cassava (Abraham, 1993), maize and lentil, oat and yam (Hoover and Vasanthan, 1994), and sweet potato (Collado and Corke, 1999; Collado et al., 2001) starches.

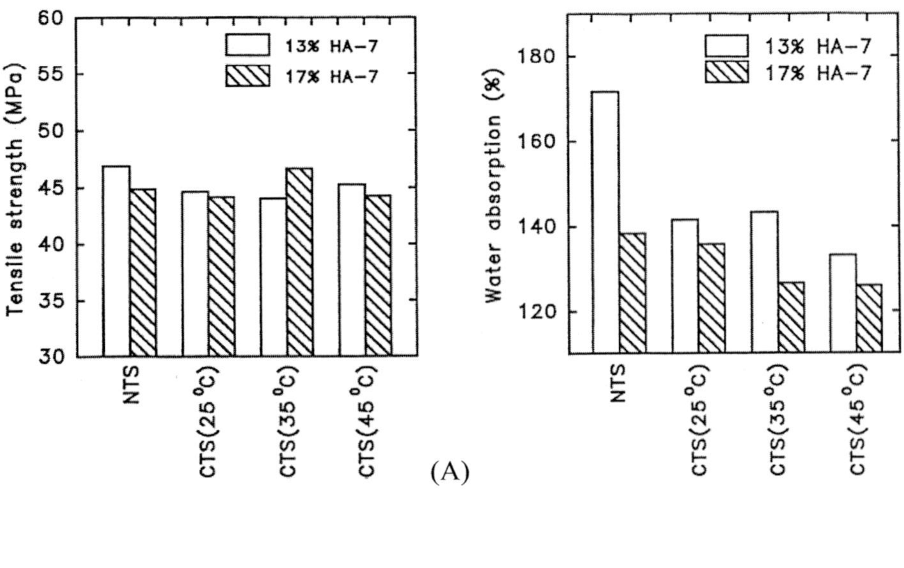

(A) (B)

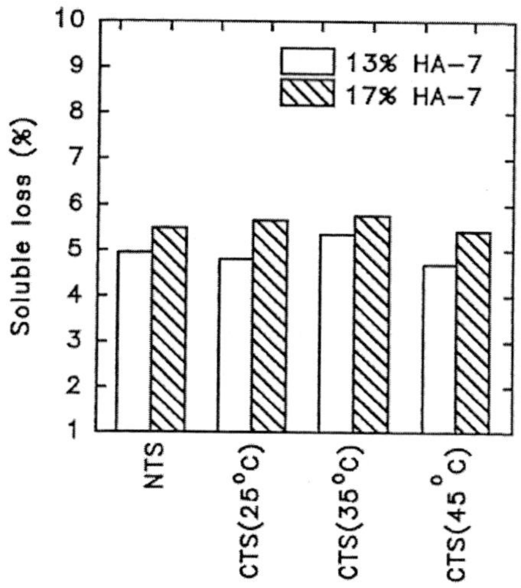

(C)

Figure 4-28. (A) Tensile strengths of noodles (54% moisture) prepared from the native tapioca starch (NTS) containing high-amylose starch (13 and 17%, dsb) and cross-linked tapioca starch (CTS) prepared at 25, 35, and 45°C reaction temperatures containing high amylose starch. (B) Water absorptions of dry noodles after being soaked in water at 25°C for 24 h. Noodles prepared from mixtures of native tapioca starch (NTS) and high-amylose starch (13 and 17%, dsb) and cross-linked tapioca starch (CTS) prepared at 25, 35, and 45°C reaction temperatures and the same amount of high-amylose starch. (C) Soluble loss of noodles prepared from mixtures of native tapioca starch (NTS) and high-amylose starch (13 and 17%, dsb), and cross-linked tapioca starch (CTS) prepared at 25, 35, and 45°C reaction temperature and high amylose starch (13 and 17%, dsb).

Collado et al. (2001) found that sweet potato starch (SPS) has limited uses, but modification of its properties can make it more suitable for use in traditional products especially starch noodles. They applied heat-moisture treatment to native sweet potato starch (HMTSPS), which was used as a substrate and composite with maize starch (MS) to produce *bihon*-type starch noodles. Their results studied indicated that noodles from SPS exposed to HMT were not sticky and were comparable to those from maize starch with regard to handling during processing. Noodle cooking time ranged from 2.5 min in 100% SPS to 3.0 min in the other samples. Yield ranged from 75% in 100% HMTSPS and 50:50 HMTSPS: MS to 78% of dry weight of raw starch in 100% MS. Cooking loss ranged from 2.5% for the commercial sample to 4.0% for 100% native SPS. The rehydration rate was lowest with the native SPS noodles at 234% (W/W), and highest for 100% HMTSPS with 262% (W/W). The hardness was highest for 100% HMTSPS with 289g and lowest for native SP noodles with 156g, while there were minimal differences in stickiness (ranging from 4.0 to 5.2g) (Collado et al., 2001).

Starch noodles with HMTSPS (100% and 50%) had higher color scores and were significantly more yellow than commercial sample and 100% MS. HMTSPS (100%) noodles were significantly less clear than the commercial samples but not significantly diffetent from the commercial sample and 100% MS. 100% HMTSPS had highest smoothness score and was significantly different from 100% MS. Preliminary quality scoring showed that acceptability scores of raw starch noodles, plain boiled, and sautéed noodles made from 100% HMTSPS and 50% HMTSPS: 50% MS were not significantly different from the commercial *bihon*. However, consumer testing is recommended to further validate acceptability to the sweet potato for *bihon* (Collado et al., 2001). Still other possibilities include the used of additives and of blends with other locally produced starches to determine their comparative advantage to the used of HMT to modify SPS for use in noodle production (Collado et al., 2001).

6.4. Biologically Treating Starches

Using corn in starch noodle making will be a good trial, but the traditional production experience showed that crude corn is not suitable for starch noodle making. Generally, a favorable mouth-feel for starchy noodles can be achieved or enhanced by adding sodium alginate, alum and other food additives as well as by modifying the starch by means of chemical and physical treatment such as oxidation and cross-linking. However, these chemicals are unpopular with consumers because of the health hazards associated with them. Spontaneous lactic acid fermentation is an important process in improving the texture of rice noodles (Lu et al., 2005). It was found that fermentation may change the amorphous region of the starch granule as well as the chemical components and thereby modify both physical properties of rice flour and texture of rice noodle. In the other hand, the method using sour liquid to extrude the starch is a traditional way in China, and is also widely used to product starch noodle. The ingredient sour liquid is an aqueous acidic fermented liquid extracted from mung bean starch slurry, which had abundant streptococcus lactics. The noodle prepared by this method is more transparent and flexile than the starch from centrifugation. It could be a practical way to introduce spontaneous lactic acid fermentation to corn starch to improve the texture of corn starch noodle.

Yuan et al (2008) study the effect of spontaneous fermentation on physical properties of corn starch and rheological characteristics of corn starch noodle, and to compare sensory characteristics of fermented corn starch noodle with those of mung bean starch noodle in order to study the feasibility of spontaneous fermentation on improvement of corn starch noodle quality. They found that maximum tensile stress was lowest for control starch noodle, and gradually increased with fermentation time, indicating that fermented corn starch noodles are harder than control sample. The reason may be that fermentation can hydrolyze short chains of amylopectin in the amorphous regions, leading to higher ratio of long-to-short chains in amylopectin and higher tendency for long chains to gel, thus the more rigid gel forms and the noodles become harder. Maximum tensile strain was obtained from the ratio of maximum extension to the original length of starch noodle. It also increased with fermentation time till the 19th day, but after that, the strain decreased from 39.0% at19th day to 33.2% at 21st day (Yuan et al., 2008).

Fermented corn starch noodle scored significantly higher than control corn starch noodle for hardness, which was consistent with the result of tensile experiment. Besides, fermented samples also had higher scores for all the other four sensory attributes, indicating fermentation significantly improved the eating quality of corn starch noodle. When compared with mung bean starch noodle, fermented corn starch noodle had lower hardness score. In fact, cooked starch noodles should be neither too hard nor too soft. These data do not indicate, however, whether the higher or lower hardness was most preferred by panelists. Fermented corn starch noodle scored significantly higher for elasticity than control corn starch noodle, but not significantly different from mung bean starch noodle. The result of overall acceptability was similar to elasticity (Yuan et al., 2008).

The above results indicated that fermentation can greatly improve the eating quality of corn starch noodle, and the quality of fermented corn starch noodle was comparable to that of mung bean starch noodle. Spontaneous fermentation is an effective and safe way to produce corn starch noodle with satisfactory quality. It will contribute to promoting the utilization of corn starch (Yuan et al., 2008).

6.5. Using Additives

Generally, the quality problem of non-mung bean starch noodles is normally partially overcome by using additives and other treatments. The formulation of non-mung bean starch noodles often includes the use of potash alum or polysaccharide gums to improve the quality of the noodle produced from it in China. Many researchers are looking for some new additives to improve the quality of non-mung bean starch noodles.

6.5.1. Adding Soybean Protein

To discuss the effect of the addition of isolated soybean protein (ISP) on physical properties of starch noodles (Harusame), Takahashi et al. (1986) made the noodles from potato, mung bean or broad bean starch mixed with ISP using a pressure extruder at 80°C. The extruded starch was heated in boiling water for 3 min, washed with water, and dried immediately, or frozen, thawed, drained and dried. The noodles were cooked in boiling water for 3 min, washed and their transparency, swelling power, solubility and texture by tensipresser were measured for comparison and they were also subjected to an organoleptic test. Their results indicated that the noodles from potato starch added by 5% ISP is transparent, has a higher tensile strength and elongation elastic modulus, less adhesiveness, being non-sticky, and lower solubility than the noodles from potato starch only. According to the organoleptic test ISP was effective in making it more acceptable, ranking next to noodles made in China. By raising the extrusion temperature from 80°C to 120°C, the cooked noodle from potato starch only was extremely hard and sticky, being difficult to separate from each other, however, the addition of ISP made the noodle more elastic and chewy, and even in the case of ordinary drying, the noodles were not sticky and easy to separate from each other. The effect of freezing of noodles before drying on physical properties of cooked noodles were evident in potato starch showing higher value in compression and tension test, and not in mung bean starch and mixture of potato and sweet potato starch (1:1). However by adding ISP, the physical properties of noodles from these starches were improved, being better than frozen noodles from potato starch (Takahashi et al., 1986).

6.5.2. Adding Fatty Acid Esters

Fatty acid esters (abbreviated as FAE) had a facilitating effect on the separation of frozen starch noodle (abbreviated as FSN). The separating effect of FAE increased in proportion to the length of the alkyl chain of FAE. Further, the separating effect was found to be closely related to the HLB (Hydrophilic Lypophilic Balance) value of FAE. Mohri (1980) studied the relation between the separating effect of FAE on FSN and the interaction of FAE with starches to verify the connection with complex-formation, syneresis, iodineaffinity, viscosity and adhesive force of starch in the presence of FAE and also the adsorption of FAE on starch surfaces.

The amount of FAE adsorbed on starch was highly dependent on the molecular weights of FAE—the greater the adsorption amount of FAE, the longer the alkyl chain length of FAE. Further, as the degree of esterification increases, the adsorption ability of glycerin fatty acid ester decreases. Because of the steric hindrance, the trioleate will not easily adsorb on starch surface. The influence of FAE on syneresis of starch gel was that as the water-solubility of FAE, namely, HLB-value increased, the starch gel containing FAE showed remarkable syneresis. This syneresis is caused by the decrease of hydration power of starch owing to the occurrence of a strong hydrogen bond between FAE and starch. FAE was also effective in lowering the viscosity of starch paste and in decreasing the adhesive force of starch. The higher the HLB of FAE, the higher the viscosity of starch paste containing them. The adhesive force decreasing effect of FAE heavily depends on the molecular weight of FAE and, in general, increases with the molecular weight. This showed that it was closely related to their ability to separate FSN. In fact, weakening the adhesive force of starch disturbing the separation of FSN was an important factor to promote the separation of FSN (Mohri, 1980).

From the above discussion, Mohri (1980) understand that separating effect of FAE has a close relationship to the interaction of FAE with starch. The following factors are at leat important in the separating effect of FAE: (a) adsorption of FAE on starch surface and (b) the subsequent action of FAE to reduce the viscosity and adhesive force of starch paste. The reason for FAE, especially sorbitan stearate or glycerin stearate, being effective in the separating action of FSN probably lies in the FAE ability to satisfy the above requirements.

6.5.3. Adding Glycerol Monostearate

Kaur et al. (2005) studied the effects of glycerol monostearate (GMS) on the physico-chemical, thermal, rheological, textural and noodle making properties of corn starch and potato starches from four different cultivars. The presence of lipids in the corn starch may be another influencing factor that delayed the swelling of individual starch granules within the noodle strands (Singh et al., 2002). The cooking time of the noodles made from different potato starch sources also differed to a considerable extent (Table 4-29). The addition of GMS increased the cooking time of corn and potato starch noodles. This may be due to the restricted supply of water to the starch granules present in the noodle strands, which delayed the swelling of the granules. The helical inclusion complexes formed between the GMS and the amylose may have possibly affected the cooking time of the noodles. After the addition of GMS, potato starch noodles showed the maximum increment in cooking time, with K. Jyoti and K. Sindhuri at the highest. The addition of GMS reduced the cooked weight as well as cooking loss for the corn and potato starch noodles. The presence of GMS may have prevented the swelling of starch granules to their full extent and transport of water that resulted in lower cooking weight. The lower cooking losses with GMS indicated its complexation with amylose within the cooked noodles.

Table 4-29. Cooking properties of corn and potato starch noodles (Kaur et al., 2005)

Starch source	Without GMS*			With GMS		
	Cooking time (min)	Cooked weight(10g)	Cooking loss(10g)	Cooking time (min)	Cooked weight(10g)	Cooking loss(10g)
Potato (kufri chandermukhi)	4.50^{b}	43.9^{b}	0.446^{bc}	5.00^{c}	42.4^{b}	0.426^{d}
Potato (kufri Sutlej)	4.0^{b}	45.6^{b}	0.470^{c}	4.50^{b}	44.6^{c}	0.398^{c}
Potato (kufri Jyoti)	3.0^{a}	50.8^{c}	0.405^{b}	4.00^{a}	42.5^{b}	0.355^{b}
Potato (kufri Sindhuri)	3.3^{a}	47.4^{c}	0.425^{b}	4.00^{a}	45.8^{c}	0.362^{b}
Corn	5.5^{c}	31.8^{a}	0.335^{a}	5.25^{c}	30.1^{a}	0.301^{a}

GMS: glycerol monostearate. Values with similar superscripts in column did not differ significantly ($p<0.05$).

Table 4-30. Texture profile analysis—textural properties of corn and potato starch noodles(Kaur et al., 2005)

Starch source	Without GMS*				
	Hardness (N)	Cohesiveness (N cm)	Springiness (m)	Gumminess (N)	Chewiness (J)
Potato (Kufri chandermukhi)	35.8^{a}	0.416^{b}	0.488^{b}	14.89^{b}	7.27^{b}
Potato (Kufri Sutlej)	48.0^{b}	0.535^{c}	0.482^{b}	25.68^{c}	12.38^{c}
Potato (Kufri Jyoti)	62.4^{d}	0.604^{d}	0.625^{c}	37.68^{d}	23.55^{d}
Potato (Kufri Sindhuri)	59.5^{c}	0.582^{d}	0.683^{c}	34.63^{cd}	23.65^{d}
Corn	31.6^{a}	0.335^{a}	0.403^{a}	10.59^{a}	4.26^{a}

GMS: glycerol monostearate. Values with similar superscripts in column did not differ significantly ($p<0.05$).

Table 4-31. Texture profile analysis—textural properties of corn and potato starch noodles (Kaur et al., 2005)

Starch source	With GMS				
	Hardness (N)	Cohesiveness (N cm)	Springiness (m)	Gumminess (N)	Chewiness (J)
Potato (Kufri chandermukhi)	34.5^{a}	0.408^{b}	0.426^{b}	14.07^{b}	5.99^{b}
Potato (Kufri Sutlej)	45.8^{b}	0.468^{b}	0.435^{b}	21.43^{c}	9.32^{c}
Potato (Kufri Jyoti)	57.6^{c}	0.452^{b}	0.514^{c}	26.033^{d}	13.37^{d}
Potato (Kufri Sindhuri)	55.4^{c}	0.422^{b}	0.575^{c}	23.37^{c}	13.43^{d}
Corn	32.2^{a}	0.292^{a}	0.381^{a}	9.4^{a}	3.58^{a}

GMS: glycerol monostearate. Values with similar superscripts in column did not differ significantly ($p < 0.05$).

The addition of GMS decreased the hardness values of the corn and potato starch noodles however, the effect was slightly pronounced (Table 4-30). The delayed swelling of the starch granules in the presence of GMS may have affected the hardness values of the noodles. Cohesiveness values were significantly higher for the potato starch noodles than corn starch noodles. The greater leaching of amylose from the potato starch granules present in the noodles may be responsible for the higher cohesiveness values. The lower cohesiveness values of the corn starch noodles may be assigned to less leaching of amylose due to the presence of lipids and strong internal bonds. The limited swelling and solubility of the corn starch may also be responsible for the lower cohesiveness of corn starch noodles. The cohesiveness values decreased with the addition of GMS in the corn and potato starch noodles (Table 4-31). The GMS may have provided the stability to the starch granules that reduced the cohesiveness by decreasing granule interaction and association within the starch noodles. The less amylose leaching due to complex formation may also have decreased the cohesiveness. The potato starch noodles from K. Jyoti and K. Sindhuri starches showed higher decrease in their cohesiveness values. Adhesiveness values, which represent the work necessary to pull the compressing plunger away from sample, were not obtained as the noodles had very low stickiness. The gumminess, chewiness and springiness values of the cooked starch noodles containing GMS were also lower than those without GMS (Table 4-31).

In conclusion, the addition of GMS brought substantial changes in the physico-chemical, thermal, rheological, textural and noodle properties. The presence of GMS decreased the swelling power and solubility of starch, while the gelatinization temperatures and enthalpy of gelatinization were observed to be increased. The cooked starch noodles containing GMS showed lower values for texture profile analysis parameters like hardness, cohesiveness, gumminess, chewiness and springiness. The change in different properties of starch and noodles with the addition of GMS was found to depend on the granule morphology. The starches with large granule populations than the potato starches with small granules and corn starch and subsequently brought greater changes in their starch and noodle properties.

6.5.4. Adding Chitosan

Chitosan is a linear polysaccharide of anhydrous β-D-glucosamine units joined by (1→4) linkages. It is obtained by deacetylation of chitin, a natural polymer, manufactured from shrimp or crab shells. Chitosan has three types of functional groups in a monomeric unit, an amino group as well as primary and secondary hydroxyl groups (C-6, C-2, and C-3, respectively). Chitosan has already been used as a functional ingredient to improve food functionality and quality in the food processing industries (Baek et al., 2001). While alum has strong ionic properties in an aqueous solution and increases starch gelatinization temperature, starch dough strength, and bleaching effect, since it is a chemical ingredient, use of natural substances is more favored. Baek et al. (2001) investigated the effect of chitosan addition on starch noodle quality, and to evaluate the possibility of chitosan as an alum replacement for starch noodle preparation.

They found that solubility of starch noodle increased with increasing cooking time and decreased as the amount of added chitosan increased. When the starch noodle containing alum (0.3%) was cooked for 6 and 12 min, the solubility values of starch noodle were 0.80 and 0.98%, respectively, which were less than the values for chitosan-added noodles (0.81–0.90%, and 1.13–1.29%, respectively). This indicates that alum was more effective in holding the noodle structure stable during cooking. At 6 min of cooking, the difference between the alum- and chitosan content increased, the noodle tended to become more stable (Baek et al., 2001).

Solubility is related to the degree of intermolecular associations. In the case of alum, it acted as a chelating agent, resulting in a more rigid and stable network formation compared to chitosan, but solubility was not significantly different with increasing amounts of alum addition. On the other hand, in the case of chitosan, the ionic-dipole interaction between chitosan ($-NH_3^+$) and starch hydroxyl groups (-OH) might facilitate intermolecular interactions that resulted in a more rigid and stable network formation. Consequently, the increased addition of chitosan caused decreasing of solubility of the starch noodle (Baek et al., 2001).

The swelling power of starch noodles increased with increasing cooking time and decreased with increasing amount of chitosan addition. Swelling power showed the same trend as solubility, but the difference in swelling power caused by different chitosan contents was smaller than that of solubility. In the case of starch noodles containing chitosan, swelling power of starch noodle with the greatest amount of chitosan (1000ppm) was 4.37 and 5.80 at 6 and 12 min of cooking, respectively. Therefore, the chitosan was not as effective as the alum addition in reducing swelling and soluble loss of the noodle. The interaction between starch and chitosan increased while water adsorption decreased as the amount of chitosan addition increased. These changes in the interactions resulted in the decreased swelling of the starch noodles with increasing amounts of the residual chitosan (Baek et al., 2001).

Hardness of the starch noodles rapidly decreased with increasing cooking time, but increased with increasing amount of chitosan addition. The starch noodles containing alum showed greater hardness than those containing chitosan. The alum-added noodles cooked for 12 min showed a similar hardness to that of starch noodles containing chitosan (1000ppm) cooked for 8 min. Therefore, the starch noodles containing chitosan readily became soft. However, as the amount of chitosan addition increased the noodles became harder. Gumminess and chewiness showed similar trends to hardness. Cohesiveness and springiness were not significantly different among the samples, but a slight decrease was observed after 10 min of cooking (Baek et al., 2001).

From a preference test of alum- or chitosan-added starch noodles cooked for 10 to 12 min, starch noodles containing 750 ppm chitosan or 0.3% alum showed the highest preference with similar scores. There was no significant difference between the two starch noodles added with alum and with chitosan in the texture properties and acceptability. However, the color of the cooked noodle appeared stronger for the alum-noodle (Baek et al., 2001).

Chitosan is widely used in food industries as an additive texture controlling agent, food mimetic, thickening and stabilizing agent, and a nutritional quality enhancer (dietary fiber, hypocholesterolemic effect) (Shahidi et al., 1999). From the sensory evaluation, it can be concluded that chitosan can be used as a replacement of alum for starch noodle quality (Baek et al., 2001).

6.5.5. Adding Polysaccharide Gums

Tan (2007) studied the effect of additives (including polysaccharide gums and alum) on the short-term and long-term retrogradation of sweet potato starch (SPS) in order to provide a theory foundation on improving the quality of its starch noodles. She measured the variety trend of viscosity of a mixture of SPS and additives during heating and cooling using Rapid

Viscosity Analysis (RVA) and its gel firmness using Texture Analysis and the T_o, T_p, T_c, $\triangle$H of the mixture system after crystallization using Differential Scanning Calorimetry (DSC), evaluated the quality of sweet potato starch noodles (SPSN) after added additives. The results indicated the RVA parameters of sweet potato starch paste and texture quality indexes of its noodles were increased with adding additives in a certain extent, especially alum, *Artemisia sphaerocephala* Krasch (ASK), konjak glucomannan (KGM), and xanthan in terms of the increased extent. The DSC thermogram of retrograded sweet potato starch with polysaccharide gums showed two peaks, one was melted amylopectin, another one was a radiative peak (87°C~105°C), which resulted from recombination of polysaccharide gums and SPS. The melting temperature of retrograded SPS increased or decreased with adding polysaccharide gums, but the enthalpy increased. There were radiative peaks owing to the conjecture that polysaccharide gums re-melted and became free at this temperature, and then competed with a few lipids in SPS, preferentially recombined with amylose in SPS, formed a firm system. The quality of SPSN added 1% (on the basis of total starch weight) of KGM and ASK (0.95:0.05) was similar to that of mung bean starch noodles and SPSN containing alum.

Combining the measurement of the glass transition temperature using differential scanning calorimetry, leached amylose, interaction test and infrared spectrum analysis of mixture systems, Tan (2007) also analyzed the mechanism of interaction between additives and sweet potato starch. The results illuminated that these mixture systems contained CMC, carrageenan and SPS was inconsistent, while those mixture systems contained alum, salt, soybean protein, glycerol, other polysaccharide gums and SPS was consistent. The mixture systems contained KGM, ASK and SPS was more steady than others. Additives combined with amylose in SPS. These interactions were strong or weak. The NaCl can weaken faintly the interactions between polysaccharide gums and SPS, while carbamid can weaken strongly their interactions. The –OH peaks in these infrared spectrum peaks of SPSN contained additives were displaced and other peaks were not displaced comparing to the original SPSN in these infrared spectrum figures. She confirmed that the mechanism of interaction between KGM, ASK and SPS was described as follows: the mixture system contained KGM, ASK and SPS was consistent. There was no new functional group in this mixture system. Their amylose and exterior chains in amylopectin juxtaposed each other by hydrogen bond, which existed inner and exterior chains, and then form minicrystal zone, which acted as junctures. The net in mixture system was held together by countless junctures so that sweet potato starch gel contained KGM and ASK had a strong texture. The interaction between SPS and alum mainly was static electricity.

Lee et al. (2002) also compared nine polysaccharide gums (sodium alginate, carboxymethyl cellulose, curdllan, gellan, guar gum, gum Arabic, k-carrageenan, locust bean, and xanthan) for their stabilizing effects in sweet potato starch gel against repeated freeze-thawing (FT) treatments. They found that the gums were added in starch gel at 0.3 or 0.6% (w/w, based on total gel weight), and total solid content in the gel was adjusted to 7% (w/w) with starch. The gels containing starch and gum were repeatedly freeze-thawed up to five times by storing at -18°C for 20 h and then at 25 for 4 h. Water release (syneresis) was measured by vacuum-filtering the freeze-thawed gels. Among the gums tested, alginate, guar gum, and xanthan were highly effective in reducing the syneresis. For example, guar gum, at 0.6%, showed the least syneresis (33.0%, w/w based on initial water content) after five FT cycles, which was less than half that of pure starch gel. At 0.3%, however, xanthan was more effective than guar gum in reducing syneresis. Xanthan reduced paste viscosity significantly, whereas guar gum and alginate increased the viscosity, but there was little relation between pasting viscosity and syneresis. The gums remained in the gel matrix during the syneresis without a significant loss. Recrystallization of starch (retrogradation) induced by FT treatment was also retarded by the presence of gums, and sodium alginate was more effective in retarding the retrogradation than xanthan or guar gum (Lee et al., 2002).

Funami et al. (2005, 2b) studied the retrogradation behavior of corn starch in an aqueous system in the presence or absence of various guar gum samples with different molecular weights. Dynamic mechanical loss tangent for starch system with 26% amylose (5 w/v%) was increased by the addition of guar (0.5%) after storage at 4 °C for 24 h, which indicated the reduction of gelled fraction in the system, leading to the retardation of short-term retrogradation of starch. This rheological change of the system related to the amount of amylose leached out the starch granules during gelatinization. The higher the molecular weight of guar, the lower the amount of amylose leached, but this effect of guar became less dependent on its molecular weight at above 15.0×10^5 g/mol. The rate constant determined from the relationship between storage time (for 14 days at 4°C) and creep compliance for the starch system (15% starch) was decreased in the presence of guar (0.5%), suggesting the retardation of long-term retrogradation of starch. This effect of guar became marked at above 30.0×10^5 g/mol, which was apparently higher than the critical molecular weight value determined from short-term retrogradation. Syneresis for the starch system (5% starch) was increased adversely by the addition of guar (0.5%) with relatively low molecular weight values (e.g., 5.0×10^5 g/mol) after storage at 4°C for 14 days, suggesting the promotion of long-term retrogradation. Functions of guar on the retrogradation behavior of starch were hypothesized considering interactions between guar and starch components; amylose and amylopectin (Funami et al., 2005, 2b).

Funami et al. (2005) studied the gelatinization and retrogradation behavior of wheat starch in an aqueous system by rheological and thermal techniques in the presence or absence of non-ionic polysaccharides, including guar gum, tara gum, locust bean gum, and konjac glucomannan. Macromolecular characteristics of each polysaccharide, including weight-average molecular weight M_w and radius of gyration R_g; were determined by static light-scattering, resulting in (1.0–3.2) ×106 g/mol for Mw and 104-217 nm for Rg; respectively. During gelatinization, addition of each polysaccharide (0.5–1% w/v) increased peak viscosity for the starch system (13%): 163–231 units larger than the control at 0.5%, whereas 230–437 units larger at 1%. Among the galactomannans tested, the order of this effect (locust>tara>guar) was contrary to that of the molecular size (guar>tara>locust). During short-term retrogradation, addition of each polysaccharide (0.5%) increased dynamic mechanical loss tangent (tan σ) for the starch system (5%) after storage at 4°C for 24 h: (16.5–26.9) ×10-2 unit larger than the control. Among the galactomannans tested, the larger the molecular size, the greater the effect to increase tan d; and this effect of polysaccharide was not explained simply by the difference in the amount of amylose leached during gelatinization. During long-term retrogradation, addition of each polysaccharide (0.5%) decreased the rate constant expressing the relationship between storage time (for 14 days at 4°C) and creep compliance for the starch system (15%): (0.9–1.5) $\times 10^{-2}$ unit smaller than the control. Among the galactomannans tested, the larger the molecular size, the greater the effect to decrease the rate constant. Functions of polysaccharide to starch were hypothesized considering structural compatibility and molecular interactions between polysaccharide and starch components; amylose and amylopectin.

7. The Future of Starch Noodles

The starch noodles on the market today use other starch materials like broad bean, and other starches besides mung bean, or just plain starch only (corn starch, tapioca starch, or potato starch). Making high-quality starch noodles involves intensive labor and liquid waste disposal, and these are bottlenecks in the process. The liquid waste is fairly rich in nutrients as it contains all the vitamins, minerals, and proteins in the starchy materials and can be used as animal feed. Attempts have also been made to recover the protein from this liquid waste.

For example, a great deal of wastewater was produced during the sweet potato starch production in China. The wastewater will pollute the environment because there are a large number of organics in it. Most of the organics of wastewater are glycoprotein, which have excellent immunodulating and antitumor biological activity. If the glycoprotein can be extracted from the waste water, not only the polluting problem can be resolved, but also can obtain a kind of biological activity product (Cheng, 2005).

Ultrafiltration was used for extracting glycoprotein from wastewater of sweet potato starch production and the structure, biological activities of the glycoprotein were studied by Cheng (2005). He found that although hollow ultrafiltrator can deal with large number of wastewater at a low cost, its membrane was polluted badly and difficult to clean, and lost of glycoprotein was serious compared with ceramic ultrafiltrator. So ceramic ultrafiltrator was more suitable for concentrating glycoprotein from wastewater of sweet potato starch production than hollow fabric ultrafiltrator. Glycoprotein of sweet potato has no acute toxicity certified by mice acute toxicity experiment, but has excellent immunomodulating activity certified by carbon clearance ratio of mice and highest clearance index (4.52) at the dose of 50mg /kg.d. When used together with Cycolphosphamide, glycoprotein also can reduce damage of body by Cycolphosphamide. Glycoprotein of Sweet Potato has excellent antitumor activity certified by mice against Sarcoma 180 implanted tumor experiment. Glycoprotein has the highest tumor inhibition rate (62%) at the dose of 50mg/kg.d. Glycoprotein also can enhance antitumor effect when used together with Cycolphosphamide. The antitumor effect of glycoprotein is due to enhance immunity of body. This is proved by the effects of proliferation of active splenocyte by ConA and S_{180} inhibition in vitro experiment. Relationship between structure and biological activity illuminated that there was no immunomodulating activity when using protein or sugar chain along; there was immunomodulating activity only when protein and sugar chains linked together (Cheng, 2005).

Exploring new products unceasingly also is very important for the evolution of the starch noodles industry. For example, Zhou et al. (2004) designed audaciously and successfully the recipe of starch noodles with vegetables as supplements based on the alum-free one. Raw materials were selected for nutrients and colors from 7 vegetables. The application amounts of the vegetables and the relationship between coagulating time and pliability were also studied. The results showed that amaranth, spinach and China squash could be used as supplement a raw materials in vegetable glass noodle processing and their application amount were 7%, 5% and 5% respectively. In addition, ascorbic acid could be used as color---

protecting agent with the optimum dosage of 1% and the optimum coagulating time was 1.0 to 1.5 h.

In conclusion, starch noodles became an important and special food and their consumers are not only in Asia but spread around world. There are different ways of consuming starch noodles as well as different recipes available in different countries. Also, there are numerous types that depend on the raw materials, product shapes, processing methods and the way of preparation and serving. However, they have also undergone changes driven by technical innovations and consumer demands. Most noodles today are produced by machine. While the actual process for manufacturing a particular type of starch noodle may differ from country to country to meet local needs, the basic principles involved are practically the same. The processing technology, such as vacuum mixing, auto-extruding, intelligentized freezing, in the production of starch noodles during the 1990s, had rapidly been developing. Although noodles are traditional foods, the technical and technological innovations are continuously evolving to adapt them to the global consumers of all ages. Recipes have been modified and continuously adjusted to suit the taste of the consumers in many countries. Production equipment has been improved, modernized, and up-graded to guaranty efficient productivity. Production costs have been optimized to make it affordable also for people in the developing nations. Product size has been modified and continuously being adjusted to the tradition of the western world. New packaging designs, new recipes and new ways of preparation continuously appear on the market to satisfy the eternal desire of consumers, particularly the younger generation, for something new, something fulfilling and something good.

REFERENCES

Abraham T E (1993), 'Stabilization of paste viscosity of cassava by heat moisture treatment', *Starch/ Stärke,* 45, 131-135.

Aktan B and Khan K (1992), 'Effect of high-temperature drying of pasta on quality parameters and on solubility, gel electrophoresis, and reversed-phase high-performance liquid chromatography of protein components', *Cereal Chemistry,* 69, 288-295.

Apolonio V-T, Perla O-D, José J-H, Juscelino T, Octavio P-L and Luis A B-P (2004), 'Starch digestibility of five cooked black bean (Phaseolus vulgaris L.) varieties', *Journal of Food Composition and Analysis,* 17(5), 605-612.

Baek M H, Cha D S and Lim S T (2001), 'Effect of chitosan addition on textural properties of sweet potato starch noodle', *Food Science Biotechnology,* 10(2), 161-165.

Banks W and Greenwood C T (1975), 'Starch and Its Components', Halsted Press: New York.

Bemiller J N and Whistler R L (1996), 'Carbohydrates'. In: Fennema OR, editor. Food chemistry, New York: Marcel Dekker, Inc. p.195.

Biliaderis C G, Grant D R and VOSE J R (1981), 'Structural characterization of legume starches. □. Studies on acid-treated starches', *Cereal Chemistry,* 58(6), 502-507.

Biliaderis C G, Grant D R and VOSE J R (1981a), 'Structural characterization of legume starches,. studies on amylose, amylopectin, and beta-limit dextrins'. *Cereal Chemisty,* 58(6), 496-502.

Cairns P, Leloup V, Miles M J, Ring S G and Morris V J (1990), 'Resistant starch: an X-ray diffraction study into the effect of enzymatic hydrolysis on amylose gels *in vitro*', *Journal of Cereal Science,* 12, 203-206.

Chang L -T (1993), 'The starch silk specialty food in China', *Starch/Stärke,* 45(2), 73.

Chang Y-H, Lin C-L and Chen J -C (2006), 'Characteristics of mung bean starch isolated by using lactic acid fermentation solution as the steeping liquor', *Food Chemistry*, 99, 794-802.

Chen Z, Sagis L, Legger A, Linssen J P H, Schols H A and Voragen A G J (2002), 'Evaluation of starch noodles made from three typical Chinese sweet potato starches', *Journal of Food Science,* 67(9), 3342-3347.

Chen Z, Schols H A, and Voragen A G J (2003), 'Physicochemical properties of starches obtained from three varieties Chinese sweetpotatoes', *Journal of Food Science,* 68(2), 431-437.

Chen Z, Schols H A and Voragen A G J (2003, b), 'Starch granule size strongly determines starch noodle processing and noodle quality', *Journal of Food Science,* 68(5), 1584-1589.

Cheng K W (2005), 'Studies on extracting of glycoprotein from wastewater of Sweet potato starch production, its biological activities and structure', Southern Yangtze University. PhD. Dissertation. 2005, 3.

Cheng Y L and Shuh M C (1981), 'Characterization of red bean (*Phaseolus radiatus var. Aurea*) starch and its Noodle quality', *Journal of Food Science,* 46, 78-81.

Christianson D D and BAGLEY E B (1983), 'Apparent viscosities of dispersions of swollen corn starch granules', *Cereal Chemistry*, 60, 116-121.

Chung H J, Jeong H Y and Lim S T (2003), 'Effects of acid hydrolysis and defatting on crystallinity and pasting properties of freeze-thawed high amylose corn starch', *Carbohydrate polymers,* 54, 449-455.

Cluskey J E, Knutson C A and Inglett G E (1980), 'Fractionation and characterization of dent corn amylomaize starch granules', *Starch/Staerke*, 32, 105–109.

Collado L S and Corke H (1999), 'Heat-moisture treatment effects on sweetpotato starches differing in amylase content', *Food Chemistry,* 65, 339-346.

Collado L S, Mabesa L B, Oates C G and Corke H (2001), 'Bihon-type noodles from heat-moisture-treated sweetpotato starch', *Journal of Food Science,* 66(4), 604-609.

Collado L S and Corke H (1997), Properties of starch noodles as affected by sweetpotato genotype, *Cereal Chemistry,* 74(2), 182-187.

Eerlingen R C, Jacobs H, Van Win H and Delcour J A (1996), 'Effect of hydrothermal treatment on the gelatinization properties of potato starch as measured by differential scanning calorimetry', *Journal of Thermal Analysis,* 47: 1229-1246.

Eliasson A-C (2004), 'Starch in food: Structure, function and applications'. CRC Press, Boca Raton, Boston, New York, Washington, DC. Page 334-336. (Moorthy S.N. Tropical sources of starch).

Eliasson A-C (2004), 'Starch in food: Structure, function and applications'. CRC Press, Boca Raton, Boston, New York, Washington, DC. P 64-71, P295-300, 578.

Eliasson A-C and Kim H R (1992), 'Changes in rheological properties of hydroxypropyl potato starch pastes during freeze-thaw treatments,□. A rheological approach for evaluation of freeze-thaw stability', *Journal of Texture Studies,* 23, 279-293.

Englyst H N, Kingman S M and Cummings J H (1992), 'Classification and measurement of nutritionally important starch fractions', *European Journal of Clinical Nutrition,* 46 (Suppl. 2), 33.

Fannon J E and Bemiller J N (1992), 'Structure of corn starch paste and granule remnants revealed by low temperature scanning electron microscopy after cryopreparation', *Cereal Chemistry,* 69, 456–460.

Food and Agriculture Organization: http://www.fao.org/

Faulks R M and Bailey A L (1990), 'Digestion of cooked starches from different food sources by porcine a-amylase', *Food Chemistry*, 36, 191–203.

Funami T, Kataoka Y, Omoto T, Goto Y, ASAI I and NISHINARI K (2005), 'Effects of non-ionic polysaccharides on the gelatinization and retrogradation behavior of wheat starch', *Food Hydrocolloids,* 19, 1–13.

Funami T, Kataoka Y, Omoto T, Goto Y, Asai I and Nishinari K (2005), 'Food hydrocolloids control the gelatinization and retrogradation behavior of starch. 2b. Functions of guar gums with different molecular weights on the retrogradation behavior of corn starch', *Food Hydrocolloids,* 19, 25-36.

Galvez F C F (1992), 'Modelling and optimization of the physical properties, sensory quality, and consumer acceptance of mungbean(*Vigna radiata L.Wilezee*) noodles using a two-stage constrained optimization procedure by response surface methodology', Ph.D.dissertation, University of Georgia, Athens, 12-14.

Galvez F C F and Resurrection A V A (1992), 'Reliability of the focus group technique in determining the quanlity characteristics of mung bean [Vigna radiata (L.) Wilczec] noodles', *Journal of Sensory Studies,* 7, 315-326.

Galvez F C F, Resurreccion A V A and WARE G O (1994), 'Process variables, gelatinized starch and moisture effects on physical properties of mungbean noodles', *Journal of Food Science,* 59(2): 378-386.

Galvez F C F, Resurreccion A V A and Ware G O (1995), 'Formulation and process optimization of mungbean noodles', *Journal of Food Processing and Preservation,* 19, 191-205.

Garcia A M and Walter W M (1998), 'Physicochemical characterization of starch from Peruvian sweet potato starches', *Starch/Stärke,* 50, 331-337.

Gidley M J, Cook D, Darke A H, Hoffmann R A, Russell A L and Greenwell P (1995), 'Molecular order and structure in enzyme-resistant retrograded starch', *Carbohydrate Polymers,* 28, 23-31.

Godet M C, Bizot H and Buléon A (1995), 'Crystallization of amylose-fatty acid complexes prepared with different amylose chain lengths', *Carbohydrate polymers*, 27, 47-52.

Guraya H S, Toledo R T and KAYS S J (1998), 'Effect of cultivar and root storage on pasting characteristics of sweet potato starch', *Tropical Science*, 38, 57-61.

Härröd M (1989), 'Time-dependent flow behavior of starch pastes, with food process applications', *Journal of Food Process Engineering,* 11, 297-309.

Hermann M (1996), 'Starch noodles from edible canna. p. 507-508. In: J. Janick (ed.), Progress in new crops'. ASHS Press, Arlington, VA.

Hizukuri S (1969), 'Effect of environment temperature of plants on the physicochemical properties of their starches', *Journal of Japanese Society of Starch Sciences (Denpun Kagaku),* 17, 73-88.

Hizukuri S (1986), 'Polymodal distribution of the chain lengths ofamylopectin, and its significance', *Carbohydrate Research*, 147(3), 342-347.

Hoover R and Sosulski F W (1985), 'A comparative study of the effect of acetylation on starches of Phaseolus vulgaris biotypes', *Starch/Stärke,* 37, 397-403.

Hoover R and Sosulski F (1991), 'Composition, structure, functionality, and chemical modification of legume starches: a review', *Can. J. Physicol Pharmacol,* 69, 79-92.

Hoover R and Vasanthan T (1994), 'Effect of heat-moisture treatment on the structure and physicochemical properties of cereal, legume, and tuber starches', *Carbohydrate Research,* 252, 33-53.

Hoover R and Zhou Y (2003), 'In vitro and in vivo hydrolysis of legume starches by a-amylase and resistant starch formation in legumes—a review', *Carbohydrate Polymers,* 54, 401–417.

http://en.wikipedia.org/wiki/Chinese_noodles

Hui Y H (2007), 'Handbook of Food Products Manufacturing. Principles, Bakery, Beverages, Cereals, Cheese, Confectionary, Fats, Fruits, and Functional Foods'. Wiley-Interscience, A John Wiley & Sons, Inc., Publication. pp 559~561.

Inouchi N, Osaka D V, Lafayette G W and Osaka H F (1987), 'Chain length distribution of amylopectins of several single mutants and the normal counterpart, and sugary-1 phytoglycogen in maize (Zea mays L.)', *Starch/Stärke,* 39, 259-266.

International Statistical Yearbook (ISY 2006/2007). *'Compiled by National Bureau of Statistics of China'* (p. 230). Beijing, China: China Financial & Economic Publishing House.

Jacobson M R and Bemiller J N (1998), 'Method for determining of the rate and extent of accelerated starch retrogradation', *Cereal Chemistry,* 75, 22-29.

Jacobson M R, Obanni M and Bemiller J N (1997), 'Retrogradation of starches from different botanical sources', *Cereal Chemistry,* 74, 571–578.

Jane J -L and Chen J -F (1992), 'Effect of amylose molecular size and amylopectin branch chain length on paste properties of starch', *Cereal Chemistry,* 69(1), 60-65.

Jane J, Chen Y Y, Lee L F, Mcpherson A E, Wong K S, Radosavljevic M, ET AL. (1999), 'Effects of amylopectin branch chain length and amylose content on the gelatinization and pasting properties of starch', *Cereal Chemistry,* 76(5), 629-637.

Jin M, Wu J and Wu X (1994), 'A study on properties of starches used for starch-noodle making', Pages 488-496 in: *Proceedings 1994 International Symposium and Exhibition on New Approaches in the Production of Food Stuffs and Intermediate Products from Cereal Grains and Oil Seeds.*G. Xie and Z.Ma, eds. CCOA/ICC/AACC Meeting. CCOA: Beijing.

Kallabinski J and Balagopalan C (1991), 'Enzymatic starch extraction from tropical root and tuber crops', *in Proc. Ninth Symp. Internat. Soc. Trop. Root Crops.* Ofori F. and Hahn S.K. (eds) 1991.

Kasemsuwan T, Bailey T and Jane J (1998), 'Preparation of clear noodles with mixtures of tapioca and high-amylose starches', *Carbohydrate polymers,* 32, 301-312.

Kaur L, Singh J and SINGH N (2005), 'Effect of glycerol monostearate on the physico-chemical, thermal, rheological and noodle making properties of corn and potato starches', *Food Hydrocolloids,* 19, 839–849.

Kaur L, Singh N and Sodhi N S (2002), 'Some properties of potatoes and their starches II. Morphological, thermal and rheological properties of starches', *Food Chemistry,* 79, 183–192.

Khader V and Rao S V (1996), 'Studies on protein quality of green gram (Phaseolus aureus)', *Plant Foods for Human Nutrition,* 49, 127-132.

Kim K E, Lim J K and Lim S -T (1999), 'Comparison between Korean and Japanese starch noodles (Korean) ', *Food Sci. Ind.* 32, 56-64.

Kim Y S, Wiesenborn D P, Lorenzen J H and Berglund P (1996), 'Suitability of edible bean and potato starches for starch noodles', *Cereal Chemistry,* 73(3), 302-308.

Kim Y S and Wiesenborn D P (1996), 'Starch noodle quality as related to potato genotypes', *Journal of Food Science,* 61, 248-252.

Krueger B R, Knutson C A, Inglett G E and Walker C E (1987), 'A differential scanning calorimetry study on the effect of annealing on gelatinization behavior of corn starch', *Journal of Food Science*, 52, 715–718.

Labell F, Eastern Editor. (1990), 'Potato starch improves oriental noodle texture', *Food Processing,* 118-119.

Lee M H, Baek M H, Cha D S, Park H J and LIM S T (2002), 'Freeze-thaw stabilization of sweet potato starch gel by polysaccharide gums', *Food Hydrocolloids,* 16, 345-352.

Lee S -Y, WOO K -S, Lim J -K, Kim H I and LIM S-T (2005), 'Effect of processing ariables on texture of sweet potato starch noodles prepared in a nonfreezing process', *Cereal Chemistry,* 82(4), 475-478.

Li J H and Vasanthan T (2003), 'Hypochlorite oxidation of field pea starch and its suitability for noodle making using an extrusion cooker', *Food Research International,* 36, 381-386.

Li J, Berke T G and Glover D V (1994), 'Variation for thermal starch properties of starch in tropical maize germplasm', *Cereal Chemistry,* 71, 87–90.

Lii C Y and Chang S M (1981), 'Characterization of red bean (*Phaseolus radiatus var. Aurea*) starch and its noodle quality', *Journal of Food Science*, 46, 78-81.

Lii C Y, Tsai M L and Tseng K H (1996), 'Effect of amylose content on the rheological properties of rice starch', *Cereal Chemistry,* 73, 415-420.

Liu W J and Shen Q (2007), 'Studies on the physicochemical properties of mung bean starch from sour liquid processing and centrifugation', *Journal of Food Engineering,* 79, 358-363.

Longton J and Legrys G A (1981), 'Differential scanning calorimetry studies on the crystallinity of ageing wheat starch gels', *Starch/Stärke,* 33, 410-415.

Lu Z H, Li L T, Min W H, Wang F and Tatsumi E (2005), 'The effects of spontaneous fermentation on the physical properties of rice flour and the rheological characteristics of rice noodles', *International Journal of Food Science and Technology,* 40, 985–992.

Madhusudhan B and Tharanathan R N (1996), 'Structural studies of linear and branched fractions of chick pea and finger millet starches', *Carboydrate Research*, 284, 101-109.

Madhusudhan B, Gowda L R and Tharanathan R N (1996), 'Studies on sweet potatoes ---- Distribution of unit chains of branched and unbranched molecules of starch', *Food Chemistry,* 57(2), 201-204.

Mestres C, Colonna P and Buleon A (1988), 'Characterization of starch networks within rice flour noodles and mung bean starch vermicelli', *Journal of Food Science*, 53, 1809-1812.

Mohri Z -i (1980), 'Interaction between starch and fatty acid esters in frozen starch noodles', *Agricultural and Biological Chemistry,* 44(7), 1455-1459.

Moorthy S N and Balagopalan C (1999), 'Physicochemical properties of enzymatically sepatated starch from sweet potato', *Tropical Science,* 39, 23-27.

Morrison W R, Tester R F, Gidley M J and KARKALAS J (1993), 'Resistance to acid hydrolysis of lipid-complexed amylose and lipid-Free amylose in lintnerized waxy, and non-waxy barley starches', *Carbohydrate Research,* 245, 289-302.

Muhammad K, Kusnandar F, Hashim D M and Rahman R A (1999), 'Application of native and phosphorylated tapioca starches in potato starch noodle', *International Journal of Food Science and Technology*, 34, 275-280.

Nguyen Q D, Jensen C T B and Kristensen P G (1998), 'Experimental and modeling studies of the flow properties of maize and waxy maize starch pastes', *Chemical Engineering Journal*, 70, 165-171.

Noda T, Takahata Y, Sato T, Ikoma H and Mochida H (1996), 'Physicochemical properties of starch from purple and orange fleshed sweet potato roots at two levels of fertilizer', *Starch/Stärke*, 48, 395-399.

Panlasigui L N, Thomson L U, Jenkins D J A, Juliano B O, Perez CO and YIU S (1990), 'Starch digestibility and glycemic response to extruded high amylose and rice noodles', *Trans. Nat. Acad. Sci. Technol. (Manila).* 12, 109-127.

Pavan G (1979), 'High temperature drying of paste products', *Lebens-mittelindustrie*, 26, 319-323.

Perera C and Hoover R (1999), 'Influence of hydroxypropylation on retrogradation properties of native, defatted and heat–moisture treated potato starches', *Food Chemistry,* 64, 361-375.

Perera C, LU Z, Sell J and JANE J (2001), 'Comparison of physicochemical properties and sugary-2 corn starch with normal and waxy cultivars', *Cereal Chemistry,* 78, 249-256.

Prakash S and Kokini J L (2000), Estimation and prediction of shear rate distribution as a model mixer, *Journal of Food Engineering*, 44, 135-148.

Purwani E Y, Widaningrum Thahir R and Muslich (2006), 'Eeffect of heat moisture treatment of sago starch on its noodle quality', *Indonesian Journal of Agricultural Science,* 7(1), 8-14.

Rask H (2003), *'Research overview on grain legume fractions'.* Pulse Crops Canada Association Saskatoon, SK.

Rask H (2004), *'Evaluation of Canadian pea/ lentil starch extraction and noodle preparation'.* Pulse Crops Canada Association Saskatoon, SK.

Ratnayake W S and Hoover R (2002), 'Pea starch: composition, structure and properties: A review', *Starch,* 54 (6), 217-234.

Redy I A and Seib P A (2000), 'Modified waxy wheat starch compared to modified waxy corn starch', *Journal of Cereal Science*, 31, 25–39.

Ring S G, Colonna P, Orford P D and Kalichevski M T (1987), 'The gelation and retrogradation of amylopectin', *Carbohydrate Research,* 162, 277-287.

Sandhu K S, SINGH N and Kaur M (2004), 'Characteristics of the different corn types and their grain fractions: physicochemical, thermal, morphological, and rheological properties of starches', *Journal of Food Engineering*, 64, 119–127.

Sandhua K S and Lim S -T (2008), 'Digestibility of legume starches as influenced by their physical and structural properties', *Carbohydrate Polymers,* 71(2), 245-252.

Sayar S, Turhan M and Gunasekaran S (2001), 'Analysis of chickpea soaking by simultaneous water transfer and water-starch reaction', *Journal of Food Engineering,* 50, 91-98.

Schoch T J and Maywald E C (1968), 'Preparation and properties of various legume starches', *Cereal Chemistry,* 45(6), 564-573.

Semeijn C, Hen Z H, Schols H A and Buwalda P L (WO/2004/095943) *'Translucent Food: Translucent food the invention relates to a translucent food product and to a method for preparing said product'.* 2004.11, 11.

Seog H M, Park Y K, Nam D Y J, Shin H and Kim J P (1987), 'Physicochemical properties of several sweet potato starches', *Hanguk Nanghaw Hakhoechi*, 30: 179-185.

Shahidi F, Arachchi J K V and Jeon Y J (1999), 'Food applications of chitin and chitosans', *Trends of Food Science and Technology,* 10: 37-51.

Shi Y-C and Seib P A (1992), 'The structure of four waxy starches related to gelatinization and retrogradation', *Carbohydrate Research*, 227,131-137.

Sievert D, Czuchajowska Z and Pomeranz Y (1991), 'Enzyme-resistant starch.□.X-ray diffraction of autoclaved amylomaize VII starch and enzyme-resistant starch residues', *Cereal Chemistry,* 68, 86-91.

Singh J and Singh N (2001), 'Studies on the morphological, thermal and rheological properties of starch separated from some Indian potato cultivars', *Food Chemistry,* 75, 67-77.

Singh N, Singh J and Sodhi N S (2002), 'Morphological, thermal, rheological and noodle-making properties of potato and corn starch', *Journal of the Science of Food and Agriculture,* 82, 1376-1383.

Singh N, Singh J, Kaur L, Sodhi N S and Gill B S (2003), 'Morphological, thermal and rheological properties of starches from different botanical sources', *Food Chemistry,* 81,219-231.

Singh J and Singh N (2003), 'Studies on the morphological and rheological properties of granular cold water soluble corn and potato starches', *Food Hydrocolloids,* 17, 63-72.

Singh U, Voraputhaporn W, Rao P V and Jambunathan R (1989), 'Physicochemical characteristics of pigeon pea and mungbean starches and their noodle quality', *Journal of Food Science,* 54, 1293-1297.

Sisawad S and Chatket I (1989), 'Transparent noodles from other bean starches'. In Trends in Food Biotechnology, *Proceedings of the 7th Word Congress of Food Science and Technology, Singapore,* October 1987. A.H. Ghee, Ed-in-chief; N.B. Hen and L.K. Kong (Ed.) p.309-318.

Steffe J F (1992), 'Rheological methods in food process engineering', East Lancing, Freeman Press.

Steffe J F (1996), *'Rheological Methods in Food Process Engineering* (2nd)'. Michigan: Freeman Press, 1–63.

Stute R (1992), 'Hydrothermal modification of starches: the difference between annealing and heat moisture treatment', *Starch/ Stärke,* 6, 205-214.

Svegmark K and Hermansson A M (1993), 'Microstructure and rheological properties of composites of potato starch granules and amylose: a comparison of observed and predicted structure', *Food Structure,* 12, 181–193.

Takeda Y, Tokunaga N, Takeda C and Hizukuri S (1986), 'Physicochemical properties of sweet potato starches', *Starch/Stärke*, 38, 345-350.

Takeda Y and Hizukuri S (1987), 'Structures of branched molecules of amyloses of various origins, and molar fractions of branched and unbranched molecules', *Carbohydrate Research,* 165, 139-145.

Takahashi S, Hirao K and D Watanabe T (1986), 'Effect of added soybean protein on physico-chemical properties of starch noodles (harusame)'. *Journal of Japanese Society of Starch Sciences (Depun Kagaku)*, 33(1), 15-24.

Takahashi S, Hirao K, Kobayashi R, Kawabata A and Nakamura M (1987), 'The degree of gelatinization and texture during the preparation of harusame noodles. (Properties and cooking quality of starches. Part 8)'. *Journal of Japanese Society of Starch Sciences (Depun Kagaku)*, 34(1), 21-30.

Takashi S, Kobayashi R, Kainuma K and Nakamura M (1985), 'Physico-chemical properties of starches from mung bean, Broad bean and commercial harusame noodles', *Nippon Shokuhin Koayo Gakkaishi,* 32(3), 181-187.

Tam L M, Corke H, Tan W T, LI J S and Collado L S (2004), 'Production of bihon-type noodles from maize starch differing in amylose content', *Cereal Chemistry,* 81(4), 475-480.

Tan H -Z (2007), 'Rheological, Thermal Properties, Molecular Structure of Sweet Potato Starch and Its Application in Starch Noodle Production'. Southern Yangtze University. *PhD. Dissertation.* 2007, 3.

Tan H -Z, Gu W -Y, Zhou J -P, Wu W -G and XIE Y -L (2006), 'Comparative study on the starch noodle structure of sweet potato and mung bean', *Journal of Food Science,* 71(8), 447-455.

Tan H -Z, Tan B, Gao H and Gu W -Y (2007), 'Rheological behaviour of mung bean starch dough', *Food Science and Technology Research,* 13(2), 103-110.

The Chinese Agriculture Trade Standards for starch noodles, (NY 5188-2002).

Tian S J, Rickard JE and Blanshard J M V (1991), 'Physicochemical properties of sweet potato starch', *Journal of the Science of Food and Agriculture,* 57, 459-491.

Tovar J, Melito C, Herrera E, Rasco´N A and Pe´Rez E (2002), 'Resistant starch formation does not parallel syneresis tendency in different starch gels', *Food Chemistry*, 76, 455-459.

Turhan M and Gunasekaran S (2002), 'Kinetics of in situ and in vitro gelatinization of hard and soft wheat starches during cooking in water', *Journal of Food Engineering,* 52, 1-7.

United States Patent 5916616----'Process for producing starch noodles by extrusion', Issued on June 29, 1999. from: http://www.patentstorm.us/patents/5916616-fulltext.html

US Department of Agriculture (2001), *'Nutrient database for standard reference',* Washington, DC: US Department of Agriculture, pp. 14.

Wang L, Wang Y -J and PORTER R (2002), 'Structures and physicochemical properties of six wild rice starches', *Journal of Agricultural and Food Chemistry*, 50(9), 2659-2699.

Wang S, Song B and Zhang L (1995), 'Status and prospects of sweet potato processing in *China'. Paper presented at the SAPPRAD Sweetpotato Processing Study Tour and Workshop*, Pingyin, Shandong Province, China. International potato Center (CIP): Bogor, Indonesia.

White P J, Abbas I R and Johnson L A (1989), 'Freeze-thaw stability and refrigerated-storage retrogradation of starches', *Starch/Stärke,* 1, 176-181.

White P J, Abbas I R, Pollak L and Johnson L J (1990), 'Intraand interpopulation variability of thermal properties in maize starch', *Cereal Chemistry*, 67, 70–73.

Wiesenborn D P, Orr P H, Casper H H and Tacke B K (1994), 'Potato starch paste behaviour as related to some physical/chemical properties', *Journal of Food Science,* 59, 644–648.

Xu A and Seib P A (1993), 'Structure of tapioca pearls compared to starch noodles from mung beans', *Cereal Chemistry,* 70(4), 463-470.

Yang J -M, Guo L -T and YU Z -J (2007), '*The production process and formula of Longkou starch noodles',* China Light Industry Press. p2.

Yang C C, Change S M and LII C Y (1980), 'Studies on the starches in Taiwain'. 4. Kidney bean. *Bull. Inst. Chem. Acad.* Sinica. 27, 37.

Yuan M -L, Lu Z -H, Cheng Y -Q and LI L -T (2008), 'Effect of spontaneous fermentation on the physical properties of corn starch and rheological characteristics of corn starch noodle', *Journal of Food Engineering,* 85, 12-17.

Yuan R C, Thompson D B and Boyer C D (1993), 'Fine structure of amylopectin in relation to gelatinization and retrogradation behavior of maize starches from three *wx*-containing genotypes in two inbred lines', *Cereal Chemistry*, 70(1), 81-89.

Zhang J -M and Chi X (2001), 'Production of Starch Noodles', China Light Industry Press, p1.

Zhang T and Oates C G (1999), 'Relationship between α-amylase degradation and physico-chemical properties of sweetpotato starches', *Food Chemistry,* 65, 157-163.

Zhou B L, Wu D T, Lu J M, Shi L and Li L (2004), 'Study on the processing variables and prescription of starch noodle made from vegetables without alum', *Science and technology of food industry*, 25(1), 73-75.

Chapter 5

TOFU

1. INTRODUCTION

1.1. Definition and Etymology

Tofu, also known as bean/soybean curd, is a soft cheese-like food made by coagulating bean/soybean milk with coagulant, and the resulted curds are formed into blocks.

Tofu is high in protein, fat and calcium and well known for its ability to absorb new flavors through spices and marinades. Due to its multiple texture, qualities and excellent nutritional value, tofu, a staple of Asian cuisines for hundreds of years, has recently become popular in Western vegetarian cooking. It is used in many different diets, including vegan and vegetarian eating plans for its eupeptic protein (above 92%) comparing with soybean (about 60%). Tofu is also a staple in many low fat, low cholesterol diets, and is regarded as a healthy part of high protein low-carb lifestyles.

The English word "tofu" comes from the Japanese tōfu, which itself derives from the Chinese dòu fu because both large-scale producing of tofu and research reports on tofu were all carried out by the Japanese. In Chinese, the characters together could be translated literally as "bean curd". The first character "dòu" refers to leguminous plants (especially soybean), which are the materials of tofu. And the second character "fu" designates the curdled state of the product with a medium state of solid and liquid phase. That means tofu is neither a solid state food nor a liquid state food. It also refers to food offered to spirits. This might imply that it might be used as a sacrifice in ancient China.

1.2. Origin and History

It is interesting that there were many mentions of tofu in the Song Dynasty (960–1279) and a little in the Wudai Dynasty (907–960), but there is not any mention before (Yang, 1994) even though it is commonly agreed that tofu had been introduced to Japan in the Tang Dynasty (618–907) (Watanabe, 1996). It is thought that for hundreds years the historiographer may have considered tofu processing a simple and natural thing which was not important enough to mention, so very little is known about the exact historic origins of tofu and its method of production. While there are many theories regarding tofu's origins in folktales, the historical information is so scarce that the status of most theories was relegated to either speculation or legend. Like the origins of cheese and butter, the exact origin of tofu production may never be known or proven.

What is known is that tofu production is an ancient technique. Tofu was widely consumed in ancient China, and techniques for its production and preparation were eventually spread to many other parts of Asia.

1.2.1. Three Theories of Origin

There are three kinds of opinions on tofu's origin. The most commonly held one maintains that tofu was invented in Northern China around 164 BC by Lord Liu An, a prince of Huainan city during the Han Dynasty. Although this is possible, the paucity of concrete information about this period makes it difficult to conclusively determine whether or not Liu An invented the method for making tofu. Furthermore, in Chinese history, important inventions were often attributed to important leaders and figures of the time. But it is also a fact that for a long time the Huainan city has celebrated a tofu festival every year and tofu making is very popular here while the Huainan tofu is famous in other parts of China, too. Yang pointed out that Huainan is an important producing area of soybean and the underground water contains many minerals so that it has been used as a coagulant (Yang, 1994).

Another theory states that the production method for tofu was discovered accidentally when boiled, ground soybean slurry was mixed with impure sea salt. Such sea salt would likely have contained calcium and magnesium salts, allowing the soy mixture to curdle and produce a tofu-like gel. This may have possibly been the way that tofu was discovered, since soy milk has been eaten as a savory soup in ancient as well as modern times. Despite its technical plausibility, there is little evidence to prove that tofu production originated in this way. But in our opinion, it is incredible because there is not any famous tofu producing area along the seaboard and brine is more difficult to use than other coagulants in tofu processing.

The last group of theories maintains that the ancient Chinese learned the method for the curdling of soy milk by emulating the milk curdling techniques of the Mongolians or East Indians. For, despite their advancement, no technology or knowledge of culturing and processing milk products existed within ancient Chinese society. The primary evidence for this theory lies with the etymological similarity between the Chinese term for Mongolian fermented milk (rufu, which literally means "milk spoiled") and the term dòufu or tofu. There is no evidence to substantiate this theory, too, beyond the point of academic speculation.

Although its development likely preceded Liu An, tofu is known to have been a commonly produced and consumed food item in China by the 2nd century BC. Although the varieties of tofu produced in ancient times may not have been identical to those of today, descriptions from writings and poetry of the Song and Yuan Dynasty showed that the production technique for tofu had already been standardized by then, to the extent that they would be similar to tofu of contemporary times.

1.2.2. In Asia and Other Countries

In China, tofu is traditionally used as a food offering when visiting the graves of deceased relatives. It is claimed that the spirits (or ghosts) have long lost their chins and jaws, and that only tofu is soft enough for them to eat. Before refrigeration was available in China, tofu was often only sold during the winter time, due to the tofu not spoiling in the colder weather. During the warmer months, any leftover tofu would be spoiled if left for more than one day.

Tofu and its production technique were subsequently introduced into Japan in the Nara period (645–794) as well as other parts of East Asia. This spread likely coincided with the spread of Buddhism as tofu is an important source of proteins in the religion's vegetarian diet. Since then, tofu has become a staple in many countries, including Vietnam, Thailand, and Korea, with subtle regional variations in production methods, texture, flavour, and usage.

The earliest document of tofu in Japan shows that the dish was served as an offering at the Kasuga Shrine in Nara in 1183. The book "Tofu Hyakuchin" (one hundred kinds of tofu dish), published in the Edo period, lists 100 recipes for cooking tofu.

Tofu is so highly esteemed in the Korean culture that the menus of many Korean restaurants are based almost entirely on tofu.

In Malaysia, Singapore, Thailand and Indonesia, tofu or tahu is widely available and used in many Malay dishes. Even the Malaysian and Singaporean Indians use tofu in their cuisine. The makers of tofu in these countries were originally the Chinese but tofu now is made by non-Chinese as well.

Tofu was not well known to most Westerners before the middle of the 20th century. With increased cultural contact and an interest in health, tofu has become almost universally known in the west. However, due largely to ongoing attempts to use tofu as a substitute for traditional western meat-foods, it is often regarded as unappetizing for its light white color and taste. In Brazil, one of the most important soybean producers, tofu is still a new food entirely.

1.3. Categories

Tofu can be categorized to many kinds according to the main raw materials, the hardness of tofu, the kinds of coagulant, the type of product and so on. The varieties of tofu and processed tofu products may be over one hundred, but what is most usually seen are south tofu, north tofu and GDL tofu.

1.3.1. North Tofu

North tofu is a kind of well-drained and pressed tofu though it still contains a great amount of moisture and can be picked up easily with chopsticks. The skin of this tofu has the pattern of the muslin used to drain it and is slightly more resilient to damage than its inside. Brine is usually used as coagulant in processing of north tofu. The moisture of north tofu may be 80~90% which is similar to "momen tofu" in Japan.

1.3.2. South Tofu

South tofu is processed with draining and pressing similar with north tofu, but the draining and pressing are light compared with north tofu. South tofu is usually prepared with Gypsum-calcium sulfate as coagulant and the moisture of south tofu may be 90~92% that is similar with soft tofu in Japan. There are some other differences except for the coagulant and draining, pressing.

1. Concentration of soymilk for south tofu is a little higher than that for north tofu. Moreover, the mesh of filter must be larger for south tofu processing so that decreasing the size and content of okara.
2. The temperatures for adding coagulant are different. Brine is a little speedy compared with calcium sulfate to form gel so it is added when soymilk is cooled to 70~80° C while the latter is used in 75~85° C.
3. The time for curd formation of south tofu is longer than 30 min usually while that of north is about 15~20 min.
4. Stirring is necessary to boost draining for north tofu before molding while it is unnecessary for south tofu.

1.3.3. GDL Tofu

GDL (glucono-delta-lactone) tofu is a kind of undrained soft tofu. From the name, we know the coagulant is glucono-delta-lactone. Because a container was filled with soymilk with GDL and then was warmed up to form the curd, GDL tofu is also known as filling tofu. It is sealed and then heated to form a curd so the spoilage bacterium contents are much lower than that of the other two kinds of tofu and the shelf life may be more than 20 days.

GDL tofu has some sourness so some can not accept it. It is also difficult to use in many Chinese dishes such as famous “Mapo tofu” because its soft texture can not endure stir-fry.

Processing of GDL tofu is needed that is suitable to mechanization production so it is usually produced in large factories near a city and is seldom done in the country. The difference in processing will be introduced later.

1.3.4. Byproducts of Tofu Production

Tofu production creates some edible byproducts. Protein-lipid film, or "tofu skin" is formed over the surface of boiling soy milk in an open pan. The films could be collected and frozen or dried into yellowish sheets known as tofu skin or soy milk skin. It is also known as yuba in Japanese and sold as a fancy food material for its unique texture and nutrition. Dried protein-lipid film contains 50~55% protein, 24~26% lipids, 12% carbohydrate, 3% ash, and 9% moisture.

The leftover solid of soymilk extracting is called okara. Okara, sometimes known in the west as soy pulp, is rich in fibre. Although mainly used as animal feed in most tofu-producing cultures, it is sometimes processed and used in cuisines, too.

1.4. Production and Consumption

The total production value of soybean food processing was 12.3 billion RMB (about 1.8 billion US$ which was 0.75% of the total value of food processing. Tofu processing is the main part of soybean food processing; it is said that about 5 million tons of soybeans are processed into tofu or related tofu foods every year.

In China, tofu may be produced locally by relatively small vendors or distributed widely by large national brands. There are innumerable family tofu hand-makers with 2~3 persons in counties all around the country and about 4000 larger factories. Some of the factories can process above 3000 tons of soybeans each year, and the sales may amount to over 100 million

RMB. Though there are no details and believable data on the numbers of tofu makers, we can speculate it would be very enormous because most soybean food enterprises are small ones. The situation in Japan is similar. There are about 16000 tofu makers with 300 thousand tons of soybean processing capacity each year in Japan (Wu Yuefang, 2006)(Table 5-1). Even though the numbers decreased in recent years, the amount of soybeans processed is still just over 200 tons yearly.

Table 5-1. Variation of Numbers of Soybean Food Factories in Japan during 2000~2004 (Wu 2006)

Year	*2000*	*2001*	*2002*	*2003*	*2004*
Number of factories	15994	15600	15028	14487	14016

The menu of tofu is very colorful. The most famous tofu dish is Mapo tofu showed in Figure 5-1. it is oriented from Sichuan province of China and now you can find it in almost all Chinese restaurants. The most important flavorings include pericarpium zanthoxyli and hot capsicum. But piquancy is also usually adjusted to different flavors.

There is little data about the consumption in China but it no question that tofu is a commonly consumed food. Chen et al. (2001) investigated tofu consumptions of a total of 1155 subjects in two districts of Shenyang and found 36.2% of subjects consumed about 250~499g tofu/week (Table 5-2) while near half of subjects consumed tofu over 500g/week..

But Ma Guansheng (2008) reported the consumption of tofu in China too. In a national survey, 55211 subjects were asked about the consumption of tofu and other soybean foods. They found about 90% of informants had a tofu dish each week and the frequency was about 2.7 times. The average quantity of tofu consumption was about 480g/week (including processed tofu food). They also found that the consumption of tofu and soybean foods for each subject had increased a little from 1982 to 2002.

Figure 5-1. Famous Mapo tofu

Table 5-2. Tofu consumption in Shenyang city, Liaonin province (Chen et al. 2001)

Tofu consumption (g/week)	*Datong district (%)*	*Tiexi district (%)*	*Total (%)*
< 250	18.2	17.2	17.9
250~ 499	30.8	39.9	36.2
500~749	25.8	25.1	25.4
≥750	24.5	17.8	20.5

It's hard to exaggerate the health benefits of tofu. It's high in calcium and vitamins, but low in fat and sodium. Because tofu is so easy to digest, it's an excellent meat substitute for individuals who have trouble digesting meat, or medical conditions such as chronic heartburn for it has no cholesterol.

2. Materials for Tofu Production

The main materials for tofu production are soybean, water, coagulant, defoamer and preservative. Defoamer and preservative may be unnecessary in some cases. Of course, for increasing the kinds of tofu, many other materials could be used such as vegetable juice, peanut milk and so on.

2.1. Soybean

Soybean is the most important material in tofu processing for it does not only decide the yield of tofu but also affects the quality of tofu.

Tofu gel is formed from a network of denatured protein and water; lipid and sugar are filled in the network so protein and lipid of soybeans have an obvious influence on the yield and texture of tofu. In general, the higher the protein of soybeans is, the higher the yield of tofu. The lipid of soybeans planted in American or South African countries such as Argentina and Brazil are usually high but protein are relatively low so it is not suitable for tofu processing. In China, there are hundreds kinds of soybeans and the protein contents may vary by a large scale. It is also necessary to pretest and confirm the suitability of soybeans for tofu processing.

In 1960, the soybean yield of China was about half the production in the world. Though China had been a top producer of soybeans for hundreds of years and was the most important exporter of soybeans, the yield of soybeans has stayed at 15 million tons while consumption of soybeans increased quickly for the huge need for soybean oil. So China had become the largest importer of soybeans just as shown in Table 5-3.

Imported soybeans mainly from American, Brazil and Argentina are all transgenic ones and are used as raw materials of oil production so most tofu is still made from non-genic soybeans in China.

Table 5-3. Production of Soybeans in the World and China

Year	*Production area (1000ha)*	*Yield (1000 ton)*	*Output (1000 ton)*	*Input (1000 ton)*	*Total of consumption (1000 ton)*	*Consumption for food (1000 ton)*	*Usage for oil (1000 ton)*	*Self-sufficiency rate (%)*
1965	8593	6140	550	0	5590	3866	966	109.8
1970	7985	8710	460	0	8250	6039	1510	105.6
1975	6999	7240	178	25	7087	5188	1297	102.2
1980	7226	7940	143	540	8337	6092	1523	95.2
1985	7718	10509	1260	280	9529	6109	2685	110.3
1990	7560	11000	1288	1	9713	5060	3903	113.3
1991	7041	9710	1090	136	8756	4617	3389	110.9
1992	7221	10300	300	150	10150	4850	4486	101.5
1993	9454	15310	1100	125	14335	5585	7605	106.8
1994	9222	16000	394	155	15761	6015	8590	101.5
1995	8127	13500	222	795	14073	5570	7470	95.9
1996	7470	13220	195	2274	14309	5750	7500	92.4
1997	8346	14728	168	2940	15472	5912	8450	95.2
1998	8500	15152	187	3850	19929	6212	12607	76.0
1999	8000	14290	230	10100	22894	6180	15070	62.4
2000	9300	15400	208	13245	26697	6222	18900	57.7
2001	9480	15410	300	10385	28310	6500	20250	54.4
2002	9546	16510	265	21417	35290	7000	26540	46.8
2003	9313	15394	319	16933	34375	7210	25439	44.8
2004	9590	17400	390	25802	40212	8000	30362	43.3
2005	9591	16350	354	28317	44440	8200	34500	36.8
2006	9100	15200	446	28726	45397	8320	35477	33.5
2007	8700	13500	350	35400	48650	8450	38600	27.7
2008	9400	16000	400	36000	51270	8600	41020	31.2

Data Sources: USDA: *PS&D Online* July 2008; USBC: *International Data Base*, August 2006.

It is said that not only proteins, lipids and other compositions related with cultivars of soybeans affect properties of tofu processing obviously, drying conditions and duration in storage may be also important in the processing of desirable tofu products.

High temperature and moisture conditions of drying or storage may result in the increasing of indiscerptible protein. The decrease of soluble protein in soymilk would lead to a decrease in the yield of tofu. Cracked and disease contaminated soybeans will affect the color and shelf life of tofu, too.

Stone soybean is a kind of kernel which can not absorb water even soaked and will have an obvious effect on the processing of soymilk. The reason for stone soybeans is still unclear but some special cultivars and extreme low temperatures during harvest may be related with it.

Soybean harvested at 3~9 months is best for tofu processing. Old soybeans would affect the yield, texture and taste of tofu. But 3 months is necessary for after-ripening (Li et al. 2003).

2.2. Water

Water is so important that the taste of tofu is decided mostly by the quality of water. Water is necessary in clearing, soaking, milling and cooling of tofu processing and affects the quality and yield of tofu. Bi Haiyan et al. (2007) said the calcium content of water changed the water absorption of soybeans as well as properties of the gel and flavour of the tofu. It is said 45~55μg/ml calcium in water is most suitable for tofu making.

According to Watanabe (1996), water with hardness below 50, iron content below 0.3ppm and alkalescence pH7~7.3 is good for tofu making. Hardness of water is usually decided by total content of calcium and magnesium. But higher calcium content is preferred because magnesium may debase the flavor of the tofu. In a word, delicious water is also suitable to tofu making.

Watanabe (1996) reported that electrolytic water can improve the processing properties of soybeans and increase the yield of tofu and may prolong the shelf life of tofu. It is useful to adjust different processing characteristics resulting from the mixing of soybeans.

2.3. Coagulant

Many kinds of coagulant are used, including calcium sulfate (gypsum, $CaSO_4 \cdot 2H_2O$), magnesium chloride (nigari, $MgCl_2 \cdot 6H_2O$), calcium chloride ($CaCl_2 \cdot 2H_2O$), magnesium sulfate ($MgSO_4 \cdot 7H_2O$), Glucono delta-lactone (GDL, $C_6H_{10}O_6$). The most popular coagulant is calcium sulfate which is especially suitable to south tofu with tender but slightly brittle texture. Nigari is a natural sea salt extract or calcium chloride derived from a mineral ore and is mainly for north tofu with a smooth and tender texture. GDL is a kind of coagulant differed from gypsum and nigari in both method of use and coagulating mechanism. The type of coagulant used has an effect on the texture and flavour of the plain tofu. Sometimes, other acidic water such as underground water or fermented soy whey with low pH can also be used in tofu.

Solubility and dissolution speed of calcium sulfate is low so it is easy to use in tofu making. Tofu made with calcium sulfate is good for softness and water retention. The dissolution of calcium sulfate is affected by the crystal and size so that the size and solubility needs to be analyzed. Calcium sulfate is added to soymilk heated to about 80° C and the Ca^{2+} ion will act as bridges among the protein molecule to form a protein network. The speed of bridge formation depends on concentration of Ca^{2+} ions but the solubility of calcium sulfate in 70~80 °C is just 0.25% so the coagulating speed is not affected significantly by an added quantity of calcium sulfate. Calcium sulfate must be cracked to fine particles so as to increase its dissolution speed in soymilk.

Nigari is mainly magnesium chloride when sodium chloride is removed. Nigari has been the most common coagulant in Japan for thousands of years. Solubility of nigari even in the solid state is very high and the action of coagulating is very fast. It is a hygroscopic salt so one must pay attention to the storage. Nigari must be added very slowly and the soymilk stirred adequately. It is difficult to use nigari directly in large-scale processing so it is usually mixed with some additions which can delay the coagulating speed.

GDL shows no coagulating reaction before it hydrolyzed to gluconic acid so cooled soymilk with GDL can be transferred to the individual plastic containers or tetrapaks. Because hydrolysis speed of GDL in low temperature is very low, it can be mixed enough and the quality of tofu may be uniform. Processing with GDL is continuous and can be fully automated. GDL tofu is characterized by a fine porosity, which retains more moisture because it is not pressed so higher production yields are obtained. GDL lowers the pH of soymilk near to the isoelectric point of soybean protein, so the tofu production has a little sourness and some consumers need to accommodate it.

Coagulant types and their properties are shown in Table 5-4. It seems tofu made with calcium sulfate has many advantages and magnesium chloride can give a special smell to tofu which some like, while GDL can be used easily in large-scale processing (Watanabe 1996).

Table 5-4. Comparing Different Kinds of Coagulants (Watanabe, 1996)

Coagulant Type	*Soymilk Temperature (° C)*	*Solubility*	*Advantages*	*Disadvantages*
Calcium Sulflate ($CaSO_4 \cdot 2H_2O$) ($2CaSO_4 \cdot H_2O$)	80-85	Low	Good color and soft texture; High yield of water holding capacity of tofu; Fast coagulating; Cheap	Hard to dissolve in water
Calcium Chloride ($CaCl_2 \cdot 2H_2O$)	75-80	High	Fast coagulating time; Easy to drain water	Crude texture; Low tofu yield
Magnesium Chloride ($MgCl_2 \cdot 6H_2O$)	75-80	High	Good flavour; High coagulating strength	Crude texture; Low tofu yield
Glucono Delta Lactone (GDL, $C_6H_{10}O_6$)	85-90	High	High yield; Easy to dissolve uniformly	Sour taste; Hard texture but easy to break

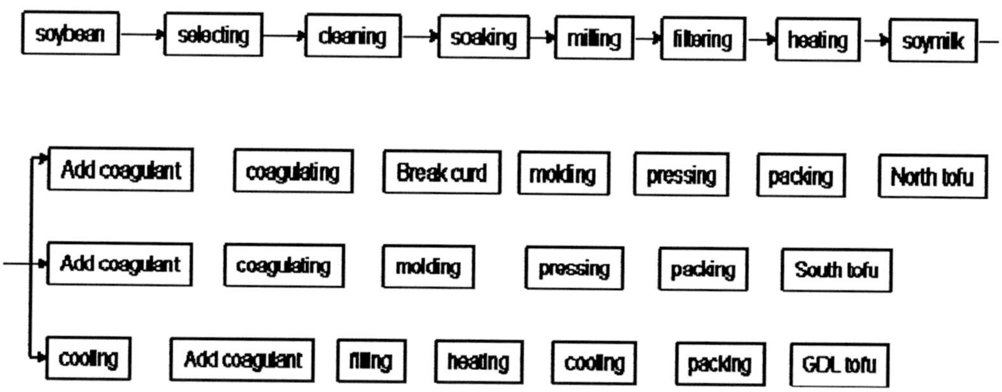

Figure 5-2. Processing technology of tofu.

2.4. Defoaming Agent

A defoaming agent is used to decrease the foam formed during soymilk boiling. The protein of soybeans has a typical construct of amphiphilic molecule so it has a high interfacial activity. During grinding and cooking, a great amount of foam results and is very difficult to control in overflow of soymilk when soymilk is near boiling. Foam will decrease the efficiency of heat transfer, prolong boiling time, but also affect the extraction of protein, increase the foam content in tofu, and finally, result in the crude construct and bad taste of tofu. There are many kinds of defoamers such as foots oil defoamer, organosilicon defoamer and mixed defoamer.

3. Processing Technology of Tofu

Tofu processing can be divided into soymilk preparation from soybeans and tofu making from the soymilk. Processing methods for varied types of tofu may be somewhat different as shown in Figure 5-2. Soymilk preparation in China and Japan is a little different in that filtering is done after the heating in Japan. The main processing steps are introduced later. Li et al. (2003) explained the processing technology of tofu and it is mentioned detailedly in this section.

3.1. Cleaning Soybeans

Soybean kernels with similar size, without cracks and impurities are ideal for tofu processing. So it is necessary to select and clean raw soybeans.

Soybean cleaning methods include dry separation and wet separation.. Dry separation mainly uses wind power or gravity screening to eliminate impurities such as straw stalks, bits of grass, stone, etc., which have quite a different specific gravity than the soybeans. However, a dry separation method occupies a large area and influences the operating environment, and most of all, it can not clean thoroughly and is unable to guarantee the sanitary quality in the following production. Therefore, dry separation has not been used in large-scale production.

The wet separation method makes use of different moving speeds in water of material with different specific gravities to separate impurities. The wet separation line includes flumes, vibrating washers, hydrocyclones and so on. In this way, the soybean is rinsed during impurities elimination, which reduces bacterium's persistence and is helpful to prolong the shelf life of tofu (Li et al. 2003).

3.2. Soaking Soybeans

After cleaning, soybeans need immersion to absorb water, before they are sent to the milling system. An appropriate duration of immersion will help protein extraction during the following steps of milling into soybean milk and will increase the quantity of soymilk.

The duration of immersion influences both quantity and quality of production. When the immersion time is appropriate, the soybean skin turns brittle, which helps to crash protein coalitions and increase the protein extraction ratio. However, insufficient or excessive immersion will cause the membrane of protein coalitions to not be soft enough or to be too soft, either of which is not suitable for the breaking of protein coalitions and extraction of protein, then reducing the protein for production and the yield of tofu. Moreover, after excessive immersion, especially at higher temperature, certain ingredients of soybean dissolves as it happens when soybeans pre-germinate. The increase of these ingredients in the soaking water will influence not only the yield but also the quality of products, and finally cause deterioration.

After sufficient water absorption, the volume of soybeans would increase by 1~2 times. Therefore, the immersion vessel should be 3~4 times soybean volume. The variety of soybean, the temperature and quality of water all influence the speed of water absorption. Temperature of water is almost directly related with needed soaking time. When the temperatures of the water are 5° C,10° C,18° C and 27°C, the needed soaking times are 24h, 18h, 12h and 8h respectively. Of course, it is also reported that soaking with hot water of 60° C, 1h increased the yield of protein-lipid film. It should be noted that too high an immersing temperature is not suitable for soybeans, as it will not only increase the respiration of soybeans and reduce nutrition, but also accelerate microorganism reproduction and cause deterioration of products. Most of the time, room temperature is applied in practice, and thus the immersion time should vary with the season or local weather. After soaking, the soybean surface is smooth without wrinkles, and the skin is coriaceous and not easy to separate. A simple way to estimate whether the immersion is sufficient is to break kernels with your hands into two hemispheres—that is easy to nip off with fingers—and if there is no white centre there has been insufficient water absorption. The volume of soaking water is commonly 2~3 times that of soybeans to ensure sufficient water absorption (Li et al., 2003).

3.3. Grinding of Soybeans

Though different in production techniques, traditional soybean products all belong to soya protein gelatin. The production of different bean products is essentially to obtain different protein colloids. The soya protein exists in the cells of cotyledon (soybean's storage tissue), enclosed in a membrane tissue mainly consisting of hemicellulose, pectin substance and so on. In mature soybean seeds, the protein epithelium is quite hard. While soaking in water, the membrane of the protein body absorbs water and swells like other tissues, turning from hard to brittle, and then to soft. It is easy to scrunch soybeans when the protein body membrane is in the brittle state, followed by dissolvation of the protein. This extracted protein will dissolve in water and then form protein sol, namely raw soybean milk. This course of using an auto-separating grinder to mill soaked beans into soybean milk is called grinding, during which, the two points below require our attention:

1. Water should be added while grinding. This can not only reduce the power consumption of the grinder, but also help to prevent over grinding of the soybean skin, which may even cause difficulties in separation of the soybean milk and residue (okara). Generally, the amount of water added is about 3–4 times the volume of the dry soybeans.

2. The soymilk grinder has two grinding wheels and the particle size of grinding can be controlled by adjusting the gap between the two wheels. When the granularity is too big, residual protein in okara will increase while the yield and quality of soybean curd might both be influenced. But if the granularity is too small, the temperature of grinder and power consumption will increase. Meanwhile it will be more difficult to separate the soy milk and residues, and even influence the taste of the tofu (Li et al., 2003).

3.4. Filtering

Filtration is mainly for the elimination of okara and adjustment of concentration. The amount of water addition during filtration differed with the original concentration and products. Okara can influence the formation of gelatin and the texture of tofu. The filtration can be conducted either before or after boiling of the soybean milk. In China, it is usually performed before boiling, while in Japan it is always after boiling.

The method of filtration after boiling is called cooked soymilk processing, while the method of filtration before boiling is called raw soymilk processing. The former can sterilize in time and prevent deterioration of the soymilk, and the tofu is elastic, coriaceous and chewy. However, boiled soy milk has higher viscosity and is hard to filtrate, which results in a higher protein residue in okara (generally over 3%); if the pressure of separating is not enough, that results in a relatively lower protein extraction and an increase of power consumption. Moreover, the water retention ability of production will be influenced as well. The increase of water isolated (after standing for a period, part of the water in bean curd will separate, and the water is called isolation water) might influence the sensory evaluation and purchase desirability. Thus, the boiled soy milk method is only used for the production of dry tofu with lower water content in China. Contrarily, as for raw soymilk processing, the sanitary condition request is higher, because soymilk is easily polluted by microorganisms, resulting in rancidity and spoilage. Due to its convenience in operation and low protein residual rate, which can be less than 2% if grinding granularity and filtration techniques are appropriate, the raw soy milk method is excessively applied in production of south tofu in Southeast China. It is also said that if the separating pressure is large enough, the extracting of protein using cooked soymilk method is 1% higher than that of raw soymilk method. But there is no report of a comparison test.

Filtration methods can mainly be divided into two kinds: traditional manual filtration and mechanical filtration. Families and small manual workshops still adopt the former one, which include hanging filtration and heavy pressing filtration. This low cost method does not need any devices, though it asks for greater labor intensity and a longer time, and leaves higher residual protein content. However, in large factories, the horizontal centrifugal screens, horizontal screens, cylinder screens and so on are used for filtration. Among which, the horizontal centrifugal screen filtration is applied most extensively, due to its high speed, low noise and complete separation of soybean milk and okara. In addition, setting an interior filtration screen in the soybean grinder itself is also applied to separate the soybean milk and okara during grinding. Although the power consumption in grinding increases somewhat, there will be only a small amount of fine granularity of okara that needs further separation.

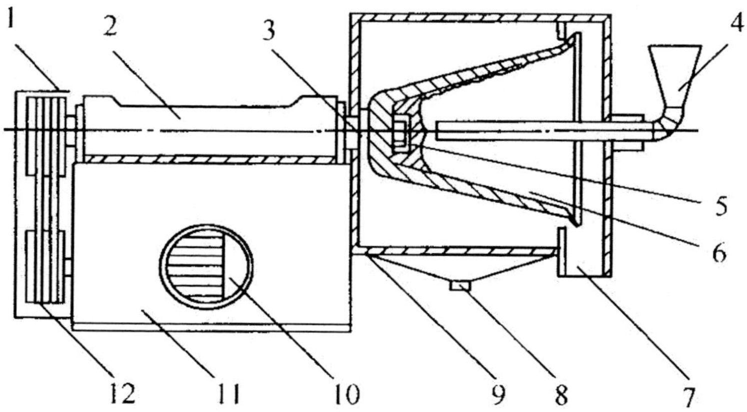

1. Strap cover, 2. Bearing case, 3. Spindle, 4. Feeding tube, 5. Separation unit, 6. Centrifuge rotor, 7. Slag discharge, 8. Soymilk exit, 9. Cover, 10. Moto,r 11. Frame, 12. Transmission unit

Figure 5-3. The structure of horizontal-type centrifuge (Li et al., 2003).

The structure of a horizontal centrifuge is shown in Figure 5-3. Soybean slurry is put into a taper rotating drum with nylon screen inside, and by the strong centrifugal force, the soybean milk and okara are draw off from the exit of liquid and residue, respectively. The whole process is continuously performed. The horizontal centrifuge is used alone or together with one another in combination to increase the protein extraction. As for the latter way, soy residue drawn off from previous centrifuge is sent to the next one with water addition, until the okara is obtained by the separation of the last (often the third) centrifuge. Generally speaking, soybean milk from the first two centrifuges is high in protein content and can be used directly in production. For example, soybean milk drawn off from the first centrifuge can be used in production of north tofu or dried tofu which prefers a relatively low water content. That from the second centrifuge can be used in south tofu or tofu film, or together with the first one. Soybean milk from the third centrifuge is low in protein content and is often mixed with the first soybean residue to make the most use of soy protein. After all separations, okara is supplied as feedstuff or sources of soy fiber food (such as okara production by fermentation). In order to obtain ideal separation efficiency, the points below demand our attention:

1. In the separation process, water of fixed quantity should be added in steps, followed by a sufficient mix to promote dissolvation of protein.
2. The appropriate water temperature is 55-60 ° C, and this is helpful to protein separation.
3. To ensure the continuous operation, temporary pause should be as less as possible to guarantees the production the stability and the soybean milk density.
4. Sieving screen should be appropriate varied with the number of centrifuge. For example, 80 mesh screan for the first separation, while the following with 100 mesh sieve.

3.5. Boiling (Cooking) of Soymilk

Boiling is a necessary step determined by the physicochemical properties of soy protein to make raw soy milk into tofu gel. Protein keeps its relatively stable sol state in raw soy milk, based on the specific molecular structure of soy protein. Hydrophobic groups in the natural protein chain distributes in the inside, while hydrophilic groups do so in the outside. Hydrophilic groups contain large amounts of oxygen and nitrogen atoms, which have unshared electron pairs. These electron pairs attract hydrogen atoms in water molecules and form hydrogen bonds, by which abundant water molecules enclose the protein colloidal particles and then a hydration shell forms, namely hydration of protein colloidal particles. At the same time, the ionization of hydrophilic groups outside generates static electrons, which absorb hydrated ions and then the absorption layer of static electricity forms, namely electric double layer outside the surface of protein colloidal particles. It is the protection of the hydration shell and electric double layer that prevents the aggregation of protein colloidal particles. However, this is a metastable system which can be disturbed by exogenous forces. Once the protection of the hydration shell and electric double layer is disturbed, this relatively stable state will disappear.

Cooking raw soy milk increases the internal energy of molecules, while the movement of protein is accelerated. That means the range and frequency of vibration of certain groups in protein molecules increases, followed by the breaking of the second bond, and the change of the special structure of protein. The polypeptide chains stretch out, which results in a reduction of the density of static electricity on the surface of the polypeptide, and then increases the attraction among protein colloidal particles. In this way, intermolecular hydrophobic bonds and disulfide bonds are formed between the hydrophobic and sulfhydryl groups. With these reactions, protein colloidal particles aggregate. Meanwhile, the densities of static electricity and the hydrophilic group increases, while the intermolecular attraction of protein decreases. In addition, the swelling of protein colloids increases the resistance of molecular thermal motion and the speed. And due to the low protein content in soy milk, a continued accumulation of protein colloidal particles is confined. As a result, a new relatively stable system, or a pre-gel system forms. In other words, cooked soy milk is produced.

Although there are no obvious differences between raw and cooked soy milk macroscopically, the state of protein molecules is completely different according to biochemical analysis. The molecular weights of proteins in raw and cooked soy milk are 600 kDa and 3,000 kDa, respectively. In fact, the soy protein in raw and cooked soy milk belongs to natural and denatured protein, respectively, as well. During the course of changing natural soy protein into its pre-gel state, protein combines with a small amount of lipid to form lipoprotein. And the lipoprotein, the amount of which increases with the prolongation of boiling, is the source of aroma in cooked soy milk.

Boiling is the most important step in the production of tofu or other soybean curd. On one side, due to the complex components of soy protein, the boiling temperature and time should ensure most of the proteins are denatured. On the other side, boiling can inactivate anti-biological actives and deconstruct the substance generating a bean smell (soybean fishy smell). Therefore, it should be ensured that the soymilk be boiled at over 100°C for 3~5 min, according to the traditional experiences. However, it is indicated that the heating process modified can effectively improve the yield and physical properties of bean curd, according to recent research. Relevant results will be introduced in detail in following chapters.

The concentration of soy milk should be adjusted before boiling by adding water. Generally, the lower the concentration, the higher the yield obtained. However, if the concentration is excessively low, a perfect net-structure of tofu gel is difficult to form, which will cause a faster isolation of water and sugar content, resulting in a decrease of yield. In addition, water addition should also take the type of tofu and consumers' tastes into consideration. Figure 5-4 shows the correlation between concentration of soymilk and the rates of gelatinization, extraction of solid contents and yield of tofu. Herein, gelatinization rate refers to the ratio of solid contents reserved in tofu after the course of gelatinization, and a higher gelatinization rate means a lower loss of solid contents. Extraction rate refers to the ratio of solid contents of soybean that dissolved into the soymilk, while the yield is the product of gelatinization and extraction rate. Gelatinization rate increases while extraction rate decreases with the increase of concentration of soy milk. In this way, the yield keeps quite stable in a reasonable range, except that an excessively high concentration may result in the yield decreasing (Li et al., 2007).

Generally, the extraction rate of soybean solid contents is between 50–70%, and it can be influenced by water addition, grinding granularity and filtration efficiency. With the increase of water addition or addition of cosolvent to help the dissolvation of soy components, solid extraction rate can be increased. However, if the soybeans have not been milled sufficiently or too much liquid content is left in the okara, the solid extraction rate will drop. The solid contents mainly consist of protein, lipid and carbohydrate. The extraction rate of protein is about 80%, and the remaining 20% is mainly soymilk in okara and the protein which can not be dissolved in hot water completely. The extraction rate of lipid is about 75%, and it can be increased by modification of milling method or addition of additives to emulsify lipid. The protein and lipid extraction are both higher than that of the total solid contents, and this indicated that the carbohydrate extraction rate is relatively lower. Calcium extraction rate is the lowest in minerals, the rates of magnesium, phosphorus and calcium extraction are 71, 74~80 and 42~47 %, respectively.

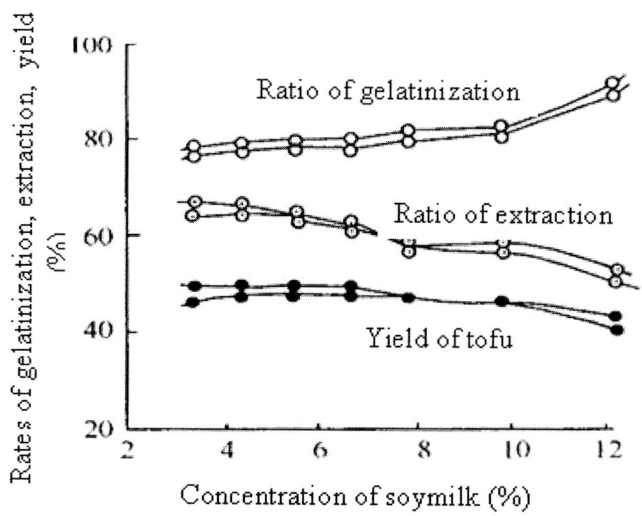

Figure 5-4. The correlation between concentration of soymilk and the rates of gelatinization, extraction of solid contents and yield of tofu (Li et al. 2007).

Some extracted solids are lost during gelatinization and, therefore, the gelatinization rate of soybean solid is generally about 76~84%. The gelatinization rate of lipid and protein are over 95% and about 90% respectively. The carbohydrate of soybeans is mainly soluble in water, the majority of which will lost with soybean whey.

Boiling the soymilk with steam in an open can is extensively applied by small or medium scale tofu makers, as the production scale can be adjusted easily. The open tank is a soy milk container with a steam pipeline on the bottom. Steam is imported directly through the pipeline into the tank during the course of boiling and cut off for 2~3 min when soy milk is boiling to prevent milk spillover, followed by the second import of steam until the milk boils once again. This two-step boiling is commonly applied to ensure a sufficient heating. When a big tank is in use, the temperature of milk decreases from up to down, as the steam goes upwards directly through the soymilk, which is of low thermal conductivity. Therefore, the first boil is a surface boil, and then letting it stand for a while can improve thermal distribution for a second and complete boil.

Closed continuous overflow heating by steam is another method for boiling soymilk, and its production line consists of 5-step tanks linked by pipelines (Figure 5-5). Each tank has a heat preservation interlayer, with a liquid entrance on the bottom and an exit on top. During production, the exit of the fifth tank is closed, and raw soymilk is imported from the entrance of the first tank until all the five tanks are fully filled. After this, steam is inputted to heat the soymilk until the temperature in the fifth tank arrives at 98~100°C, and then the boiled soy milk in the fifth tank is put out. Raw soy milk is then added again from the first tank. In this way, the input of raw material and output of boiled soymilk performs continuously by the steps of heating and rise of temperature in five tanks. Temperatures of the 5 tanks from the first to the 5th are controlled as 40, 60, 80, 90 and 98~100°C, respectively. And the gap between the heights of each nearby tank is about 8 cm. Making use of gravity overflow, it costs only 2~3 min to finish the course of flow from the entrance of raw soy milk to the exit of boiled soy milk, and the flow can be adjusted by steam pressure according to the scale of production.

Figure 5-5. Large-scale soymilk boiling line.

The continuous electric heating line is excessively applied in soymilk boiling in Japan. Generally, shallow cubic containers with an electrode board on two sides are used. Soymilk is heated while flowing by in the line of containers and the exit temperature can just satisfy the requests. This method has priority in continuous large-scale production for its high automatization, convenience of control, sanitation and cleanness.

3.6. Coagulation

Coagulation, as one of the most important steps of tofu making, is the course of protein thermal denaturization by addition of coagulant, turning the soybean milk from sol into gel, which mainly contains coagulant addition and curd forming two steps. Here, the coagulation of south and north tofu was introduced in detail, while that of filling tofu is completely different and will be introduced latter.

Coagulant addition is a critical procedure of tofu making. With the addition of coagulant, protein of soybean milk turns from sol into gel, namely tofu jelly (tofu flower), which consists of soy protein, lipid and water filled in the protein network. The water content in tofu jelly can be divided into combined or free water. The combined water mainly combines with the residual hydrophilic groups in the protein network by hydrogen bonds and often, each gram of protein contains 0.3~0.4g combined water. Even suffering exogenic forces during formation, it is stable and not easy to outflow. However, the free water is easy to outflow, as it is kept in the network by sorption of surface energy of the capillary. Water retention refers to the water holding capacity of tofu jelly. The structure, as well as the water retention, softness and elasticity, can be influenced by the coagulation conditions. Generally, tofu jelly with a large network of protein, firm structure and good water retention can produce soft, tender tofu and result in a higher yield. Conversely, tofu jelly with a small network of protein, loose structure and low water retention will influence the softness, elasticity and even yield. Therefore, the water loss ratio (refers to the ratio of water content isolated from bean curd during the storage) is also influenced by the gelatin network structure.

Mechanization of the coagulating processing is the most difficult one in tofu making for the coagulant must be added evenly and the coagulate speed must be controlled suitably. Figure 5-6 shows a large-scale equipment which can add coagulant automatically. Cooked soymilk adjusted to a suitable temperature is poured into tofu trays in a conveyer and coagulant is injected into the soymilk by a rotational filler. And then the tray is oscillated or stirred to decentralize the coagulant. However, because the gypsum is difficult to dissolve in water, it is still hard to add automatically with a machine.

It is reported that, varieties and quality of soybean, water quality, types and addition amount of coagulants, boiling temperature, concentration and pH of soybean milk, coagulation time, as well as stirring methods and so on will influence the coagulation process. Among these factors, boiling temperature, the concentration and the pH of soybean milk, the coagulation time, and stirring methods, etc., have significant influence on the quality of tofu.

Figure 5-6. Equipment which can add coagulant automatically.

While adding coagulants, the coagulation speed of protein and the temperature of the soymilk are closely related. An excessively high temperature will increase the internal energy and accelerate the aggregation speed, resulting in the shrinkage of framework, reduction of elasticity and lowering of water retention. Meanwhile, it will cause an uneven distribution of coagulants and result in poor quality easily, even under highly skilled operation. On the contrary, an excessively low temperature will greatly slow down the coagulation speed and make it hard to form curd, or result in production with a high water content but a lack of elasticity. Therefore, the temperature of soymilk should be adjusted according to the characters of production, the type and addition ratio of coagulants and even the method of adding. Generally speaking, the higher the temperature is, the harder and coarser the texture of curd will form. Often, the temperature of soymilk for south and north tofu is about 70~75°C, and an even higher temperature of 80~85°C is suitable for production of products such as dried tofu which need lower water retention. Commonly, a relatively higher temperature is suitable for gypsum and a lower temperature of soymilk is needed for nigari respectively. And as for filling tofu, boiled soymilk should be cooled before coagulant (GDL) is added. Figure 5-7 shows the correlation between coagulation temperature and hardness of tofu using 0.6% of gypsum as coagulant when the solid content and concentration of soymilk are 11.3 and 5.3% respectively. Just as shown, the hardness of tofu almost linearly increased with the increase of temperature.

Coagulation time has much to do with the properties of the tofu gel. Figure 5-8 shows the different hardness of tofu formed at 70°C for different times, with a coagulant addition of 0.6% into the soybean milk with 5.3% protein content. The hardness of tofu changes fastest at the beginning of 40 min, during which the coagulation is almost finished. However, it kept increasing even after 2 h from the beginning of coagulation. Therefore, soymilk should be allowed to stand for at least 40 min to ensure a complete coagulation. It is also important to keep the temperature during the period of coagulation for the following procedure of molding as well.

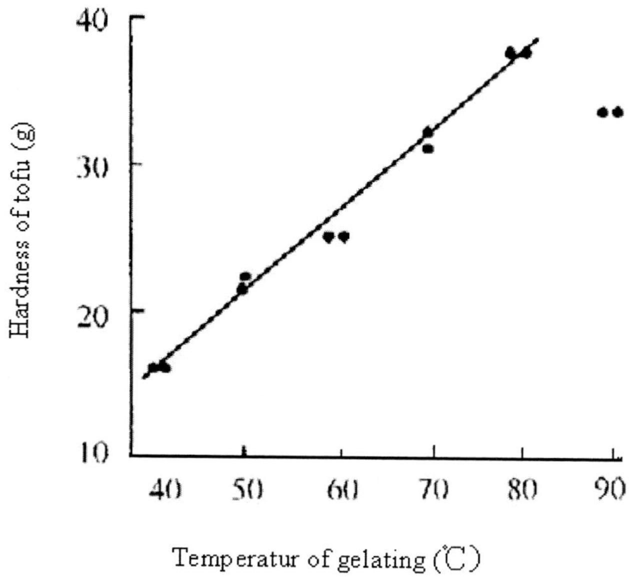

Figure 5-7. The correlation between temperature of gelating and hardness of tofu (Li et al., 2007).

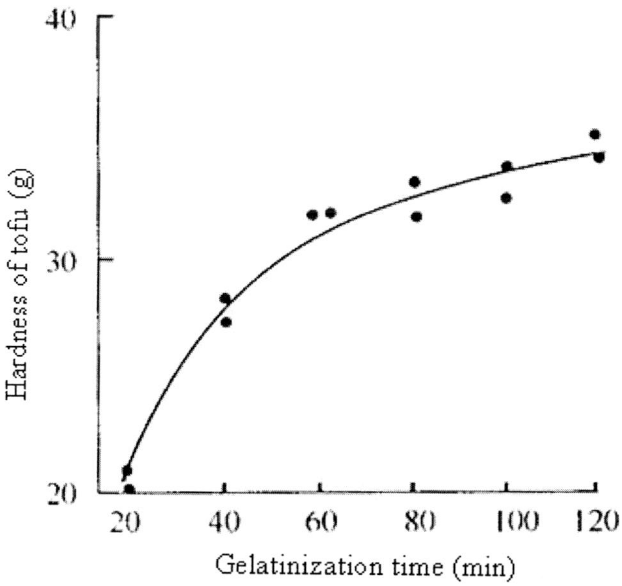

Figure 5-8. The correlation between the hardness of tofu and gelatinization time (Li et al. 2007).

The addition ratio of coagulants has significant influences on the quality of curding, too. But the addition ratio depends on the protein content and temperature of soymilk. Generally, a too low addition ratio of coagulants will cause insufficient coagulation and lower the hardness, while a too high addition ratio will cause uneven coagulation, increase of water isolation and lowered yield. Figure 5-9 shows the variation of tofu hardness with different addition of coagulants.

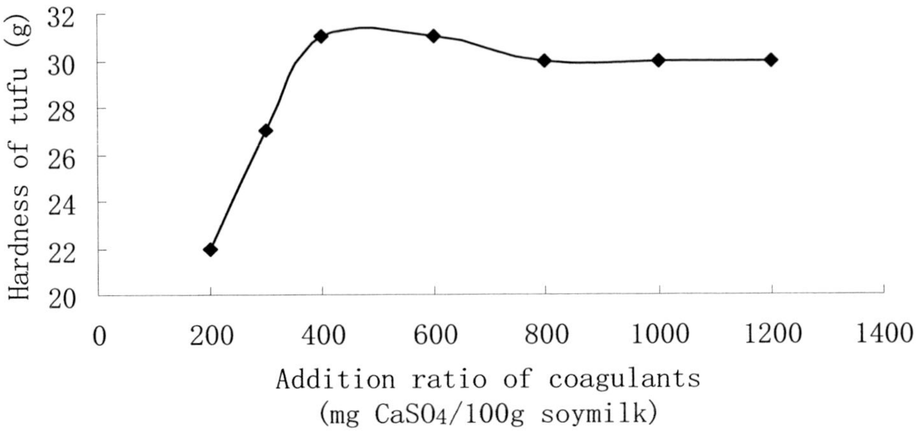

Figure 5-9. The correlation between the hardness of tofu and the addition ratio of coagulants.

The concentration of soymilk, which mainly refers to the protein content of soymilk, is another important factor that influences the coagulation. As Figure 5-10 shows, an excessively low concentration, that is, low addition ratio of water to the weight of soybeans will result in a small volume and low yield of tofu with a texture that is too hard. If protein content is too low, curd formation would be difficult or the formed curd would have a low water retention. While an excessively high concentration will result in uneven coagulation and white slurry generation, because coagulation is too fast once the coagulants are added into the soymilk. Generally speaking, the appropriate concentration for north and south tofu is about 3.2 and 4.5% respectively. And a practical way to control the concentration is to add 5~6 and 8~9 times water of the weight of dry soybean for south and north tofu respectively.

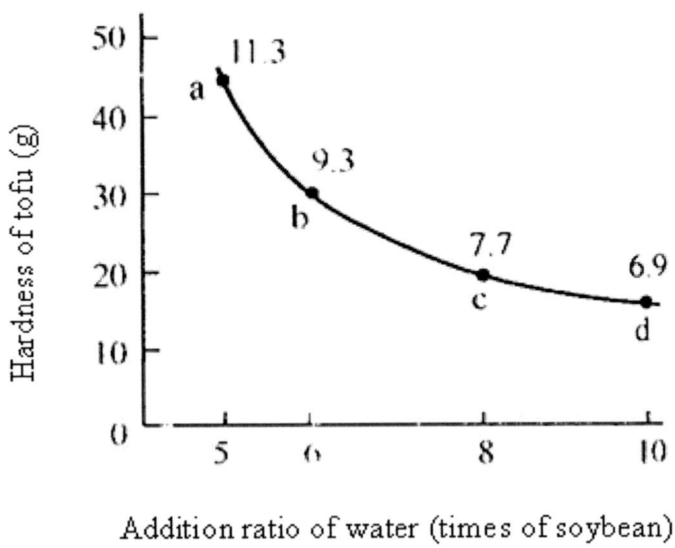

Figure 5-10. The correlation between tofu hardness and protein content of soymilk (Li et al., 2007).

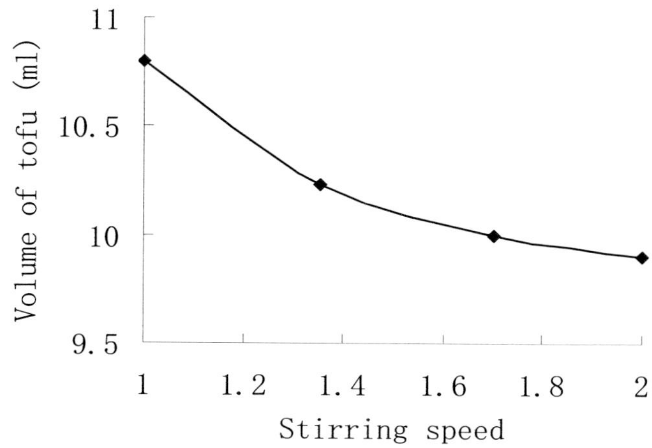

Figure 5-11. The correlation between stirring speed and tofu volume (Li et al., 2007).

Stirring during coagulant addition can help to evenly distribute them, and thus it can be mixed sufficiently with protein before a gel forms. Therefore, the speed of stirring and efficiency of coagulation is directly related. Figure 5-11 shows the relationship between the speed of stirring and the volume of tofu produced. The higher the stirring speed, the lower the addition of coagulants and the coagulation is faster, finally resulting in a small volume, firm texture and network, and vice versa. The speed and time of stirring depends on the type of production and the extent of coagulation, respectively. Stirring should be stopped once the coagulation is finished to avoid destroying the tofu jelly. Suitable stirring is necessary to obtain soft, tender and elastic products and high tofu yields. Otherwise, a further stirring will destroy the texture of the tofu jelly and influence the water retention, resulting in coarse and poor texture and low yields of tofu. However, if the stirring time is not enough, the network has not completely formed, resulting in an incomplete, soft but not elastic texture with white slurry mixed in it and low yield as well. And the stirring methods should also ensure the sufficient and complete mixture of soymilk and coagulant to gain a homogenous texture of tofu products at the end.

After adding the coagulant, standing is necessary, which allows the process of gelatinizing to continue and form firm enough protein networks, just as shown in Figure 5-8. It is better to keep the tofu standing statically to firm the structure further. Otherwise, a gel-network flimsily formed might be destroyed easily, and this will cause crevices in the inner tissue and distort the shape of the tofu, especially for filling tofu. However, letting it stand too long is not good for maintaining the temperature and for the proceeding of the steps that follow.

3.7. Molding

Molding is the procedure in which coagulated tofu jelly is put into a mold container (tray) and the redundant whey is extruded by adding forces. This is helpful to firm further the network and to increase the elastic and chewy properties of soy products. Except for soft tofu, such as south tofu, it is necessary to extrude part of the water. However, the water in the network is hard to extrude in a short time only by pressing, so we need to cut the tofu flower to destroy its formed network to satisfy the different requirements of tofu products, as well as the techniques of molding. South tofu contains a high water content and does not need to be broken up. As for north tofu, the curd cut into cubes about 8~10 cm in size will be suitable. While for dry tofu or tofu sheets which have a low water content, it is better to cut the curd into cubes about 0.5~0.8 cm in size or completely destroyed so as to extrude more water or even almost all the water.

Molding includes pouring into molds, pressing and molding, removing tofu from molds and chilling.

Curd enclosed in mesh cloth goes through pressing and molding in a cubic wooden mold (tofu box) with gaps to drain the residual water, and the curd structure becomes even firmer. The mesh size of cloth influences the drainage speed and, usually, a cloth with a bigger mesh for draining water freely is suitable for production of north tofu and will press its pattern onto the surface of the tofu. While only cloth with a fine mesh, which can slow down the drainage speed, can be used for south tofu.

Molding is necessary to make protein gel tighter after the tofu jelly is poured into molds. The temperature of the tofu flower, the pressure and duration of press all also influence the products. Pressure that is not high enough can not firm the gel to a perfect structure, but pressure that is too high will unfortunately break the network of protein gel. The pressure is about 1~3 kPa depending on the products and, often, the pressure for north is a little bit higher than that of south tofu in production.

Temperature maintenance and appropriate duration of pressing help to firm the protein gel as well, in addition to an appropriate pressure. If the beginning temperature is too low, the structure can not be firmed even with very high pressure, and water will not be drained, ending up with a loose texture of curd. The appropriate temperature and duration of pressing are about 65~70 °C and 15~25 min. Moreover, the curd should be reshaped during molding for north tofu production. After pressing, the water content should be about 90 and 80~85 % for south and north tofu, respectively.

After molding and pressing, the tofu should be removed from the mold in a water trough to reduce dehydration, to prevent sticking to the cloth, to keep the product neat and sanitary and to prolong the shelf life somewhat.

Table 5-5. Variation of main composites during tofu processing (per 100 g martial) (Watanabe, 1996)

	Solid contents		Protein		Fat		Ash		Ca		P		Mg	
	Moisture (%)	Solids (%)	(g)	(%)	(g)	(%)	(g)	(%)	(g)	(%)	(g)	(%)	(g)	(%)
Soybean	11.3	100	35.7	100	18.8	100	5.0	100	22.4	100	5.9	100	232	100
Soaking water	—	—	—	—	—	—	0.0	0.6	0.4	0.6	2.2	1.2	2.2	3.0
Okara	76.0	32.4	5.5	18.2	4.3	27.0	1.0	24.8	87.0	43.7	99.0	20.2	44.0	24.2
Soy milk	93.9	66.1	3.0	78.4	1.4	73.0	0.4	74.6	13.8	55.7	48.3	78.6	16.0	72.8
Soybean whey	98.1	11.0	0.4	5.3	0.0	0.9	0.4	39.2	19.4	42.9	7.4	6.7	12.1	27.4
Tofu	88.93	56.5	6.0	73.1	3.0	72.1	0.6	60.8	83.6	163	91.0	71.9	22.5	45.4

The nutrition of solid contents including protein, lipid, ash and mineral will change with the different key processings during tofu making just as in Table 5-5. As shown, the protein content is very low in the soaking water, and thus the immersion has little influence on the nutrient content.

3.8. Production of Filling Tofu

GDL tofu is coagulated in a sealed LDPE or PVC container so the tofu is entirely separated from the environment. So the sanitation of the tofu is good and the shelf life is much longer than other types of tofu. Because GDL coagulant which has good dissolution is mixed with cooled soymilk so it is easy to be decentralized in soymilk and suitable for large scale production.

3.8.1. Coagulating Mechanism of GDL Tofu

Coagulating of filling tofu includes both hydrolyzation of GDL and gelation of soybean protein,. GDL has not the function to form a protein gel but its hydrolysis product–gluconic acid—has the function. GDL hydrolyzes very slowly at room temperature (about 30°C) but its hydrolysis speed increases rapidly with the increasing of temperature.

Processing of filling tofu includes also cleaning, soaking, grinding of soybeans and cooking, coagulating. But the protein concentration of soymilk for filling tofu is much higher than that for south or north tofu processing. The protein concentration of soymilk is 4.5 % or more and about 5 kg soymilk may be produced from 1 kg soybean. There is not isolation of whey during the coagulating, so if the concentration is too low, the tofu will be frangible or even not able to be molded.

Because of special hydrolyzation of GDL, the adding of coagulant must be under 30°C, otherwise it will cause instant coagulation when it come into contact with the soymilk.

3.8.2. Adding of GDL

GDL is added to cooled soymilk in a suitable ratio and put into a container after being stirred sufficiently. Not only is sufficient stirring necessary, but also the containers must be filled as soon as possible so as to avoid the increasing of soymilk viscidity resulting from the gelation of protein. Table 5-6 shows that if the addition of GDL is higher, the hardness of the filling tofu is also larger. But if the addition is over 0.5%, the filling tofu will have an obvious sour flavour, so the addition of GDL must be controlled to 0.2~0.3% of soymilk.

Table 5-6. The relation between addition ratio of GDL and quality of filling tofu (Li et al., 2003)

Addition ratio of GDL (%)	*Hardness of tofu (g)*	*pH of tofu or soymilk*	*Flavour of tofu*
0.1	Can not form gel	6.3	Natural
0.3	37.6	5.7	Natural
0.5	59.4	5.3	A little sour
1.0	60.4	4.6	Obviously sour

GDL must be dissolved by cooled boiled water or cooled boiled soymilk. Soymilk mixed with GDL should be filled in 15~20 min and is not suitable for storage.

3.8.3. Concentration of Soymilk for Filling Tofu

Because coagulation of GDL is very strong, if addition of GDL is enough, tofu curd could be formed in a comparatively low concentration of soymilk. But as mentioned above, adding too much GDL may result in a sour flavour of the tofu. As shown in Table 5-7, even if the addition of water is 15 times the weight of the soybeans, tofu curd is formed, at the same time, the hardness of the curd is very low even if the addition of GDL is 0.3% of soymilk. With an increasing amount of water addition, the hardness and pH of the filling tofu decreased. If the water addition is too high and the solid content of soymilk is too low, the hardness increased little with the increasing of GDL addition. Usually, the concentration of soymilk for filling tofu is higher than that for other tofu processing, and the addition of water must be controlled to 5~7 times the weight of the soybeans.

Table 5-7. The effects of concentration of soymilk on the processing of filling tofu (Li et al., 2003)

Addition of water (times of soybean)	*Solid content of soymilk (%)*	*Addition of GDL (% of soymilk)*	Clarified water (ml)[a]	*Isolated water (ml)[a]*	*pH*	*Hardness (g)*
6	10.0	0.1	0.6	0.2	6.1	10.7
		0.2	0.7	0.6	5.8	47.7
		0.3	1.1	1.0	5.5	72.0
7.5	8.9	0.1	0.5	0.8	6.0	7.7
		0.2	0.7	2.1	5.7	38.8
		0.3	1.2	3.3	5.4	43.2
10	5.9	0.1	1.0	3.9	6.0	7.5
		0.2	1.1	5.1	5.5	26.2
		0.3	1.3	7.7	5.2	32.0
15	3.8	0.1	1.1	12.9	5.8	9.5
		0.2	7.9	10.7	5.2	14.5
		0.3	9.4	11.3	4.9	19.5

[a] The quantity from 1 kg soybean.

3.8.4. Gelling Temperature

Soymilk mixed with GDL is poured into a container and then is heated to coagulate. Just as shown in Figure 5-12, when the heating temperature is 85~90 °C, soymilk would coagulate speedily and the hardness of the filling tofu would be high. If the temperature is near or over 100 °C, the soymilk would be boiled and the coagulating is too speedy. There are many negative effects on the quality of filling tofu including a large number of air holes in the tofu, shrunk gelation, isolation of water and crude structure. If the temperature is lower than 70 °C, the curd of the tofu is very weak. Usually, the gelling temperature of filling tofu is 85~90 °C and the heating time for coagulating is 15~20 min.

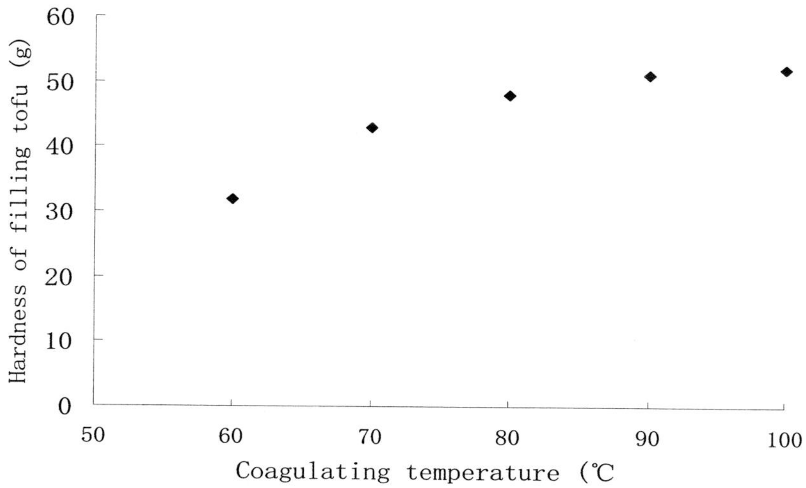

Figure 5-12. The relationship between coagulating temperature and hardness of filling tofu.

Table 5-8. Correlations of soybean quality characteristics and tofu yield from 30g soybeans (Chen et al. 2004)

Soybean cultivar	Protein (% d.b.)	Lipid (% d.b.)	Tofu yield (g)	Tofu volume (cm^3)
Zhongzuo 015	48.46	17.87	46.40	38.81
Zhongzuo 96-952	45.01	23.51	55.35	46.30
BN 1003-12	41.68	22.00	55.41	46.35
96-274	47.45	19.50	44.82	37.49
Yi-1358	45.38	22.00	48.55	40.61
99S	46.65	19.50	44.66	37.36
Jingyin No.1	45.65	21.00	60.32	50.46
BN1010	45.29	21.00	47.01	39.32
BN1003	45.16	21.50	45.38	37.96
Zhonghuang 18	44.69	20.67	66.41	55.55
Fengda 988	46.38	15.50	42.88	35.87
9901	49.27	17.00	49.34	41.57
Zhonghuang 4	47.35	21.00	45.32	37.92
Yi 75-14	47.83	18.50	37.81	31.64
Zhongzuo 947	45.33	21.00	59.19	49.52
Jingfeng No.1	45.82	20.33	50.35	42.13
Zhongzuo 975	47.81	18.00	43.68	36.55
Zhongpin 5807	44.60	20.50	43.76	36.55
Zaoshu 17	45.65	20.83	59.62	49.88
Zhongzuo 983	46.41	22.97	45.03	37.68
Average	45.99	20.21	49.56	41.48
SRD	1.76	2.01	7.37	6.17

4. PROGRESS OF STUDY ON TOFU PROCESSING

With the benefits of tofu well-known in China and spreading around the world, there are many researchers focused on the raw material, processing technology and nutrition of tofu.

4.1. Study of Soybeans for Tofu Processing

Chen et al. (2004) studied the correlations of soybean quality characteristics and bean curd yield. They planted 20 kinds of soybean cultivars in Beijing and analyzed the protein, lipid contents and the tofu yield. As shown in Table 5-8, protein contents (dried base) varied from 41.68 % of BN 1003-12 to 49.27 % of 9901. While the lipid contents of 20 cultivars varied from 15.50 % of Fengda 988 to 23.51 % of Zhongzuo 96-952. But yield of tofu made from 30 g soybeans varied from 37.81 g of Yi 75-14 to 66.41 g of Zhonghuang 18. It is understandable that the protein or lipid content of soybean cultivar is not directly related with tofu processing character.

Zhang and Wei (2006) studied the correlation between soybean varieties and texture properties of tofu gel, and concluded that the protein or lipid content of different cultivars had no significant effect on the tofu gel, too.

Rajni Mujoo et al. (2003) studied the characterization of storage proteins in different soybean varieties and their relationship to tofu yield and texture. They used 7 kinds of cultivars with different 11S (glycinin), 7S (β -conglycinin) protein contents (Table 5-9) and investigated tofu yield and firmness (Table 5-10) made from these soybeans. The results showed that cultivars of S-2020 with the lowest protein content had the lowest tofu yield and firmness, but cultivar of Vinton-81 with the highest protein content and second highest 11S/7S ratio had the largest firmness but just medium class of tofu yield. There was no clear relationship between tofu yield and soybean protein content; however, the firmness of tofu prepared from these varieties decreased as the protein content of the soybean decreased.

Table 5-9. Mean peak area percentages of 11S and 7S fractions of soybean proteins separated by RP-HPLC, and their ratio (Rajni Mujoo et al. 2003)

Soybean variety	11S	7S	11S/7S
Vinton-81	76.9	23.1	3.33
S-20F8	77.8	21.8	3.57
HP-204	75.6	24.4	3.10
IA-2034	64.9	35.1	1.85
Steyer	70.6	29.4	2.40
IA-2020	69.2	30.9	2.24
S-2020	67.3	32.7	2.05

Table 5-10. Moisture and protein contents of seven soybean varieties, and tofu yield and firmness (Rajni Mujoo et al. 2003)

Soybean variety	Moisture (%d.b.)	Protein (%d.b.)	Tofu yield (kg/kg soybean)	Tofu firmness (N)
Vinton-81	9.67	49.6	2.93	10.02
S-20F8	9.40	49.1	2.69	9.91
HP-204	9.06	48.5	3.20	8.53
IA-2034	9.66	47.9	3.22	8.19
Steyer	9.71	47.9	3.14	7.97
IA-2020	9.19	45.9	3.43	7.84
S-2020	9.04	42.9	2.90	6.93

The study indicated that the 11S protein fraction and the 11S/7S ratio are both good indicators for these properties of tofu, based on total protein analysis. The correlations of different peaks of total proteins, 11S and 7S protein fractions and their ratios to tofu yield and texture were calculated (Table 6). Peak 5, separated from total proteins, showed significant correlation ($P<0.05$) with tofu yield ($r=0.741$) and peak 6 showed correlation ($P<0.05$) with tofu firmness ($r=-0.761$). Soybean 7S content showed negative correlation ($P<0.01$) with tofu firmness, with a value of $r=-0.823$. Soybean 11S content ($P<0.05$) and the 11S/7S ratio ($P<0.01$) were also significantly correlated with tofu firmness, showing values of $r=0.820$ and $r=0.861$, respectively. However, peak 7 of the total proteins had an inverse correlation with tofu firmness ($r=-0.832$, $P<0.05$) (Table 5-11). Peak 7 comprises one of the components of the 7S fraction of soybean proteins, as identified from chromatographic separation of 11S and 7S fractions. The results indicate that 7S content, 11S content and 11S/7S ratio each appear to be associated with tofu firmness. However, peak 7 of the 7S fraction shows the most significant negative correlation. The strong negative relationships found between peak 7 of the total proteins and tofu firmness, and between peak 7 of the 7S protein fraction separated by RP-HPLC and tofu firmness indicated a role of the 7S fraction in determining tofu firmness. While many other studies pointed towards a relationship between the soybean 11S protein fraction and tofu textural properties. But the role in defining tofu texture of peak 7 fraction of β -conglycinin (7S) still needs to be studied.

Table 5-11. Correlation coefficients of different RP-HPLC peaks, protein fractions and protein ratios with tofu yield and firmness (n=7) (Rajni Mujoo et al. 2003)

Peak	Tofu yield	Tofu firmness
5	0.741*	0.634
6	0.558	0.761*
7	0.046	0.832*
9	0.812*	0.646
7S	0.518	0.823*
11S	0.507	0.820*
11S/7S	0.561	0.861**

* P<0.05.

** P<0.01.

Table 5-12. Mean squares for characteristics of seed, soymilk, and tofu of 10 soybean genotypes grown at three locations for 2 years (Poysa and Woodrow, 2002)

Trait	*Genotype*	*Location*	*Year*	*Yr_Loc*	*Yr_Geno*	*Geno_Loc*	*Residual*
Seed							
Protein	16.01***	2.42**	184.80***	2.89**	1.12*	1.05*	0.34
Oil	2.53***	1.79***	22.08***	0.53*	0.33**	0.16	0.09
Sugar	0.77***	0.69*	5.46***	0.74*	0.13	0.14	0.13
Sucrose	1.15***	0.72***	7.92***	0.1	0.1	0.04	0.04
Stachyose	0.10***	0.02	0.62 ***	0.02	0.01	0.01	0.01
Remainder	6.06***	1.86***	43.01***	1.01***	0.39*	0.52**	0.13
Seed mass	7.03*	1.45	97.79***	72.29***	3.16	2.17	2.56
Colour (L*)	3.29***	2.13*	14.54***	2.10*	0.23	0.63	0.38
Water absorption factor	0.01	0.003	0.02	0.024*	0.009	0.005	0.005
Protein/oil ratio	0.14***	0.05***	1.73***	0.03**	0.02**	0.009*	0.004
Protein/remainder ratio	0.12***	0.02**	1.20***	0.03***	0.009**	0.008**	0.002
Protein/sugar ratio	0.55***	0.25**	5.43***	0.17*	0.06	0.04	0.03
Soymilk							
Yield per kg seed DM	0.33***	0.05	2.18***	0.25**	0.04	0.04	0.03
Solids content	0.82***	0.43***	6.04***	0.31**	0.04	0.07	0.04
Colour (L*)	1.13*	0.04	0.23	4.02**	0.35	0.37	0.41
pH	0.007**	0.005	0.018**	0.004	0.001	0.002	0.002
% seed DM recovered	1.45	30.90**	4.85	35.98**	4.55	3.18	3.68
GDL-tofu							
Yield per kg seed DM	0.24***	0.02	1.15***	0.11*	0.03	0.03	0.02
Solids content	0.59***	0.21	4.38***	0.16	0.14	0.1	0.09
Colour (L*)	1.61*	2.53*	5.0**	3.71**	0.65	0.45	0.49
pH	0.004*	0.002	0.153***	0.006*	0.001	0.001	0.001
Hardness (compression)	0.14***	0.01	0.10**	0.05*	0.05**	0.01	0.01
Firmness (compression)	0.005***	0.001	0.001	0.002*	0.001**	0.001	0.001
Hardness (penetration)	0.005**	0.005*	0.033***	0.014***	0.002	0.001	0.001
Index	203.24***	118.01*	29.81	81.11	84.41**	13.16	23.06
CS-tofu							
Yield per kg seed	0.28***	0.01	0.59***	0.14**	0.03	0.04	0.02
Solids content	0.52**	0.11	7.31***	0.14	0.16	0.08	0.1
Colour (L*)	1.98**	0.08	27.41***	3.35**	0.6	0.11	0.39
pH	0.027***	0.007	0.125***	0.098***	0.004	0.002	0.002
Hardness (compression)	0.137***	0.004	0.022	0.044	0.019	0.021	0.015
Firmness (compression)	0.005***	0.001	0.001	0.002	0.001	0.001	0.001
Hardness (penetration)	0.006***	0.001	0.011**	0.008**	0.003*	0.001	0.001
Index	243.47***	0.49	64.73	93.21*	34.36	39.03	22.18

* $P<0.05$.

** $P<0.01$.

*** $P<0.0001$.

Poysa and Woodrow (2002) studied the stability of soybean seed composition and its effect on soymilk and tofu yield and quality. Five soybean cultivars planted in three places for two years were used as samples. Main compositions were analyzed and soymilk and tofu were prepared from these samples (Table 5-12). They concluded that genotype, location, and yearly effects were highly significant for most seed components measured, with year consistently having the largest and location the smallest effect. Genotype and year effects were highly significant for soymilk yield, solids levels, and pH. Location effects were much less significant, as were the interaction effects generally. Mean soymilk yield per kilogram of seed dry matter ranged from 6.99 l for CHR450 to 7.65 l for X799, while the solids level per kilogram of soymilk ranged from 95 g for X799 to 107 g for CHR450. All genotypes had similar percent DM (dried soymilk output) recovered in soymilk, averaging 74.5%. Yield of soymilk, GDL tofu, and CS (calcium sulphate)- tofu were all positively correlated with seed protein and strachyose and negatively correlated with seed oil, free sugar, sucrose and remainding contents. Seed protein was also positively correlated with tofu hardness and firmness, while seed oil, free sugar, sucrose, and remainding content were generally negatively correlated with these tofu quality parameters. They developed models from stepwise regression analyses and indicated that, when making soymilk and tofu with the above procedures, variation in seed protein and remainder contents, and their ratios, could account for a substantial percent of the variability in soymilk yield and GDL tofu yield, while seed protein and sucrose content are the most important determinants of CS-tofu yield. Protein content, per se, plays a smaller role in accounting for variability in tofu quality parameters.

Even though there were many studies on the soybeans for tofu making, because the varieties of soybeans is very large and the compositions are complex, it is still difficult to give some parameters for evaluating which kind of soybean.is suitable for tofu.

4.2. Study on Soymilk

Concentration adjusting of soymilk is one of the most important technologies in tofu processing. Protein content is the most significant component affecting the behaviors of soymilk, but for convenience, the concentration of soymilk is usually showed by solid content because it is easy to measure.

There were many researches on the effects of soymilk concentration on the quality and yield of tofu in the past 20 years, and the requests on soymilk for tofu making are well known just as explained in the former section. Recently, most of the research focused on the pretreatment of soybeans or soymilk. Tang (2007) studied the effect of thermal pretreatment of raw soymilk on the gel strength and microstructure of tofu induced by microbial transglutaminase (MTGase). The raw soymilk was heated at different heating rates (about 1.6, 6.3 and 23.8 ° C /min) from 20 to 95 ° C in a water bath, and then kept at 95 ° C for 5 min, or at similar heating rates from 20 to 75~78 ° C and kept for 5 min first, and then further heated to 95 ° C and kept at 95 ° C for another 5 min. The former method is called one-step method and the latter is called two-step.

Table 5-13 shows the influence of thermal pretreatment with different preheating method and heating rate on the viscosity of soymilk and gel hardness of MTGase-induced tofu. At the same heating rate, there were no significant ($P \leq 0.05$) differences for the viscosity of soymilk and tofu gel hardness, between one- and two-step preheating methods. However, the viscosity of soymilk and the tofu gel hardness were highly dependent upon the heating rate of the pretreatment. The viscosity of soymilk ($P>0.05$) decreased significantly with an increase in the heating rate from 1.6 to 23.8 ° C /min, irrespective of which heating method was chosen. This result suggests that the thermal denaturation and subsequent aggregation of soy proteins in soymilk are remarkably affected by the heating rate of the pretreatment, and the structure of soy proteins is more unfolded at a lower heating rate (e.g., about 1.6 ° C /min) than that at a higher one. In the case of a low heating rate, the unfolded proteins seem to be stable in the heat-treated soymilk, to a certain extent.

At a high heating rate (e.g., about 23.8 ° C /min), all protein constituents in soymilk might be denatured simultaneously within a short period of time, and those denatured or completely unfolded proteins might easily aggregate each other to form large aggregates. Thus, the differences in soymilk viscosity at different heating rates may be attributed to the differences in the extent of aggregation of the denatured proteins.

The gel hardness of the tofu formed from soymilk, treated at relatively low heating rates, e.g., less than 6.3 ° C/min, was significantly ($P \leq 0.05$) higher than that at high heating rates (e.g., about 23.8 ° C/min). The data suggest that the soymilk, heat-treated at lower heating rates, is more suitable for the gel formation of tofu by means of MTGase. However, the gel hardness of tofu was insignificant at $P>0.05$ between at 6.3 and 1.6 ° C/min.

Tang (2007) also reported that at a constant enzyme concentration of MTGase, the tofus formed from different preheated soymilks exhibited significantly ($P \leq 0.05$) higher gel hardness than those from the unheated ones (control), except the tofu induced by 50 units per100 ml soymilk of MTGase from soymilk treated at 95 ° C for 5 min (Figure 5-13). These results showed that, like in the conventional tofu-making cases, the thermal denaturation of soy proteins in soymilk by heat pretreatment is indispensable for making tofu using MTGase as the coagulant.

Table 5-13. Effect of heat pretreatment with different preheating method and rate on soymilk viscosity and gel hardness of tofu, induced by MTGase (Tang, 2007)

No.	*Heating pretreatment*		*Soymilk viscosity (cP)*	*Hardness of tofu-gel (g)*
	Method	Heating rate (° C/min)		
I	One step	1.6	82.2±8.9[a]	12.2±1.8[a]
II	Two step	1.6	68.5±9.2[a]	16.8±2.7[a]
III	One step	6.3	17.5±2.0[b]	18.1±6.2[a]
IV	Two step	6.3	20.4±1.8[b]	17.0±1.7[a]
V	One step	23.8	2.7±0.6[c]	8.5±2.5[b]
VI	Two step	23.8	2.0±0.7[c]	6.2±1.4[b]

Different superscript characters (a~c) indicate significant difference ($P<0.05$) in a same column.

The viscosity values at 25 °C were obtained at a rate of 60 rpm, after equilibrating for 3 min. The tofu was induced by 100 units per 100 ml soymilk of MTGase at 37 °C for 16 h. Mean values ± SD deviations of three replicates are given.

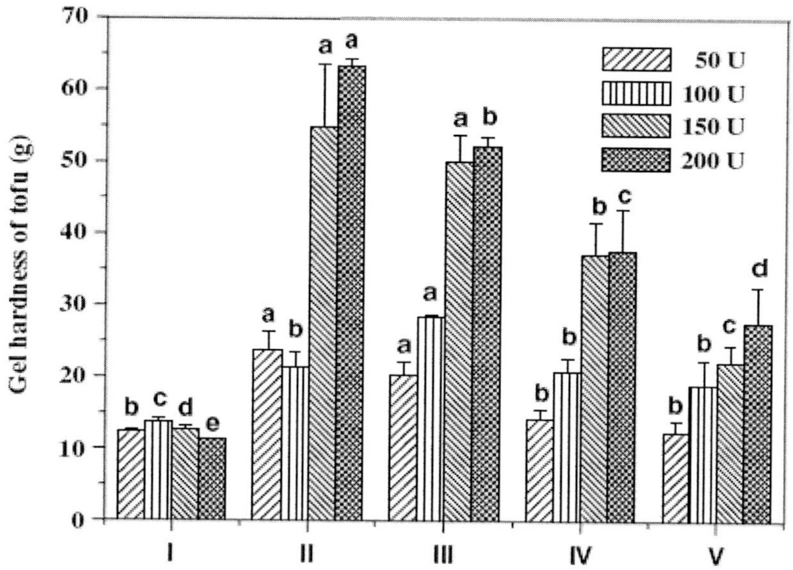

Figure 5-13. Effects of selective thermal pretreatment of soymilk and enzyme amount on gel hardness of MTGase-induced tofu (Tang, 2007).
The tofus were induced by various levels of MTGase (50, 100, 150 and 200 units per 100 ml soymilk) at 37 °C for 16 h. I, not heated (control); II, heated at 75 °C for 10 min; III, heated at 75 °C for 30 min; IV, heated at 75 °C for 30 min, and then at 95 °C for 5 min; V, heated at 95 °C for 5 min. Error bars indicate mean values 7SD deviations of three replicates, and different characters on the top of each column (a, b, c and d) indicate significant difference (P≤0.05) at a constant enzyme level.

Cai et al. (1997) researched the effect of processing method and soybean varieties on moisture and solid recovery of raw soymilk. Soymilks were prepared with bench scale (139 g soybean) and production scale (6500 g soybean). As shown in Table 5-14, the small-scale method exhibited more profound effects of soybean variety on the yield and moisture content of tofu than the production method. Variation range in yield made by the bench scale method was greater than that obtained by the production scale method. It is clear that the bench scale method can improve the solid recovery more significantly than the production scale because of sufficient extracting of residue. That is to say, sufficient extraction in soymilk processing could be improved to increase the quality and solid recovery of soymilk. Of course, different varieties of soybean had varied solid recovery and means extraction method may need to be adjusted to soybean cultivars.

Protein, lipid and ash content of soymilk and tofu were also measured with two-scale processing (Table 5-15, 5-16, 5-17). It was concluded that protein contents of soymilk and tofu varied with soybean variety. The higher the protein in soybeans, the higher the protein in the soymilk and tofu. The correlation between soybean and tofu protein contents were highly significant (R=0.84 and $p \leqslant 0.001$ for the bench scale, R=0.93 and $p \leqslant 0.001$ for the production scale).

The variations of lipid contents of soymilk or tofu made from two-scale processing differed also with the varieties of soybean. Generally, soybeans with high lipid content produced soymilk with high lipid content. Production scale yields a higher lipid content of soymilk than that of the bench scale. At the same time, lipid content of tofu was not affected obviously by processing scales. In bench scale, the lipid contents of tofu made from 13 varieties of soybeans had no significant difference while they were much different in the production scale. That is to say, if the processing technology in the production scale can be adjusted according to the character of soybean, the usage of lipid can be improved too. The difference of lipid extraction may be resulted from the separating strength that is larger in bench scale than that in production scale.

Table 5-14. Effect of processing method and soybean varieties on moisture and solid recovery of soymilk (Cai et al. 1997)

Variety	*Moisture (%)*		*Solid recovery (%)*	
	Bench	Production	Bench	Production
Proto	89.9±0.1[a]	89.3±0.2[a]	79.4±0.6[a]	57.8±1.2[bcd]
T5	90.2±0.3[a]	88.7±0.1[ab]	75.8±2.4[a]	60.6±0.7[ab]
Corsoy-97	90.2±0.4[a]	88.9±0.2[f]	75.6±2.9[a]	59.5±1.2[abcd]
Vinton	90.1±0.1[a]	88.3±0.1[abcd]	77.5±0.4[a]	63.0±0.6[a]
Kato	90.1±0.2[a]	89.5±0.4[a]	77.5±1.3[a]	57.1±2.3[bcd]
Hardin	90.5±0.6[a]	89.5±0.4[abc]	72.7±4.7[a]	57.8±2.4[d]
Sturdy	90.0±0.0[a]	89.0±0.0[def]	78.6±0.3[a]	69.8±0.2[abc]
SBB100ND	89.9±0.0[a]	89.5±0.1[ab]	77.7±0.0[a]	56.0±0.4[cd]
SBB100SD	90.2±0.2[a]	89.0±0.4[bcde]	76.0±1.2[a]	58.7±2.4[bcd]
Stine 2220	90.1±0.0[a]	89.4±0.1[ef]	77.1±1.1[a]	56.9±0.5[bcd]
Stine 1590	90.1±0.2[a]	89.3±0.2[ef]	77.2±1.9[a]	57.8±1.4[bcd]
Stine 0380	90.2±0.1[a]	89.2±0.0[cde]	76.6±0.9[a]	58.5±0.0[bcd]
Stine 1570	90.7±0.6[a]	88.9±0.8[abc]	72.0±4.6[a]	60.1±4.5[abc]

Data are means ±SD of two replicates (two determinations) on the wet weight basis. Means within the same column followed by different letters are significantly different ($p \leq 0.05$).

Table 5-15. Protein content of soymilk and tofu produced from 13 soybean varieties with two-scale processing (Cai et al. 1997)

Variety	*Soymilk (%)*		*Tofu (%)*	
	Bench	Production	Bench	Production
Proto	51.3±0.1[a]	52.7±0.6[a]	57.8±0.5[a]	57.6±0.1[a]
T5	50.6±0.5[a]	51.2±0.7[b]	54.2±0.6[b]	53.7±0.2[b]
Corsoy-97	46.4±2.3[b]	47.2±0.6[d]	53.6±1.8[bc]	49.8±0.8[c]
Vinton	51.4±0.8[a]	51.7±0.0[ab]	53.1±1.0[bc]	52.4±0.0[b]
Kato	46.0±0.9[b]	49.3±0.2[c]	50.8±2.2[cde]	52.6±0.1[b]
Hardin	45.3±2.8[b]	45.6±0.8[ef]	51.8±3.5[bcd]	47.3±1.3[d]
Sturdy	43.9±0.2[b]	45.0±0.5[f]	48.4±0.5[e]	47.3±1.3[d]
SBB100ND	45.5±1.2[b]	46.8±1.2[de]	49.7±0.1[dc]	48.4±2.4[cd]
SBB100SD	49.1±1.3[a]	52.3±0.5[ab]	54.4±0.9[b]	53.7±1.6[b]
Stine 2220	43.7±0.2[b]	45.3±0.4[ef]	48.1±0.0[e]	49.5±0.4[cd]
Stine 1590	45.6±0.2[b]	46.0±0.9[def]	49.0±0.4[de]	48.7±0.3[cd]
Stine 0380	46.2±0.1[b]	47.1±0.6[d]	50.3±0.3[cde]	52.6±0.1[b]
Stine 1570	44.4±0.5[b]	46.6±0.5[de]	48.5±1.1[de]	48.2±0.0[cd]

Data are means ± SD of two replicates (one determination per replicate except that tofu on production scale had four determinations per replicate) on the dry weight basis. Means within the same column followed by different letters are significantly different ($p \leq 0.05$).

Table 5-16. Lipid content of soymilk and tofu produced from 13 soybean varieties with two scales (Cai et al. 1997)

Variety	*Soymilk (%)*		*Tofu (%)*	
	Bench	Production	Bench	Production
Proto	17.8 ± 0.1^{f}	19.1 ± 0.0^{e}	19.8 ± 0.5^{a}	21.3 ± 0.1^{aa}
T5	19.9 ± 0.4^{bcde}	20.0 ± 0.5^{bcde}	20.5 ± 0.7^{a}	21.0 ± 0.5^{bc}
Corsoy-97	22.0 ± 1.0^{a}	21.5 ± 0.6^{a}	22.0 ± 1.6^{a}	22.0 ± 0.0^{ab}
Vinton	19.5 ± 0.1^{bcd}	19.8 ± 0.9^{cde}	21.0 ± 1.3^{a}	20.0 ± 0.1^{c}
Kato	19.1 ± 0.1^{bcde}	20.9 ± 0.4^{abcd}	21.6 ± 0.3^{a}	22.0 ± 1.3^{ab}
Hardin	18.4 ± 0.6^{def}	20.4 ± 0.5^{abcde}	22.1 ± 1.1^{a}	21.0 ± 0.4^{bc}
Sturdy	19.5 ± 0.2^{bc}	21.2 ± 0.9^{ab}	21.6 ± 0.0^{a}	21.6 ± 0.3^{ab}
SBB100ND	18.3 ± 0.5^{ef}	19.6 ± 1.0^{de}	20.0 ± 0.1^{a}	19.9 ± 1.0^{c}
SBB100SD	17.5 ± 0.5^{f}	19.6 ± 0.1^{de}	19.9 ± 0.0^{a}	19.7 ± 0.6^{c}
Stine 2220	19.0 ± 0.1^{bcde}	21.2 ± 0.4^{ab}	21.5 ± 0.5^{a}	22.6 ± 0.3^{a}
Stine 1590	19.6 ± 0.6^{b}	21.3 ± 0.5^{ab}	21.6 ± 0.4^{a}	21.7 ± 0.1^{ab}
Stine 0380	18.4 ± 0.1^{cdef}	20.2 ± 0.6^{abcde}	21.6 ± 0.4^{a}	22.0 ± 0.5^{ab}
Stine 1570	19.1 ± 0.7^{bcde}	21.1 ± 0.4^{abc}	21.4 ± 0.4^{a}	21.7 ± 0.1^{ab}

Data are means ± SD of two replicates (one determination per replicate except that tofu on production scale had four determinations per replicate) on the dry weight basis. Means within the same column followed by different letters are significantly different ($p\leq0.05$).

Table 5-17. Ash content of soymilk and tofu produced from 13 soybean varieties with two scales (Cai et al. 1997)

Variety	*Soymilk (%)*		*Tofu (%)*	
	Bench	Production	Bench	Production
Proto	6.4 ± 0.0^{a}	6.2 ± 0.1^{a}	7.8 ± 0.0^{a}	7.8 ± 0.1^{a}
T5	5.4 ± 0.0^{g}	5.2 ± 0.0^{g}	6.7 ± 0.0^{cd}	7.0 ± 0.1^{cde}
Corsoy-97	5.4 ± 0.3^{fg}	4.9 ± 0.1^{h}	4.5 ± 0.0^{f}	5.7 ± 0.1^{g}
Vinton	5.4 ± 0.0^{fg}	5.3 ± 0.0^{fg}	7.2 ± 0.2^{abc}	7.3 ± 0.0^{bc}
Kato	5.9 ± 0.3^{cde}	5.7 ± 0.0^{bc}	7.3 ± 0.5^{abc}	7.7 ± 0.0^{ab}
Hardin	5.9 ± 0.4^{cde}	5.5 ± 0.0^{def}	7.3 ± 0.5^{c}	7.1 ± 0.0^{cde}
Sturdy	5.5 ± 0.1^{efg}	5.4 ± 0.1^{fg}	6.1 ± 0.1^{de}	6.7 ± 0.0^{def}
SBB100ND	6.3 ± 0.0^{ab}	6.2 ± 0.0^{a}	7.6 ± 0.3^{ab}	7.8 ± 0.3^{a}
SBB100SD	6.0 ± 0.0^{bcd}	5.6 ± 0.0^{cd}	6.1 ± 0.3^{de}	6.8 ± 0.0^{cde}
Stine 2220	5.5 ± 0.2^{efg}	5.4 ± 0.1^{befg}	6.1 ± 0.0^{de}	6.3 ± 0.3^{f}
Stine 1590	5.6 ± 0.1^{defg}	5.6 ± 0.1^{cde}	6.2 ± 0.0^{de}	6.7 ± 0.2^{ef}
Stine 0380	5.8 ± 0.1^{defg}	5.6 ± 0.1^{cde}	5.9 ± 0.1^{e}	6.7 ± 0.3^{def}
Stine 1570	6.1 ± 0.1^{abc}	5.9 ± 0.1^{b}	7.1 ± 0.2^{bc}	7.1 ± 0.1^{cd}

Data are means ± SD of two replicates (one determination per replicate except that tofu on production scale had four determinations per replicate) on the dry weight basis. Means within the same column followed by different letters are significantly different ($p\leq0.05$).

Table 5-18. Texture profile analysis of tofu (Noh et al. 2005)

Sample	*Hardness*	*Cohesiveness*	*Springiness (mm)*	*Gumminess (g)*	*Chewiness (g mm)*
$TSN_{2,5}$[A]	511.2[aB]	0.289[a]	7.498[a]	118.5[a]	888.4[a]
$TSF_{2,5}$	852.3[b]	0.350[b]	7.906[bc]	259.2[bc]	2047.3[bc]
TSN_5	625.6[ab]	0.360[b]	7.717[b]	198.4[ab]	1533.4[ab]
TSF_5	1175.9[c]	0.344[b]	7.819[c]	324.5[c]	2537.7[c]

[A] $TSN_{2,5}$, tofu from soymilk heated for 2.5 min, made from the unfrozen soybean; TSN_5, tofu from soymilk heated for 5 min, made from the unfrozen soybean; $TSF_{2,5}$, tofu from soymilk heated for 2.5 min, made from the frozen soybean; TSF_5, tofu from soymilk heated for 5 min, made from the frozen soybean;

[B] Means within the same column followed by different letters are significantly different ($p \leq 0.05$).

Ash content of tofu was larger than that of soymilk (Table 5-17). It resulted from use of calcium sulfate because most of the coagulant was kept in the tofu. It is reported ~80% of calcium from coagulant was retained after tofu formation. In other words, about 300mg calcium would be increased for 300 g tofu in normal addition of coagulant so tofu made from calcium coagulant is an excellent source of calcium too.

4.3. Progress of Study on the Processing Methods of Tofu

Tofu processing includes cleaning, soaking and grinding of soybeans, filtering, boiling and coagulating of soymilk, molding of tofu. Almost all of the steps affect the yield and quality of tofu. Most of processing parameters are understood, for many studies had been done long ago. But the effects of boiling are complicated and need further study because of different compositions of protein subunits of soybean and variation of their changeful gelling characters. So the study on processing is mainly focused on the boiling (cooking or heating) of soymilk.

Noh et al. (2005) researched the effects of freezing of soybean on the texture, tofu yield. Soaked beans were frozen to -20 °C for 5 h by air-blast freezing. As shown in Table 5-18, with the freezing of soybeans, the hardness, gumminess and chewiness of tofu improved significantly in either of 2.5 min or 5 min heating. That is to say, freezing of soybeans can shorten requisite heating time or decrease the solid concentration of soymilk to increase tofu yield while keeping the texture. It is said the effect may result from the freezing which promotes the hydrophobic coagulation of soy protein.

The yield of tofu prepared from frozen soybeans was lower than that from unfrozen soybeans and the yield decreased with an increase of the heating time (Table 5-19). In addition, both freezing and prolonged heating also resulted in tofu with low moisture content. Decrease in the yield was reflected by the lower moisture content. The lower yield and moisture content of tofu from frozen soybeans may be ascribed to the denser and more compact structure, which made water easily release from the curd during pressing. Protein content of tofu from frozen soybeans was found to be higher than that of tofu from unfrozen soybeans. In contrast, tofu from frozen soybeans contained less fat than that from unfrozen soybeans. These results indicate that fats in the coagulum from frozen soybeans are more easily released during pressing, probably suggesting that freezing considerably decreases the fat-binding capacity of protein. In other words, freezing of soybeans enhanced the aggregation of the protein molecules during heating and led to an increased participation of the soy protein in the gel network, thereby resulting in tofu with higher protein content. It may be necessary to study further the effect of freezing on tofu processing. For example, the texture of tofu may be adjusted to be a little firm and compact by increasing water addition so as to increase the yield of tofu with frozen soybeans.

It was also reported that tofu from frozen soybeans had lower (better) scores than that from unfrozen soybeans on a 9-point hedonic scale (Table 5-20). The soymilk from frozen soybeans produced a more uniform-structured gel than that from unfrozen soybeans, suggesting that the freezing can control the coagulating process of soymilk in a positive way.

The force to extrude the soymilk from frozen soybeans was researched. The force increased faster and reached the maximum value earlier than that from unfrozen soybeans (i.e., a faster coagulation). A difference in the force between the soymilk coagulums from the frozen and unfrozen soybeans was evident, regardless of heating time.

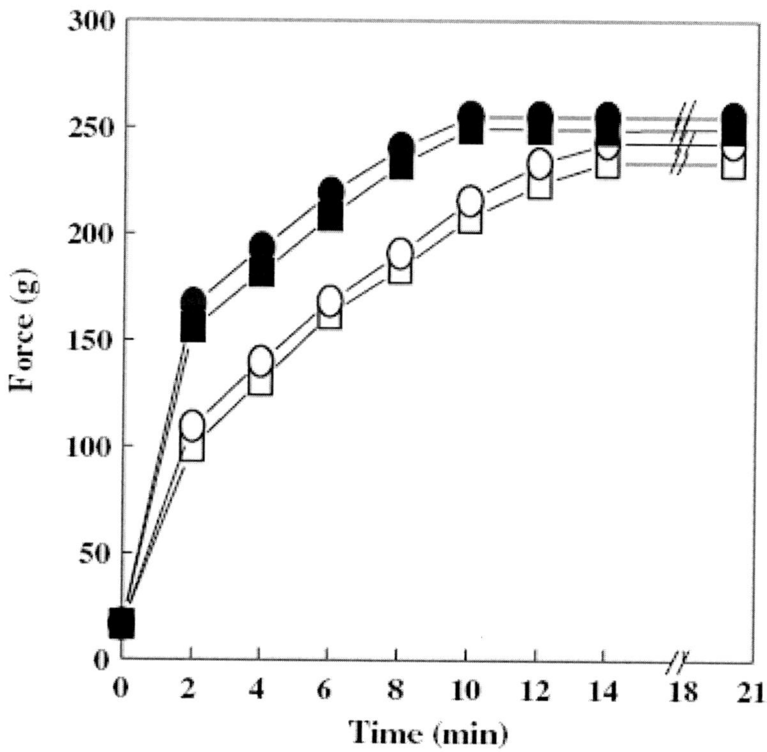

Figure 5-14. Changes in the force obtained by a back extrusion test for the soymilk coagulum (Noh et al. 2005).

The soymilk coagulum was prepared using soymilk heated for 2.5 or 5 min, made from frozen soybeans or unfrozen soybeans: □unfrozen and 2.5-min heating; ○ unfrozen and 5-min heating; ■frozen and 2.5-min heating; ●frozen and 5-min heating. The values were obtained from triplicate experiments and expressed as mean values.

Table 5-19. Synersis, yield and content of moisture, protein and fat in tofu prepared under different conditions (Noh et al. 2005)

Sample	*Yield (g)*	*Water (%)*	*Protein (%)*	*Fat (%)*	*Synersis (%)*
$TSN_{2,5}$[A]	301.8[aB]	89.6[a]	9.6[d]	2.208[a]	2.9[a]
$TSF_{2,5}$	284.9[a]	79.1[b]	16.2[b]	1.451[b]	4.4[b]
TSN_5	289.8[a]	79.2[b]	14.0[c]	2.206[a]	3.3[a]
TSF_5	279.7[a]	77.2[c]	19.3[a]	1.597[b]	4.9[b]

[A] Sample designations are the same as in Table 5-18.
[B] Means within the same column followed by different letters are significantly different ($p \leq 0.05$).

Table 5-20. Effect of freezing of soybeans on sensory characteristics of tofu (Noh et al. 2005)

Sample	Color	Flavour	Mouthfeel	Overall acceptability
$TSN_{2,5}$[A]	2.5[aB]	5.1[a]	6.2[a]	7.1[a]
$TSF_{2,5}$	2.8[a]	4.0[bc]	4.0[bc]	4.7[c]
TSN_5	2.5[a]	4.8[ab]	5.0[b]	6.0[b]
TSF_5	2.6[a]	3.5[c]	3.1[c]	3.6[d]

[A] Sample designations are the same as in Table 5-18.
[B] Means within the same column followed by different letters are significantly different ($p \leq 0.05$).

Cotyledon of soybean seeds is composed of Palisadelike cells in which spherical protein bodies adhered to lipid bodies (also known as spherosomes). Lee et al. (1992) examined the effect of freezing on the microstructure of soybeans using an optical microscope. It was found that freezing ruptured cell membranes and the lipids were pushed out of the cells when the frozen soybeans were heated, while the leakage of lipid bodies was not observed in unfrozen soybeans. Protein is easily separated from the lipids as they become frozen, and become partially insoluble. With the leakage of the lipid bodies induced by freezing, the protein molecules in the frozen soybeans may become closer to neighbouring protein molecules. This can facilitate the formation of intermolecular disulphide bonds. It is speculated that, the soymilk from the frozen soybeans coagulates faster (by enhanced hydrophobic interactions and intermolecular disulphide linkages) than that from unfrozen soybeans, resulting in a more uniform-structured tofu gel network. But in production scale, freezing of soybean is still seldom used perhaps because of the increase in cost.

Soymilk must be heated to denature before coagulated with the function of coagulant. So the heating temperature and speed are considered a key point to tofu production. The protein composition is very complex, as mentioned above, and different subunits of protein have varied denaturalization temperatures. For example, the denaturation temperature of glycinin is approximately 20 °C higher than that of β-conglycinin, Liu et al. (2004) researched the effects of one-step or two-step heating on soymilk and tofu. They found extending heating time from 5 to 10 min induced significant decrease in tofu's apparent breaking strength and Young's modulus, but did not significantly affect tofu's syneresis rate in one-step heating when GDL was coagulant. But two-step heating significantly increased the apparent Young's modulus and breaking strength, and reduced the syneresis rate of tofu (Table 5-21). The improvement of tofu physical properties was attributed to the selective thermal denaturation of soybean proteins.

Table 5-21. Effect of heating method on soymilk viscosity and tofu gel physical properties (Liu et al. 2004)

Heating method	Heating condition	Soymilk viscosity (cP)	Tofu-gel physical properties		
			Apparent breaking strength (kPa)	Apparent Young's modulus (kPa)	Syneresis rate (%)
One-step	95°C, 5 min	38±2.1[b]	18.9±0.4[b]	103.2±1.3[b]	16.0±0.4[a]
	95°C, 7min	38±2.5[b]	17.5±0.4[c]	99.1±2.1[bc]	16.0±0.3[a]
	95°C, 10min	38±1.4[b]	17.3±0.4[c]	97.5±2.4[c]	16.3±0.5[a]
Two-step	75°C, 5min and then 95°C,5 min	96±2.8[a]	20.2±0.5[a]	126.1±1.3[a]	14.4±0.3[b]

Data were expressed as means ± SD of duplicate experiments, and means in the same column with different superscripts are significantly different ($p \leq 0.05$).

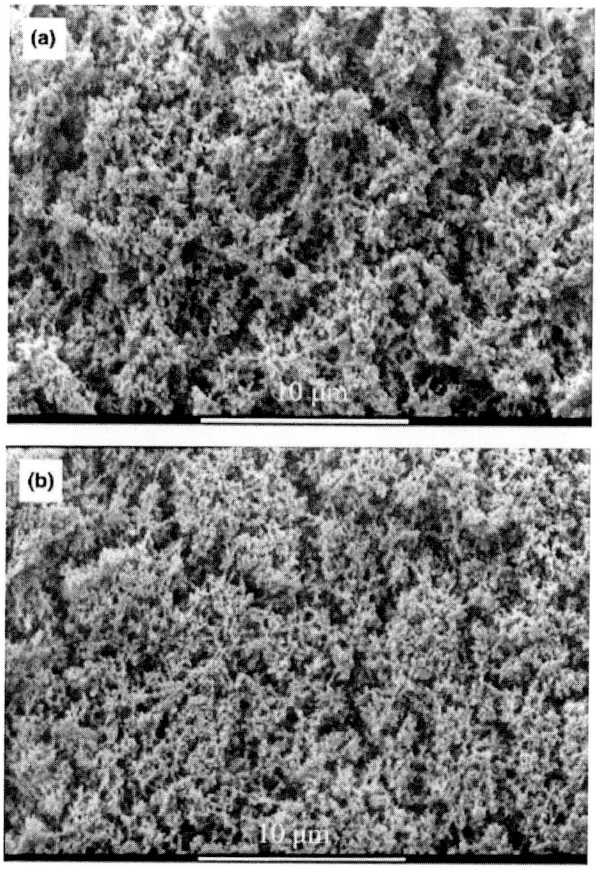

(a)

(b)

Figure 5-15. Effect of selective thermal denaturation on the microstructure of filling tofu (Liu et al. 2004). (a) one-step heating and (b) two-step heating.

Table 5-22. Effect of soymilk solid content on selective thermal denaturation [A] (Liu et al. 2004)

Soymilk solid content (%)	Heating method	Syneresis rate (%)	Apparent breaking strength (kPa)	Apparent Young's modulus (kPa)
9.3	One-step	29.1±0.4[a]	15.1±0.3[a]	67.0±2.5[b]
	Two-step	25.5±0.5[b]	15.2±0.5[a]	82.3±3.8[a]
	STD efficiency [B]	-12.4%		22.8%
10.4	One-step	23.5±0.5[a]	17.2±0.4[a]	89.3±2.3[b]
	Two-step	20.8±0.4[b]	17.7±0.3[a]	106.5±3.1[a]
	STD efficiency [B]	-11.5%		19.3%
11.7	One-step	16.0±0.4[a]	18.9±0.4[b]	103.2±3.0[b]
	Two-step	14.4±0.3[b]	20.2±0.5[a]	126.1±2.7[a]
	STD efficiency [B]	-10.0%	6.9%	22.2%

[A] Data were expressed as means ± SD of duplicate experiments, and means (in the same column for the same solid content) with different superscripts are significantly different ($p \le 0.05$).

[B] STD efficiency: the difference between two-step heating and one-step heating divided by one-step heating.

Microstructures of tofu made from two-step heating and one-step heating were observed by SEM (Liu et al., 2004). As shown in Figure 5-15, the matrix structure of the tofu network prepared with two-step heating was denser, finer and more homogeneous than that of one-step heating. It can be concluded that the differences in the microstructures induced to the differences in the physical properties of two types of tofu gels.

They confirmed that two-step heating significantly increased the tofu's apparent Young's modulus and reduced the syneresis rate for all of the test soymilk concentration but only increased the breaking strength in higher solid content soymilk (11.7%).

Research of Wang et al. (2007) also confirmed two-step heating was good for south tofu production either in yield or solid recycle (Table 5-23). They suggested that, in one-stage heating, the dissociation/association of 7S and 11S globulin occurs at the same time, whereas in two-stage heating, 7S globulin subunits dissociate/associate selectively at the first heating stage, then in the second stage, 7S sub-units and/or newly dissociated 11S sub-units associate cooperatively. These moderately associated protein molecules enabled tofu to form a tight gel after the addition of a coagulant.

Table 5-23. Effect of heating method on soft-tofu yield and solid recycle [A] (Wang et al. 2007)

Heating method	Heating conditions	Solids recovered (%)	Yield of soft-tofu (g/g soybean)
One-step	100 °C,5 min	71.80 ± 1.96 [b]	3.95 ± 0.09 [b]
Two-step	70 °C,10 min and then 100 °C 5 min	75.66 ± 1.19 [a]	4.13 ± 0.09 [a]

Values in a column with different letters were significantly different ($p < 0.05$).

[A] Values represent the mean ±SD; n=3.

There are many researches on the mechanism of tofu gel forming, too, even though it is still not clear entirely. Tang et al. (1997) researched the interaction between protein and lipid in heated soymilk. When the temperature of heated soymilk rose from 65 to 75°C, a part of lipid and almost all of α and α' subunits of β-conglycinin in the particulate fraction liberated and moved to soluble fraction. With the increasing of temperature from 75°C, the lipids in the soluble and particulate fractions began to liberate and to shift to the floating fraction. Almost all lipid (neutral lipid) shifted to the floating fraction at 90°C. The interactions of proteins with lipids occurred in soybeans, especially in soybeans stored at high temperatures and high relative humidity, and it may weaken the soy protein three-dimension network by decreasing the probability of protein-protein interactions, which induced the changes in tofu texture (Hou et al., 1997).

Wang and Damodaran (1991) stated that the texture of soy protein gels is fundamentally related to the molecular weight-average and the hydrodynamic shape of the polypeptide in the gel network rather than their chemical nature such as the amino acid composition and distribution. Kang et al. (1991) suggested that the glycinin/β-conglycinin ratio affects the texture of the gels. Nishinari et al. (1991) reported that tofu gels prepared from soybean cultivars without the A_4 subunits (A_5 in their nomenclature) were harder than those prepared from cultivars with the A_4 subunit. However, these limited results do not allow us to draw conclusions regarding which factor(s) play(s) the most important role in the rheological properties of soy protein gels. Furthermore, these studies with the exception of Nishinari et al. (1991), were based on purified soy protein systems rather than soy foods where the other constituents may affect the final texture characteristics of soy products. To determine the factors contributing to the textural properties of tofu made on a pilot plant scale, the correlation between soybean physico-chemical properties (NSI, amounts of soy proteins and subunits) with tofu texture characteristics (hardness, fracturability, and cohesiveness) must be evaluated statistically.

4.4. Study on the Quality and Nutrition of Tofu

Tofu is rich in protein and lipid because the extracting ratio of protein and lipids from soybeans are about 80% and 75% (Watanabe, 1996). But the mineral content is not so high and many studies have focused on further improving the quality and nutrition of tofu by adding ingredients.

Karim et al. (1999) studied the effects of adding carrageenan on the quality and yield of tofu made with different coagulants. As shown in Figure 5-16, the addition of carrageenan to soymilk prior to coagulation resulted in significant increases in yields of tofu ($p<0.05$). At a concentration of 1 g/l and 2 g/l, carrageenan increased the yield of CS-tofu by approximately 9.5% and 33%, respectively. While as shown in Figure 5-17, a texture analysis revealed that CS (calcium sulphate)-tofu was harder than CA (calcium acetate)- and GDL-tofu. Addition of carrageenan, at 1 g/l and 2 g/l, resulted in a significant decrease ($p<0.05$) in hardness of CS-tofu (Figure 2) by 21.2% and 55.9% and that of CA-tofu by 21.6% and 38.3%, respectively.

The effect of carrageenan on the flavour of tofu was evaluated by a panel on a 9 point scale in that same research. Colour, favour, mouthfeel and overall acceptability of tofu prepared with CA/2 g/l carrageenan, CA and CS alone (without carrageenan) were evaluated on a 9 point scale (Table 5-24). The acceptability scores ranged from 5.05 to 7.53 and the highest scores were given to CA-tofu with 2 g /l carrageenan, which had a smooth, soft but firm texture. That is to say, addition of carrageenan can improve the flavour of CA tofu.

Table 5-24. Effect of coagulants and carrageenan on sensory characteristics of tofu [a] (Karim et al. 1999)

Sensory parameters	CA+ carragenean tofu [b]	CA-tofu [c]	CS-tofu [d]
Colour	7.53	7.05	7.00
Flavour	6.74	6.26	5.84
Mouthfeel	6.68	5.95	5.05
Overall acceptability	6.63	6.21	6.05

[a] on the 9-point hedonic scale.
[b] Calcium acetate 0.1 mol/l and carrageenan at 2 g/l
[c] Calcium acetate only at 0.1 mol/l.
[d] Calcium sulphate only at 0.2 mol/l.

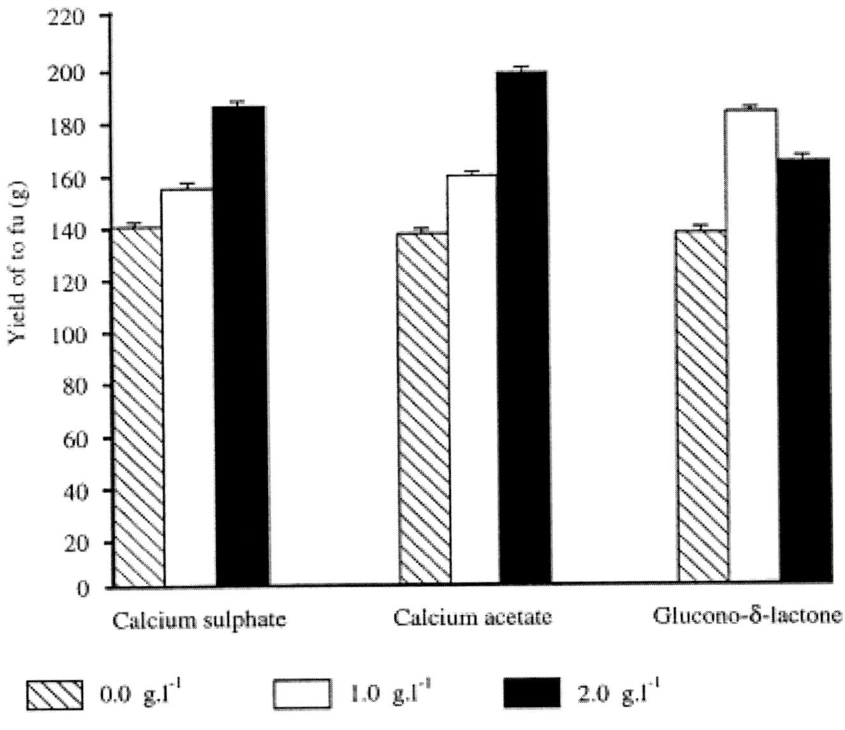

Figure 5-16. Effect of coagulants and carrageenan on yield of tofu (Karim et al. 1999).

Chang and Chen (2003) researched the effect of chitosan on the gel properties and shelf life of tofu. Table 5-25 shows the effect of 2.0% chitosan on the gel properties of tofu. Regardless of the method of preparation, the gel strength of tofu was increased (for 5~305% versus control samples) by the addition of 2% chitosan. The shelf life of tofu was improved, too, due to chitosan generally having high solubility and positive charges in an acidic environment.

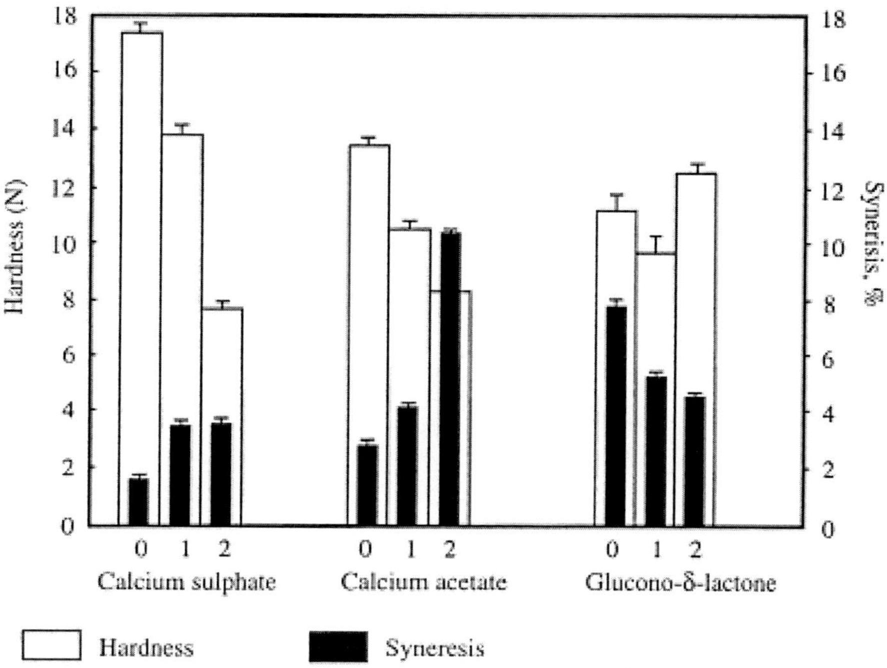

Figure 5-17. Effect of coagulants and carrageenan on hardness and syneresis of tofu. (Karim et al. 1999). The numbers on x-axis represent carrageenan concentrations (g/l).

In addition, the degree of deacetylation (DD) of chitosan is also related with the effect. The increase in gel strength decreased with the rising of DD (Figure 5-18). This indicated that the higher molecular weight (Mw) of chitosan strengthened the gel structure of tofu more than lower weight molecules did. High Mw chitosan apparently bound to a higher extent with soy protein. More entanglements occurred between the polysaccharide and protein molecules and resulted in a more stable gel network and firmer texture.

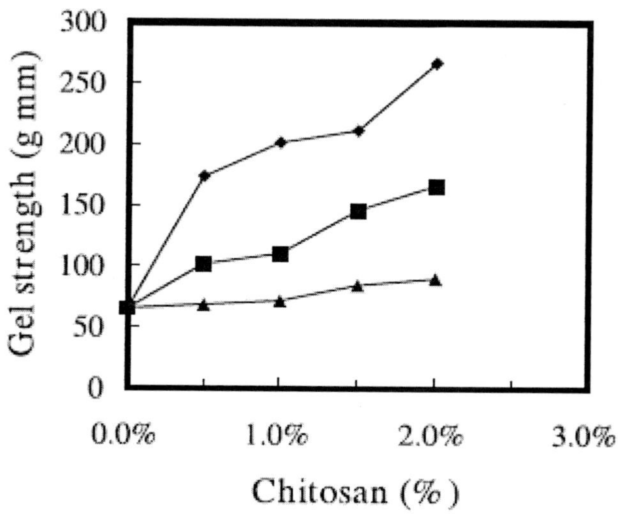

Figure 5-18. The gel strength of acetic acid tofu after adding chitosan with different degrees of deacetylation: (□) 54%; (■) 73%; (▲) 91% (Chang and Chen, 2003).

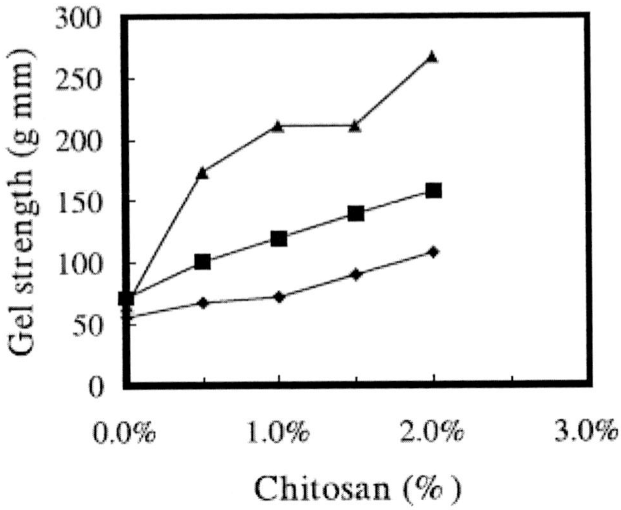

Figure 5-19. The effect of 54% DD chitosan on the gel strengths of tofu prepared by different curdling agents: (□) GDL; (■) Gypsum; (▲) Acetic acid (Chang and Chen, 2003).

Table 5-25. Changes in the gel strength, water content and shelf life of different types of tofu caused by the addition of 2 % chitosan (Chang and Chen, 2003)

		Control	*54%DD*	*73%DD*	*91%DD*
GDL tofu	Gel strength (g mm)	56±2.1	108±0.0	75±0.0	59±0.0
	Water content	80±0.30	84±0.66	81±1.40	82±0.21
	Shelf life [a] (days)	5	7	8	13
Gypsum tofu	Gel strength (g mm)	71±7.2	210±30.0	158±2.4	121±12.0
	Water content (%)	74±0.71	79±7.20	82±0.97	80±0.65
	Shelf life [a] (days)	8	13	13	18
Acetic acid tofu	Gel strength (g mm)	66±2.30	267±0.0	166±2.8	90±8.9
	Water content (%)	79±0.84	78±0.22	78±0.22	76±2.1
	Shelf life [a] (days)	12	15	20	22

[a] The total days of storage in 4 °C refrigerator until significant mold growth and staled flavour were observed.

It is also said that chitosan was more soluble in soymilk containing acetic acid and would allow more entanglements or interactions to occur between chitosan and soy protein molecules. So acetic acid tofu was more sensitive to the addition of chitosan.

Kim et al. (2007) tried to improve the quality of tofu by adding oyster shell powder. When stored at 4 °C for 24 h, syneresis of tofu prepared with shell powder was significantly ($p < 0.05$) lower than that of tofu prepared without shell powder. It is generally accepted that syneresis in the protein gel during storage is caused by an increased cross-linking among protein molecules through various interactions, making the protein gel matrix denser. With the addition of shell powder, the decrease in syneresis could result from enhanced water retention in the gel microstructure.

At the same time, shell powder affected the texture of tofu (Table 5-26). With the addition of shell powder, the hardness of tofu increased significantly and subsequently increased the gumminess and chewiness. However, no significant (p>0.05) further increase in textural properties such as hardness, gumminess and chewiness were observed when the amount of added shell powder increased from 0.05% to 0.20%. The result indicated that the presence of calcium ion in the mixed coagulant system increased the hardness of tofu.. The hardening of tofu by the addition of calcium ion can be affected by the way protein interacts with calcium and other constituents, e.g. phytic acid, in soy milk and anions to form the microstructure into a gel. Gelation of food protein involves heat denaturation followed by aggregation. If aggregation is relatively slower than denaturation, an ordered structure will be promoted, by allowing the denatured molecules to orient themselves in a systematic fashion prior to aggregation. Conditions that retard the intermolecular interaction will result in a more homogeneous and regular network and consequently a stronger tofu structure.

Table 5-26. Texture profile analysis of tofu [A] (Kim et al. 2007)

Shell powder (%)	Hardness (N)	cohesiveness	Gumminess (N)	Chewiness (N mm)	Springiness (mm)	adhesiveness
0.0	3.71±0.57[c]	0.54±0.024[a]	1.99±0.31[c]	1.71±0.26[c]	8.64±0.13[ab]	-0.12±0.04[a]
0.05	8.17±0.69[a]	0.54±0.019[a]	4.42±0.48[a]	3.78±0.39[a]	8.53±0.17[ab]	-0.13±0.07[a]
0.1	5.91±1.02[b]	0.57±0.027[a]	3.33±0.50[b]	2.95±0.46[b]	8.93±0.22[a]	-0.12±0.03[a]
0.2	7.31±0.49[a]	0.56±0.027[a]	4.08±0.37[a]	3.46±0.43[ab]	8.52±0.30[b]	-0.12±0.02[a]

[A] means ±SD of 10 replicates. For each type of tofu, the same letters in the same column are not significantly different (p>0.05).

The usage of shell powder can improve the storage ability. The changes of the viable microbial counts of tofu prepared with shell powder and a control tofu prepared with single use of $MgCl_2$ during storage at 10 °C for 11 days was observed. All tofu had initial bacterial concentrations of 208 CFU/g at the beginning of storage. These values are similar to initial concentrations (102 CFU/mL) found in the solution which immerses tofu. Tofu prepared with 0.1% and 0.2% addition of shell powder did not show any changes of viable counts by the second day and the fourth day of storage, respectively. However, viable microbial counts of tofu prepared without shell powder increased more rapidly than those of tofu prepared with shell powder during longer storage periods.

Tseng and Xiong (2008) reported the effect of inulin on the rheological properties of GDL-tofu. With the increasing of inulin addition, textural attributes of GDL- tofu were significantly influenced by the presence of inulin, but in a complex manner. Specifically, the incorporation of inulin increased ($P < 0.05$) both the hardness and the breaking strength of tofu, but in a non-dose dependent manner (Table 5-27). For instance, the tofu with 3% inulin had a hardness improvement ($P < 0.05$) by 22.3% compared to the control (inulin-free); with 4% inulin, the hardness almost did not increase further. The addition of more than 3% of inulin did not further enhance the breaking strength, too. On the other hand, inulin at 3% levels significantly decreased the tofu gels' cohesiveness and increased their deformability. Tofu gels containing 4% inulin showed, on average, a 46.6% increase ($P < 0.05$) in structure deformability and a 4.6% decrease ($P < 0.05$) in cohesiveness compared to control samples. Hardness and breaking strength are important textural parameters that are commonly used to evaluate the quality of tofu curds. GDL-tofu can be found in different degrees of consistency, ranging from soft to extra firm. Firmer GDL-tofu, generally with higher protein content, are gaining more popularity because they do not fall apart during handling and can be combined with many other ingredients to create a variety of dishes. Inulin was shown to affect the texture of silken tofu at constant protein and GDL levels, producing harder gels with a higher breaking strength. It was revealed that the addition of inulin facilitates the network formation of soy proteins by increasing the density of protein cross linking, producing SPI gels with a greater compactness and smaller pore sizes. The presence of inulin was believed to alter the aggregation pattern of soy protein and transform them into a more interactive matrix system that was more resilient to compression.

In China, many kinds of colored tofu are produced and sold by adding colorful plant extracts such as spinach juice, carrot juice and so on. The processing is almost the same as normal tofu, except for premixing the plant extract with soymilk. The color and flavour may vary by adjusting the addition of different kinds and ratios of plant extract. This kind of tofu is especially welcomed by children and it is an efficacious method to solve the problem of disliking some vegetables.

Kao et al. (2004) researched the recoveries of all the isoflavones in soybeans, soaked beans, soybean and tofu and found that it ranged from 65% to 91% based on the extraction. The concentrations of glucosides, acetylglucosides and malonylglucosides in soybeans followed a decreasing order with the increase of both soaking temperature and time of soybeans, because of conversion to aglycones by cleaving the ester or glucosidic group. However, for aglycones, the concentrations increased with the above condition. During the heating of soymilk, the concentrations of glucosides and acetylglucosides showed an increase while malonylglucosides decreased and the aglycones did not show a significant change. The highest concentration of isoflavones was obtained in tofu with 0.3% calcium sulfate when compared to the higher percentage of the same coagulant. The loss of isoflavones may be found mostly in whey after coagulating and then in the soaking water.

The effects of tofu intake on the health of consumers have also been researched (Shi et al. 2008). The study reported the association between tofu intake and anemia at the population level in Jiansu province of China (Table 5-28). The findings indicated that the anemia in the region could be related to high rates of infection and inflammatory processes. Although the region is one of the richest in China, the prevalence of anemia is the highest in China as observed by three national nutrition surveys. Taking into account the long tradition of consuming tofu in the country, its low economic burden, as well as the health benefits, it may be promising to promote tofu as part of a healthful food choice in the prevention of anemia. The results may provide useful knowledge to other countries with the tradition of eating tofu that have a high prevalence of anemia. A high intake of tofu was associated with a lower prevalence of anemia. It was reported that high intake of tofu was associated with lower risk of anemia in both men and women. The association still remained highly significant after adjusting for sociodemographic characteristics, BMI, intake of foods known to affect iron status, and iron. Tofu intake was negatively associated with having higher serum ferritin levels in women but not in men. Chen et al. (2007) also reported the association between tofu intake and serum polychlorinated dibenzo-p-dioxins and dibenzofurans (PCDD/Fs).

Table 5-27. Textural analysis of GDL-coagulated silken tofu gels containing different concentrations of inulin (n =18) (Tseng and Xiong 2008)

treatment	hardness	cohesiveness	Deformability (%)	Rupture force (g)
Control (inulin free)	110.1 ± 6.7 a	0.86 ± 0.011 a	5.19 ± 0.97 a	424.3 ± 8.4 a
1%raftiline®HP-gel	115.8 ± 6.1 ab	0.86 ± 0.015 a	6.89 ± 0.77 ab	445.2± 10.3 b
2%raftiline®HP	125.5 ± 4.9 bc	0.85 ± 0.013 ab	8.25± 0.54 c	478.6 ±14.7 cd
3%raftiline®HP	134.7 ± 4.4 cd	0.82 ± 0.019 b	7.48± 0.76 bc	475.2 ± 11.2 cd
4%raftiline®HP	137.4 ± 5.3 d	0.82 ± 0.013 b	7.61 ± 0.63 bc	487.9± 12.6 d

Values are means ± SD deviation.

[abcd] Means within the same column without a common superscript differ ($P < 0.05$).

Table 5-28. Odds ratio (OR) and 95% confidence interval (CI) for anemia according to tofu intake quartiles (Q) in adults living in Jiangsu, China(n_2,849) (Shi et al. 2008)

		Tofu intake				
		Q1 (low)	Q2	Q3	Q4 (high)	P for trend
Men	Model 1[a]	1	0.77 (0.55-1.09)	0.71 (0.46-1.09)	0.27 (0.16-0.45)	<0.001
	Model 2[b]	1	0.77 (0.55-1.10)	0.70 (0.45-1.07)	0.27 (0.16-0.45)	<0.001
	Model 3[c]	1	0.85 (0.59-1.21)	0.76 (0.49-1.18)	0.31 (0.19-0.52)	<0.001
	Model 4[d]	1	0.81 (0.57-1.16)	0.76 (0.48-1.19)	0.30 (0.17-0.50)	<0.001
	Model 5[e]	1	0.83 (0.58-1.19)	0.78 (0.50-1.22)	0.31 (0.18-0.52)	<0.001
Women	Model 1[a]	1	0.88 (0.67-1.17)	0.67 (0.50-0.89)	0.26 (0.17-0.40)	<0.001
	Model 2[b]	1	0.89 (0.67-1.18)	0.68 (0.51-0.91)	0.27 (0.18-0.41)	<0.001
	Model 3[c]	1	0.93 (0.70-1.24)	0.70 (0.52-0.94)	0.30 (0.20-0.46)	<0.001
	Model 4[d]	1	0.94 (0.70-1.26)	0.70 (0.52-0.95)	0.31 (0.20-0.47)	<0.001
	Model 5[e]	1	0.95 (0.71-1.27)	0.70 (0.52-0.94)	0.32 (0.21-0.49)	<0.001

[a] Adjusted for age.

[b] Adjusted for age, body mass index (continuous).

[c] Adjusted for age, body mass index, urban/rural, household socioeconomic status, education, south/north.

[d] Additional adjusted for intake of fruits and vegetables, and intake of pork, beef, and lamb (continuous).

[e] Additional adjusted for intake of energy and iron.

Tofu and processed tofu foods are the most healthful foods and have been accepted by more and more people recently. The colorful and unique texture of tofu foods give the convenience and possibility to use with other foods. With the spreading of soybean plants around the world, as a main source of soybean foods, tofu would be discovered and spread widely. In developed countries, it will be a wonderful health food, and in developing countries, it will be a perfect protein and lipid resource.

References

Abd Karim, G.A. Sulebele, M.E. Azhar, C.Y. Ping (1999) 'Effect of carrageenan on yield and properties of tofu'. *Food Chemistry* 66 (1999) 159-165.

Bi Haiyan, Zhao Lihong, Liu Liping. (2007) 'The Effects of Calcium Content of Water on the Quality of Tofu'. *Cereals and Oils Processing* 12:113-115.

Cai T.D., Chang K.C., Shih M.C., Hou H.J., Ji M..(1997) 'Compaarison of bench and production scale methods for making soymilk and tofu from 13 soybean varieties'. *Food Research International*, 1997, 30(9):659-668.

Chang K.L.B., Lin Y.-S., Chen R.H. (2003). 'The effect of chitosan on the gel properties of tofu (soybean curd)'. *Journal of Food Engineering*, 57 (2003) 315–319.

Chen Changzhong, Wang Xiaobin, Chen Dafang, Li Guang, Alayne Ronnenberb, Hirokatsu Watanabe, Wang Xinru, Louise Ryan, David C. Christiani, Xu Xiping. (2001) 'Tofu Consumption and Blood Lead Levels in Young Chinese Adults'. *American Journal Epideniol*, 153(12):1206-1212.

Chen Hsiu Ling, Su Huei Jen, Lee Ching Chang (2007) 'Association between tofu intake and serum polychlorinated dibenzo-p-dioxins and dibenzofurans (PCDD/Fs) in the elderly Taiwanese'. *Environment International* 33 (2007): 265–271

Chen Xuezhen,Xie Hao,Jia Haorong,Li Shuchen. (2004) 'The Studies on Correlations of Soybean Quality Characters and Bbean Curd Yield'. *Chinese Agricultural Science Bulletin,* 20(1):138-141.

Hou, H. J., Chang, K. C. and Shih, M. C. (1997) 'Yield and textural properties of soft tofu as affected by coagulation method'. *Journal of Food Science*, 62: 824-827.

Lee, D. R., Choi, Y. H., Kim, M. K., & Yun, S. E. (1992). 'Influence of freezing upon the cooking time and eating quality of beans'. *Journal of Korean Agricultural and Chemical Society*, 35, 219–226.

Li Lite, Li Zaigui, Ying Lijun (2003), 'Processing and utilization of soybean'. Chemical Industry Press, Beijing.

Liu Zhi-Sheng, Chang Sam K.C., Li Li-Te, Eizo Tatsumi (2004) 'Effect of selective thermal denaturation of soybean proteins on soymilk viscosity and tofu's physical properties Food'. *Research International* 37 (2004) 815–822

Ma Guansheng, Cui Chaohui, Zhou Qin, Hu Xiaoqi, Li Yanping, Zai Fengying, Yang Xiaoguan. (2008), 'Consumption of soybean and soybean foods in China'. *Food and nutrition in China,* 1:40-43.

Nishinari K, Kohyama K, Zhang Y, Kitamura K, Sugimoto S, Saio K and Kawamura Y (1991) 'Rheological study on the effect of the A5 subunit on the gelation characteristics of soybean protein'. Agric Biol Chem 55:351-355.

Noh E.J., Park S.Y., Pak J.I., Hong S.T., Yun S.E. (2005) 'Coagulation of soymilk and quality of tofu as affected by freeze treatment of soybeans'. *Food Chemistry* 91 (2005) 715–721.

Poysa V., Woodrow L. (2002). 'Stability of soybean seed composition and its effect on soymilk and tofu yield and quality'. *Food Research International* 35 (2002) 337–345.

Rajni Mujoo, Dianne T. Trinh, Perry K.W. Ng. (2003) Characterization of storage proteins in different soybean varieties and their relationship to tofu yield and texture. *Food Chemistry*, 82 (2003) 265-273.

Shi Zumin, Hu Xiaoshu, Yuan Baojun, Pan Xiaoqun, Dai Yue, Gerd Holmboe-ottesen, Julee Byles (2008) 'Strong Negative Association between Intake of Tofu and Anemia among Chinese Adults in Jiangsu, China'. *Journal of the American Dietetic Association.* 7:1146-1154.

Tang Chuanhe (2007) 'Effect of thermal pretreatment of raw soymilk on the gel strength and micro structure of tofu induced by microbial transglutaminase'. *LWT-Food Science and Technology*, 40(8):1403-1409.

T.H. Kao, Y.F. Lu, H.C. Hsieh, B.H. Chen (2004) 'Stability of isoflavone glucosides during processing of soymilk and tofu'. *Food Research International* 37 (2004): 891–900.

Tseng Yen-Chang, Xiong Youling L. (2008). 'Effect of inulin on the rheological properties of silken tofu coagulated with glucono-δ-lactone'. *Journal of Food Engineering, online* (2008,07).

Wang Li-Jun, Li Dong, Eizo Tatsumi, Liu Zhi-Sheng, Chen Xiao Dong, Li Li-Te. (2007) 'Application of two-stage ohmic heating to tofu processing'. *Chemical Engineering and Processing,* 46 (2007) 486–490.

Wang CH, S Damodaran, (1991) 'Thermal Gelation of Globular Proteins: Influence of Protein Conformation on Gel Strength'. *Journal of Agricultural and Food Chemistry.* 39(3):433-438.

Watanabe Tokuzi. (1996) 'Basic Science of Tofu'. *Food Journal Co.Ltd.* Osaka, Japan. 1996.03:p18.

Wu YueFang. (2006) 'Analysis on the development of China soybean food enterprises from the situation of soybean processing in Japan'. *China Food Industry*, 4:38-39.

Yang Jian. (1994). Origin and development of Chinese tofu. *Agricultual History*, 1:217-227.

Young Soon Kim, Yang Mun Choi, Dong Ouk Noh, Seung Yong Cho, Hyung Joo Suh (2007) The effect of oyster shell powder on the extension of the shelf life of tofu. *Food Chemistry* 103 (2007) 155–160.

Zhang Mingjing, Wei Yimin. (2006) 'studies on correlation between soybean vaneties and texture properties of tofu-gel'. *Soybean Science,* 25(1):32-37.

Chapter 6

SUFU

1. INTRODUCTION

Sufu is a traditional fermented soybean food originating in China. It is a cheese-like product with a spreadable creamy consistency and a pronounced flavor. Sufu is a popular side dish consumed mainly with breakfast rice or steamed bread. It has a long history and written records date back to the Wei Dynasty (220~265 AD) (Han, Rombouts, & Nout, 2001). Sufu is made by fungal solid state fermentation of tofu (soybean curd) followed by aging in brine containing salt and alcohol.

Sufu is the original name for this product (Wang and Hesseltine, 1970). Because of the numerous dialects used in China and the difficulties of the phonetic translation from Chinese into English, sufu has been mentioned in the literature under many different names, such as tosufu, fu-ru, toe-fu-ru, tou-fu-ru, teou-fu-ru, fu-ju, fu-yu, foo-yue, and fu-i (Wang and Hesseltine, 1970; Lin et al., 1982). Sufu also is known as to-fu-zu in Mandarin, tau-zu or tao-hu-yi in Taiwanese, and tofuyo, nyu-fu or funyu in Japanese (Chou, 1998; Su, 1986), chao in Vietnam, ta-huri in the Phillippines, taokaoan in Indonesia and tao-hu-yi in Thailand (Beuchat, 1995). These names confuse Western people as well as the Chinese. Officially, sufu should be named Furu (or Tofuru) in Chinese (Han, Rombouts, & Nout, 2001).

Because of its characteristic salty flavor, sufu is consumed widely by Chinese as an appetizer (Chou and Hwan, 1994). It has a relatively high protein content, can be eaten directly as a relish, or cooked with vegetables or meats. Because sufu has the texture of soft cream cheese, it would be suitable for use in Western countries as a cracker spread or as an ingredient for dips and dressing (Wang and Hesseltine, 1979). In the Western world, sufu has been called either Chinese cheese or bean cake (Wang and Hesseltine, 1979). Otherwise, it can be used in the same manner as cheese (Fukushima, 1981).

Sufu is produced both commercially and domestically in China and other Eastern countries. Indigenous celebrated sufu brands include Wanzhihe furu in Beijing, Kedong furu in Heilongjiang, Shaoxing furu in Zhejiang, Guilin furu in Guangxi, et al. Several types of sufu are available commercially—Tsao sufu, red sufu, Kwantung sufu, rose sufu, and Yunnan sufu, and they are different in taste and flavor. In China, the annual production of Sufu is estimated at over 300,000 metric tons (Han, Rombouts, & Nout, 2001). It has become a giant commercialized industry. To improve the production and quality of sufu, it has been studied extensively to determine the ideal fermentation parameters, essential microorganisms involved in the fermentation process, biochemical and chemical changes that occur during fermentation, nutritive values of the product, organoleptic characteristics of different types of sufu and possible toxicological problems that may arise throughout the fermentation.

Sufu are produced in homes, villages, small cottage industries, and even larger commercial processing plants. There are many types of sufu produced by various processes in different localities in China, therefore, sufu includes a large number of products. The focus of this report is to examine the processing methods, organoleptic characteristics, microbiological aspects, nutritional quality, biochemical and chemical changes, and practical applications iinvolved in the manufacture of different sufu.

2. THE CLASSIFICATION OF SUFU

Sufu are generally classified into the following categories according to Han et al. (2001).

(1) According to the microorganisms involved in fermentation, sufu can be classified into four types, that is, mould-fermented tofu, naturally fermented tofu and bacteria-fermented tofu. The base for all form types is tofu, a curd from soybean milk by adding Calcium salts. Four steps are normally involved in making these types of sufu: (a) preparing tofu, (b) pehtze (pizi) fermentation with a pure culture microorganism fermentation or natural fermentation, (c) salting, (4) ripening.

Pure Actinomucor elegans or Mucor wetungkiao serves as suitable organisms for preparing mold-type sufu. Because it is more suitable to a higher temperature (37°C) than the other molds, Rhizopus species are usually used for sufu production in the South of China. Among these organisms, Actinonucor elegans is stated as being the one used most in the commercial production of sufu. Pehtze is prepared by spraying mold suspension on the tofu cube and the tofu cube was incubated at room temperature until white mycelia cover on the surface of the tofu cube (about 48~72 hours). The product can be kept a good shape.

For bacteria-type sufu, the tofu is usually presalted before preparing pehtzethe with bacterial fermentation. During the pre-salting, the tofu adsorbs the salt till the salt content of tofu reaches about 6.5%, which takes about 2 days. Pehtze is prepared by pure Bacillus spp. or Mirococcus spp. at 30~38°C for about 1 week. In order to keep the shape of the final product, pehtze is dried at 50~60°C for 12 h before salting. The ripening time normally takes less than 3 months. This sufu is made in some places, such as Kedong (Heilongjiang) and Wuhan (Hubei). Among all the sufu products, the flavor and mouthfeel of this sufu is the best.

From the view of processing technologies, a noticeable sufu is enzymatically ripened sufu, only three steps are normally involved in making this type of sufu. That is, preparing tofu, salting and ripening. Non-fermentation before ripening is the most prominent feature of this type of sufu. Some koji, red kojic rice, or rice wine is added in the dressing mixture for enzymatic ripening. Because of a lack of the enzymes involved in the ripening, the ripening takes a long time (about 6~10 months), and the product tastes coarse. This product of sufu is produced in only a few areas of China, such as Taiyuan (Shanxi) and Shaoxin Qifang (Zhejiang).

(2) According to the color and flavor, sufu can be classified into four types, that is, red sufu, white sufu, grey sufu and the other type. The categories are mainly based on the different ingredients of dressing mixtures in the ripening stage.

The dressing mixture of red sufu mainly consists of salt, angkak (red kojic rice), alcoholic beverage, sugar, flour (or soybean) paste and some spices. Angkak gives the red variety of sufu its color as shown in Figure 6-1. The outside of this type of sufu is red to purple, the interior is light yellow to orange. For its attractive color and strong flavor, the red sufu is the most popular product in all of China.

Angkak or red kojic rice has a specific aroma and purple red color and is used as a natural coloring agent in red sufu and some other traditional food. This colorant originated in China and is produced by fermenting rice with Monascus purpureus. A similar colorant, anka, is produced in the Philippines (Hesseltine and Wang, 1980).

White sufu is untreated or uncolored which has similar ingredients as red sufu in the dressing mixture but without angkak. It has an even light yellow color inside and outside. White sufu is popular in the south of China because it is less salty than red sufu.

The dressing mixture of grey sufu contains the soy whey left over from making tofu, salt and some spices. Grey sufu is ripened with a special dressing mixture, which could be dominated by both bacteria and mould enzymes and results in a product with a strong, offensive odor. The preparation of this type of sufu is a top secret in the industry and is slowly becoming a lost art (Wang and Fang, 1986).

The dressing mixture of the other type of sufu includes various ingredients, such as vegetables, rice, bacon, and an even higher concentration of alcohol. For instance, a sufu type, called Zui-Fang, is made by adding high levels of ethanol to the dressing, resulting in a marked alcoholic bouquet in this product. Therefore, Zui-Fang means drunk sufu.

Figure 6-1. Red sufu.

3. PROCESSING DEVELOPMENT IN SUFU MANUFACTURE

Although a broad range of sufu making processes exist in China, sufu making consists of four major processes, which are (1) preparation of tofu; (2) pehtze fermentation; (3) salting or brining; and (4) ripening. The flavor, aroma, and texture of sufu developed during the ripening process are essentially dependent on the surface of the tofu and on the ingredients added in the aging solution. Different ingredients can result in different colors and flavors. The schematic diagram for production of sufu is shown in Figure 6-2. The product of sufu was shown in Figure 6-3.

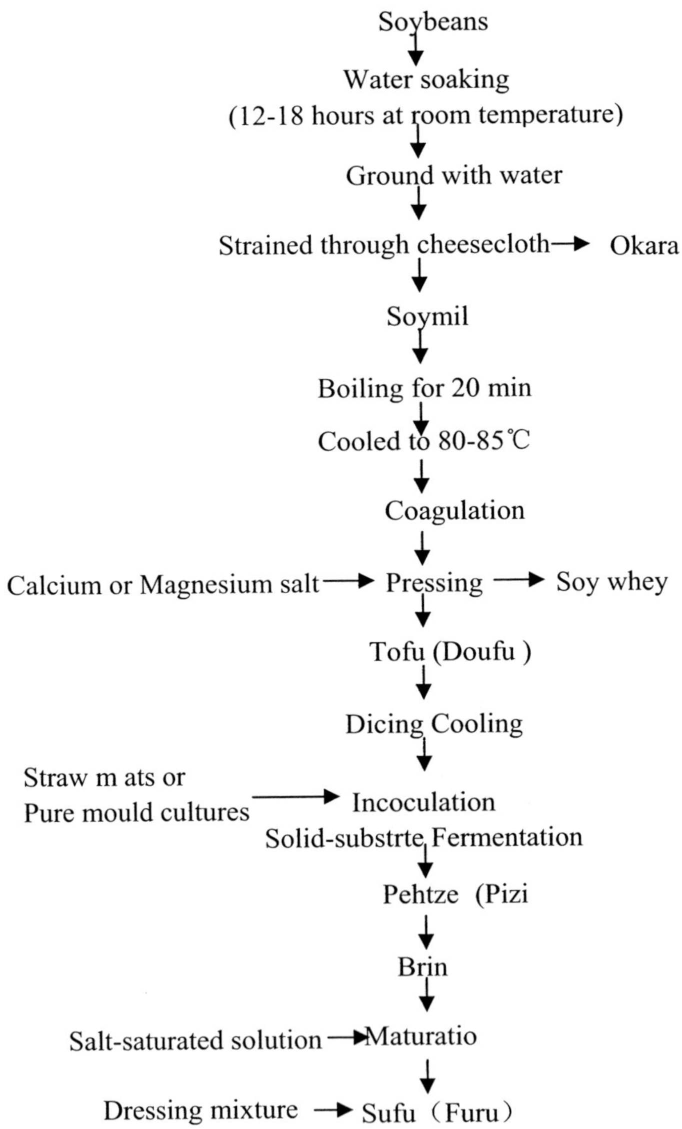

Figure 6-2. Flow diagram of sufu processing.

Figure 6-3. The production of sufu.

3.1. Cleaning

Cleaning the soybeans is the first step in preparation for the production of sufu. This step is carried out in order to remove dirt, stones, weed seeds, damaged and possibly decomposed beans, and any other foreign matter.

3.2. Soaking

Soaking is a process whereby soybeans are soaked in excess water overnight at room temperature in order to facilitate protein extraction. Soft water is beneficial for the extraction of soy protein. The ratio of water and soybeans is about 1:3.5. alkaline water was found to be more efficient than tap water and acidic water.

3.3. Preparation of Tofu

Hard tofu, with 79~87% moisture content, is generally used as the substrate to make sufu.

Traditionally, soybeans are washed and soaked overnight in water until the soaked soybeans are about two times higher than the soybeans before soaking in weight, and then ground in a stone mill into a slurry. The slurry is diluted and pressed to obtain soymilk. Coagulation is achieved by acid or by addition of salts, such as calcium sulphate and magnesium sulphate. The precipitate is pressed to remove excess water (soy whey) with cheesecloth bags using stones or wooden planks. Finally, a soft but firm cake-like tofu results, which can then be cut into cubes of desired sizes.

The production of tofu is highly mechanized nowadays. Preparation of tofu used for sufu mainly follows the processing technologies for commercial tofu except for slight differences in some steps. First, coagulation is achieved by addition of coagulants, such as calcium sulphate, magnesium sulplate, or sea salt by about 2.5-to-3.5% of dry weight of soybeans to the warm soymilk (80~85 °C). Generally, 20% (by weight) more coagulant is used to produce tofu for sufu preparation than for regular tofu. After the addition of coagulants, the mixture is agitated vigorously to facilitate mixing of the additives with protein. The agitated mixture is set aside for 10 min to complete the coagulation process. The precipitate is slowly pressed to remove excess water (soy whey) with cheesecloth bags using stones or wooden planks. Finally, soft but firm cake-like tofu results, which can then be cut into cubes of desired sizes (normally rectangular pieces, approximately 3.2*3.2*1.6cm). Typically, tofu used for sufu fermentation has about 83% moisture content, 10% protein, and 4% lipid (Wang and Hesseltine, 1970). Tofu used for sufu preparation generally is pressed harder than ordinary tofu (Su, 1986).

3.4. Preparation of Pehtze (Pizi)

Fermentation is used to make pehtze, which is freshly prepared soybean curd grown with mold but not yet processed and aged into the final sufu product (Li, 1991). The fungus responsible for sufu fermentation originated from rice straw. Actinomucor spp., Mucor spp., or Rhizopus spp. are normal contaminants in rice straw (Su, 1986). In the traditional method of preparing sufu, cubes of tofu are placed in wooden trays, the bottom of which is made of bamboo strips loosely woven together. The loaded trays are piled up and surrounded with straw for natural inoculation and fermentation. The temperature is 15~20 °C which is not favourable for bacteria, yeasts and other moulds except for Mucor spp. This step takes 5~15 days and varies depending on locality and season, till tofu cubes were overgrown by the mold mycelia. This method does not yield a high quality pehtze or sufu because of undesirable contaminating microorganisms. To avoid contamination and bacterial spoilage in traditional sufu preparation, tofu cubes are exposed to bright sunlight for several hours prior to inoculation with mold. Heat from the sun dehydrates the surface of the tofu cubes, making them less susceptible to bacterial spoilage (Su, 1986).

The traditional method of preparing sufu at home is to cut tofu into small pieces and then put them into boiling water for about one min. the tofu is placed in a bamboo tray to allow water to drain. After the tofu cools to room temperature, the tray is set in the open air for two days to allow natural mold growth. After tofu becomes pehtze, it is put in a jar, and liquor is added for the aging process. The pehtze must have white or light yellow-white mycelium to ensure that the final sufu has an attractive appearance. Before the pehtze is moved to the salting treatment, the mycelial mat of mould should be flattened by hand so that a firm film will be formed over the surface of the sufu to keep its shape.

In order to prevent the growth of contaminating bacteria, Wai (1968) suggested soaking the tofu cubes (2.5*3*3 cm) in a solution containing 6% NaCl and 2.5% citric acid for 1 h, followed by 15 min hot air treatment at 100 °C. This acidic saline solution will prevent the growth of putrefactive bacteria, while allowing mold growth. Lactic acid can be used instead of citric acid. The treated tofu cubes are mounted on sticks, separated from each other, and placed in a tray with pinholes in the bottom and top to aid air circulation, because mycelia must develop on all sides of the cubes. After cooling, the cubes are inoculated over their surface by rubbing with pure culture of an appropriate fungus grown on filter paper impregnated with a culture solution or by spraying the pure culture spore solution on each side. After inoculation, the cubes are incubated at 20 °C or lower for 3~7 days. Different molds may require different time and temperature for sufu fermentation (Fukushima, 1985). Rhizopus chinensis var. chungyuen grows well in 7 days at 12 °C, whereas Mucor hiemalis and M. silvaticus grow better in 3 days at 20 °C (Su, 1986). Lin et al. (1982) studied the growth of M. hiemalis, M. silvaticus, M.praini, and R. chinensis during sufu fermentation and reported that incubation time and temperatures varied from 3 to 7 days and 12~25 °C, respectively. At the end of the proper incubation period, the cubes are covered with a luxurious growth of white mycelium and have no disagreeable odor. The mold cube usually contains 74% water, 12.2% protein, and 4.3% lipid (Wang and Hesseltine, 1979).

Nowadays, pehtze, fresh bean curd overgrown with mycelium of moulds is produced by means of solid substrate fermentation after inoculation with pure culture moulds. The fungal genera involved (Actinomucor, Mucor and Rhizopus) all belong to the Mucoraceae. The mold used in sufu fermentation must have high proteolytic and lipolytic activities to hydrolyze protein and lipid in tofu and develop the desired flavor, texture, and consistency. The white or yellow, dense, and tenacious mycelium will form a strong film on the surface of the pehtze to protect the shape of the finished sufu from distortion. The mold growth does not produce any disagreeable odor, astringent taste, or mycotoxins (Hesseltine and Wang, 1980; Su, 1986).

According to these criteria, some Mucor spp., Actinomucor spp., and Rhizopus spp. could be used for making high quality sufu. Among them, Actinomucor elegans and Actinomucor Taiwanensis, Mucor sufu, and Mucor wutungkiao have been mentioned as popular starter cultures. The spore suspension (~10^5CFU/ml) is harvested and inoculated on the surfaces of the tofu with manually operated sprayers, comparable to those used for spraying plants. The inoculated tofu is placed, evenly spaced in wooden or plastic trays, the bottoms of which are made of bamboo or wooden strips. The loaded trays are piled up in an incubation room, where a controlled temperature (about 25 °C), a relative humidity (88~97%) and good aeration are needed for optimum growth of the mycelia. The thin white mycelia are developed in 8~12 h and a thick mycelial mat is formed after 36~40 h of incubation. Then the room temperature is decreased by aeration to prevent over-growth of mould, until a slightly yellowish white color appears, at which point formation of fresh pehtze is complete. The total cultivation time is about 48 h, which is much less than in the traditional way (5~15 days) (Han et al., 2001).

Chou et al. (1998) reported that optimum conditions for growth of A. taiwanensis were 25~30 °C at 97% relative humidity when tofu of 65% moisture content was inoculated. Under these conditions, a maximum production of protease, lipase, α-amylase and α-galactosidase was achieved.

Before pehtze is transferred to the salt treatment, the mould mycelial mat should be flattened by hand, in the same as was done in the traditional way.

3.5. Salting

In the conventional method, the pretreated pehtze is transferred into a big earthen jar and salt is spread between layers of pehtze as they pile up in the jar. During this period, the pehtze adsorbs the salt until salt content of pehtze reaches about 16%, which takes 6~12 day. The salted pehtze is removed from the jar, washed with water, and then transferred to another jar for further processing.

Freshly prepared pehtze has a bland taste. The flavour and aroma of sufu develop during the salting and ripening process. The added salt imparts a salty taste to the product and also retards the growth of mold and contaminating microorganisms (Lu, 1997). More importantly, the salt releases the mycelia-bound proteases (Chou et al., 1993). During fermentation, mold growth is limited to the surface of the cubes, and the mycelium does not penetrate into the tofu cubes. The enzymes produced by the mold, on the other hand, are not extracellular. They are bound loosely to the mycelium, possibly by ionic linkage. Salt enhances the release of these enzymes to hydrolyze the protein in tofu. Wai (1968) also demonstrated that enzyme digestion occurred mostly during the first 10 days of aging (Wang and Hesseltine, 1970).

Freshly prepared pehtze can be salted in a number of ways. Firstly, pehtze can be salted by sprinkling with a layer of salt in containers in accordance with a traditional method. This method takes longer and makes the pieces of pehtze varying widely in salt concentration. Pehtze can also be soaked in a saturated salt solution. After 4~5 days at room temperature, the salt content of the pehtze can reach over 12% and the moisture content decreased by 10~15%. Final moisture levels may vary in the range 50~65%. In one conventional method, pehtze is immersed in a solution containing 12% NaCl and 10% alcohol (distilled liquor or rice wine is used). The pehtze immersed in alcoholic saline solution can be sold without aging (Su, 1986).

3.6. Ripening

The differences between the various types of sufu are mainly caused during the ripening process since different dressing mixtures are added in salted pehtze. The ingredients of dressing mixture vary with social customs, climate, and location and so on. The most common dressing mixture used consists of angkak 2%, alcoholic beverage 8~12%, salt (final salt content 10~12%), sugar 5~10%, flour (or bean) paste 3~5%, and spices. *Monascus purpureus,* after growing on soaked and polished rice, produces a deep purplish-red compound called ang-kak. Following drying, the ang-kak is ground into a powder and used to color food red (Whitaker, 1978; Hesseltine and Wang, 1979). Ang-kak is added to the brine to give sufu a red color. Hot pepper added to the brine would make hot sufu. Additional essence can be added into the dressing mixture to supply a special flavour (Han et al., 2001). For example, rose sufu can be made by aging in brine containing rose essence (Wang and Hesseltine, 1970). Therefore, the taste and aroma of sufu, in addition to its own characteristically mild ones, can be enhanced or modified easily by the ingredients in the brine solution (Hesseltine and Wang, 1978).

For the ripening, alternate layers of pehtze and dressing mixture are packed into jars, and the ratio is about 2:1 between salted pehtze and dressing mixture in the conventional method. The mouth of the jar is wrapped with sheath leaves of bamboo and sealed with clay. The sealed jars are aged for 6 months for further maturation (Han et al., 2001), and then the product is ready for market. The exact composition of the brine and length of aging vary among manufactures.

Ripening requires much time and space. Although nowadays the ripening time is shorter than the 6 months that the traditional process took, modern processes still take about 2~3 months. Reduction of ripening times can be achieved by using smaller cubes of tofu, lowering the salt content from ~14% to ~10%, lowering alcohol content from ~10% to ~6%, keeping the ripening temperature at a higher and more constant level, and using smaller jars. The high concentration of salt is considered not only to retard the hydrolysis of protein and lipid during ripening, but also to result in health problems in consumers for its saltiness. However, high salt content could prolong the shelf life of the product. A coating of whole blocks of pehtze with paraffin (m.p. 60°C) seems to be a satisfactory solution to get out of the above dilemma (Wai, 1964). The pehtze was mixed with salt (7% of pehtze weight) and then coated with a layer of melted paraffin. The pehtze coated with solidified paraffin could be stored in a glass container for one month at room temperature. Addition of stem bromelain to soymilk as a coagulant to prepare tofu could accelerate the ripening and enhance the flavour of sufu (Han et al., 2001).

The added alcohol is believed to give a pleasant odor to the product and prevent the growth of contaminating organisms (Chou and Hwan, 1994; Shaw and Chou, 1990). The soybean lipids are degraded to some extent to fatty acids. The added alcohol reacts with the fatty acids chemically or enzymatically to form esters, providing the pleasant odor of the product (Wang and Hesseltine, 1970). However, alcohol has been found to exert an inhibitory effect on the activity of a protease produced by *A. taiwanesis* (Shaw and Chou, 1990). Chou and Hwan (1994) confirmed that the addition of alcohol to the brine solution retards the hydrolysis of protein during the aging period.

Table 6-1. Amino acids content of sufu samples

Amino acid	Red sufu[a] (g/100g sufu)	Grey sufu[a] (g/100g sufu)	Sufu[b] (g/100g sufu)	White sufu[c] (Molar ratio %)
Alanine	0.32	0.70	10.0	7.0
Arginine	0.38	0.27	2.1	2.5
Aspartic acid	1.00	0.66	5.1	13.7
Cystine	0.59	0.20	0.4	
Glutamic acid	2.15	2.08	0.6	22.0
Glycine	0.54	0.42	4.4	7.0
Histidine	0.20	0.18	1.4	1.9
Isoleucine	0.88	0.58	4.8	4.5
Leucine	0.81	0.95	8.8	7.6
Lysine	0.59	0.29	7.0	7.3
Methionine	0.51	0.14	0.7	
Phenylalanine	0.59	0.59	4.6	2.6
Proline	0.38	0.29	2.4	7.7
Serine	0.34	0.27	2.3	5.2
Threonine	0.45	0.23	2.0	4.1
Tryptophan	0.09	0.05	0.6	
Tyrosine	0.54	0.25	2.2	1.0
Valine	0.16	0.58	5.3	5.2

[a] Wang (1995) and Wang and Du (1998); [b] Su (1986): commercial sample non-specified. [c] Liu and Chou (1994).

The flavour and aroma of sufu develop during the ripening step. During this period, the enzymes produced by the mould act upon their respective substrates, and it is likely that hydrolysis of protein and lipid provide the principal compounds of the mild, characteristic flavour of sufu. The pleasant and palatable taste is considered to be related to the content of free amino acids, mainly glutamic acid, in the oriental food (Chou et al., 1993). Glutamic acid, aspartic acid, leucine/isoleucine and alanine are predominant free amino acids in sufu (Table 6-1). Fatty acids derived from soybean lipids react with the added alcohol chemically or enzymatically to form esters, also providing the pleasant odor of the product (Chou and Hwan, 1994).

4. Enzymes Produced during Fermentation

During the pehtze preparation, the mycelium of the starter culture will grow and finally cover the entire surface of tofu cubes (Wang et al., 1974). Lin et al. (1982) indicated that after 48 h of incubation, the mycelium of *Mucor* sp. was densely spread over the tofu surface and grew outward for about 2.5 to 3.0 cm, but little mycelium could grow inward from the surface. Scanning electron microscopic observation showed that the mycelium penetrated only 0.18 cm deep.

As the mycelium grow, various proteolytic enzymes, such as pepsin- and trypsin-like enzymes, are produced by the starter culture on tofu (Liu, 1932; Lu et al., 1995; Zhou et al., 1990). Liu et al. (1965) and Wang (1967b) described proteases produced by *M. sufu* and *M. hienalis* NRRI 3103, respectively. Lipase, phosphatase, amylase, α-galactosidase, glutaminase, invertase, trypsin-like proteases, pepsin-like protease, oxidase, and catalase also have been found to be produced by the sufu starter culture (Chou et al., 1988; Chou et al., 1994; Hesseltine and Wang, 1967; Liu and Chou, 1992; Su, 1986).

Liu and Chou (1992, 1994) observed that when *A. taiwanensis* and *A. elegans* were used to prepare pehtze, lipase, and amylase increased with time during the incubation period of 48 h. The most marked increase of enzyme activity was noted after 24 h of cultivation. Furthermore, cultivation temperature and humidity greatly affected enzyme production by *A. taiwanensis* on tofu. Chou et al. (1988) reported the highest yields of protease (112 U/g of dry tofu) and lipase (1448 U/g of dry tofu) after 60 h of incubation at 97% humidity and 25 °C. On the other hand, the highest yield of α-amylase (1949 U/g of dry tofu) was observed after 48 h of incubation at 96-to-97% humidity and 30 °C, and the highest amount of α-galactosidase (387 U/g of dry tofu) was observed at 35 °C and 96% humidity after 60 h of growth.

5. The Characteristics of Sufu

5.1. The Physical Properties of Sufu

The overall quality of sufu is determined in part by physical properties such as color, taste, and texture. Sufu, as seen in the market, usually is in the form of red, pale yellow, or white blocks. Pale yellow or white sufu is untreated, whereas red sufu is colored with ang-kak. According to Zhang (1997), good sufu should be fresh, soft, and light yellow in color and have a special flavor. For red sufu, the outside should be red but the inside yellow, with a special flavor and soft texture (Fu, 1997). The flavor, aroma, and texture of sufu developed during the ripening process are essentially dependent on the enzymes produced by mold grown on the surface of tofu. Except for differences in taste and flavor, the types of sufu are generally similar in composition (Chou et al., 1988). Also, the enzyme production by mold on tofu is affected greatly by incubation temperature, humidity, and cultivation time.

5.2. The Chemical Composition and Nutritional Quality of Sufu

The chemical composition of sufu affects its nutritional quality. Sufu is considered a high quality protein food. Except for water, protein is the major component in sufu. Tofu contains 50-to-55% protein and 30% lipid on a dry mass basis (Wang and Hesseltine, 1979). Pehtze is bland in taste, and the characteristic flavor, aroma, and texture of sufu develop during the aging period. These changes can be attributed to the action of the hydrolytic enzymes produced by the starter culture during pehtze preparation. Lin et al. (1982) indicated that hydrolysis occurred on tofu during the aging period, and Wang (1967a) reported that it appeared only when the membrane-bound enzymes were released from mycelium after the salting treatment. Wai (1968) reported that soybean proteins were digested into peptides and amino acids by mold protease. Contents of amino nitrogen and normal nitrogen and the dissolution ratio of sufu all increased during fermentation (Chou and Hwan, 1994; Lin, 1982). Liu and Chou (1992) found that the contents of total nitrogen and amino nitrogen increased in the brine infusion as the aging period extended. After 30 days of aging at room temperature, the total soluble nitrogen increased from 1.00~2.74% and total insoluble nitrogen decreased from 7.89~6.05%, so overall change was small; free fat acids increased from 12.8~37.1%, and total lipids remained unchanged (Wang and Hesseltine, 1979).

The amino acid content of sufu is presented in Table 1. Glutamic acid and aspartic acid were the most abundant amino acids found in red sufu and grey sufu, which are about 30% of total amino acids and are related with the delicious taste of sufu. The cystine and methionine may be lower in grey sufu than in red sufu because of their degradation or conversion to other sulfur compounds during maturation, which may contribute to the offensive odor of grey sufu (Han et al., 2001).

Yen (1986) reported that the average amino contents in 15 samples of commercial sufu from Taiwan, China were: cadaverine (0.039 mg/g), histamine (0.088 mg/g), beta-phenylethylamine (0.063 mg/g), putrecine (0.473 mg/g), tryptamine (0.150 mg/g), and tyramine (0.485 mg/g). Tyramine and putrescine were the major amines found, and these might have a potential harmful effect on human beings if levels are very high.

Lipid is the second major component of sufu. Chou and Hwan (1994) observed that lipid content fluctuated during the aging process of sufu prepared with either *A. taiwanensis* or *A. elegans*. The free fatty acid content increased then decreased during the aging period. Regardless of the starter organism used for sufu, linoleic acid (18:2) was the highest followed by oleic acid (18:1), palmitic acid (16:0), linoleic acid (18:3), and stearic acid (18:0).

The complex flavour of sufu was reported to contain 22 esters, 18 alcohols, 7 ketones, 3 aldehydes, 2 pyrazines, 2 phenols and other volatile compounds by Hwan and Chou (1999). Maturation in the presence of ethanol resulted in higher levels of volatiles. Ho et al. (1989) compared the volatile flavor compounds of red sufu and white sufu. Red sufu contains much larger amounts of alcohols, esters and acids, which may be due to the fermentation of angkak by Monascus spp. The esters give red sufu its characteristic fruity aroma. White sufu contains a large quantity of anethol, which seems to be the major contributor of its flavour. The volative compounds detected in red/white type of sufu are shown in the Table 6-2.

Table 6-2. Volatile compounds detected in red/white types of sufu

Alcohols	Ethanol, 2-butanol, Propanol, 2-Methylpropanol, Butanol, 3-Methylbutanol, Hexanol, 3-Octanol, 2-Ethylhexanol, Benzyl alcohol, Phenylethyl alcohol
Esters	Ethyl butyrate, Ethyl 2-methylbutyrate, Ethyl hexanoate, Ethyl heptanoate, Ethyl octanoate, Exthyl benzoate, Ethyl dodecanoate, Phenylethyl propanoate, Ethyl tetradecanoate, Ethkyl palmitate, Ethyl stearate, Ethyl oleate, Ethyl linoleate
Miscellaneous	Acetic acid, Phenol, 2-Nonanone, 2,6-Dimethylpyrazine, 2-Ethyl-5-methylpyrazine

Source: Ho et al. (1989) and Hwan and Chou (1999).

The chemical compositions of fresh and dried tofu, pehtze, and sufu are presented in Table 6-3 (Su, 1986). Fresh tofu contains more moisture than pehtze and sufu. Sufu has a high-fat content compared with tofu and pehtze. The changes in nitrogenous compounds of tofu, pehtze, and sufu are presented in Table 6-4 (Su, 1986). Table 3 (Su, 1986) shows the amino acid profile of sufu. The proteases from the start culture can hydrolyze the soybean proteins into peptides and amino acids (Ferng and Chiou, 1993).

Table 6-3. Compositions of tofu, pehtze and sufu

Component (%)	Tofu		Pehtze		Sufu	
	Fresh	Dried	Fresh	Dried	Fresh	Dried
Moisture	75.8	—	70.0	—	59.7	—
Protein	16.0	66.0	17.9	59.7	15.9	39.4
Fat	7.2	29.7	9.8	32.8	20.3	50.4
Carbohydrate	0.1	0.4	0.5	1.7	0.0	0.0
Fiber	0.0	0.0	0.4	1.3	1.1	3.7
Ash	0.9	3.9	1.4	4.5	3.0	7.4

Source: Su (1986).

Table 6-4. Changes in Nitrogenous compounds of tofu, pehtze, and sufu

Nitrogen compound (%)	Tofu	Pehtze	Sufu	Aging solution
Protein nitrogen	99.1[a]	64.0	83.5	5.7
Normal nitrogen	1.4	18.8	17.8	54.5
Ammonia nitrogen	0.04	7.4	0.8	10.4

[a]Percent of total nitrogen.

Source: Su (1986).

Table 6-5. Some physiological substances in sufu

Component	Source	Function	Reference
Tryp-Leu	Japanese sufu	ACE inhibitory activity	Kuba et al., 2003
Sufu water extract	Red sufu	Anti-α-glucosidase activity	Chen, 2006
Sufu water extract	Sufu fermented with Actinomucor elegans 3.118	AChE inhibition ability	Chen, 2006
SOD	White sufu	Radical scavenging activity	Rao et al.,1996
Isoflavone	Red sufu	Antioixdant activity	Zhang et al., 2006

5.3. Physiological Function of Sufu

Recently, much attention has been paid to the physiological function in foods. The angiotensin I-converting enzyme (ACE) is a dipeptidyl carboxy peptidase associated with the regulation of blood pressure. It converts angiotensin I to the potent presser peptide, angiotensin II, and also degrades depressor peptide bradykinin. ACE inhibitors from various foods have been recently studied in terms of their ability to prevent alleviate hypertension. ACE inhibitory activity was observed in a sufu extract. Some of them were isolated to homogeneity from the extract, and one was identified to be Tryp-Leu (IC50 value, 29.9 μM). The inhibitory activity of the peptide was completely preserved after a treatment with pepsin, chymotrypsin or trypsin (Kuba et al., 2003). Wang et al. (2003) compared the ACE inhibitory activity of 15 sufu, and found that the ACE inhibitory activity of these sufu were ranged from 0.71~1.94 mg/mL.

Oxidative injury to the living body by reactive oxygen or free radicals has been shown to play a role in many lifestyle-related diseases (Osawa et al. 1995; Wanasundara et al. 1997; Niki 1998). Several papers (Zhang et al., 2006; Wang et al., 2003) reported the antioxidant activity of sufu. Quan et al. (2006) found that sufu could scavenge 1,1-diphenyl-2-picrydrazyl (DPPH) radical and sufu obtained by Actinomucor elegans had higher antioxidant activity than that obtained by Rhizopus arrhizus. Isoflavones was thought to be the major functional material in sufu (Zhang et al., 2006). Table 6-5 shows some physiological components in sufu.

6. Microbiological Aspects of Sufu

During processing of tofu, which includes boiling the soymilk, the initial vegetative microflora is effectively eliminated (Kovats et al., 1984). However, the postboiling pressing of the curd to form cakes and the handling of the cakes before packing allow possible microbial contamination. Szabo et al. (1989) analyzed 346 samples of tofu and found *Staphylococcus aureus*, psychrotrophs, coliforms, and *Yersina enterocolitica.* Tuitemwong and Fung (1991) studied bacterial populations of commercial tofu and found that the most common Gram-positive organisms were *Streptococcus* sp., *Pediococcus* sp., and *Lactobacillius* sp., and the most common Gram-negative bacteria were *Pseudomonas putida*, *P. aeruginosa*, *Enterobacter agglomerans*, and *E. cloacae.* Other studies also obtained similar results (Rehberger et al., 1984). Two food poisoning outbreaks related to contaminated tofu were caused by *Yersinia enterocolitica* and *Shigella sonnei* (Jackson, 1990).

Much fewer data on microbial flora of sufu have been reported. Pao (1995) studied the relationship between brine composition and microbiological quality of commercial sufu and found high levels (>5 log cfu/g) of nonhalophiles and moderate levels (3~4 log cfu/g) of halophiles in 60% of all brands tested. The predominate halophile was *Pediococcus halophilus*, which helped the overall sensory development of sufu at early stages of the fermentation. This result indicates that the presence of *P. halophilus* in sufu fermentation appears to be desirable. He also found that the brine composition, especially ethanol percentage, influenced the growth or survival of microorganisms in sufu. Tang (1977) conducted an extensive survey on the toxicity and safety of fermented foods. She collected 80 samples of sufu from the markets throughout Taiwan, and only a small percentage (1.3%) was found to have bacteria, but they were not pathogenic species.

Shi and Fung (2000) reported that sufu fermentation and aging can control common foodborne pathogens, such as *Escherichia coli* O157:H7, *Salmonella typhimurium*, *Staphylococcus aureus*, and *Listeria monocytogenes*, so sufu is a safe product even though its preparation does not include pasteurization. Before fermentation, pathogens were inoculated onto tofu (substrate for sufu) at 5 log cfu/g or 3 log cfu/g, and starter culture (*Actinomucor elegans*) was inoculated at 3 log cfu/g. After 2 days of fermentation at 30 °C, the four pathogens reached 7 to 9 log cfu/g, and the mold count reached 6~7 log cfu/g. After fermentation, sufu samples were aged in a solution of 10% alcohol + 12% NaCl. After 1 month of aging, the total bacterial count was 6~7 log cfu/g, but all foodborne pathogens and mold were reduced to nondetectable levels. The total bacterial count decreased after aging for 2 months and 3 months, but the differences were not significant ($P > 0.05$) compared with the count after 1 month. Microorganism in experimental sufu from different aging periods and in commercial sufu were compared. A total of 270 isolates were purified and identified by the BBL Crystal Identification System. From the experimental sufu samples, 49 *Bacillus* spp. (20.4%), 167 *Enterococcus* spp. (69.6%), 6 *Shewanella putrefaciens* (2.4%), and 18 miscellaneous Gram-negative bacilli (7.5%) were identified. From commercial sufu samples, 17 *Bacillus* spp. (56.7%), 2 *Enterococcus durans* (6.7%), 5 miscellaneous Gramnegative bacilli (16.7%), 5 *Corynbacterium aquaticum* (16.7%), and 1 *Shewanella putrefaciens* (3.3%) were obtained. Although the longer aging period did not significantly decrease the total bacterial count, it may help in the development of sufu flavor.

REFERENCES

Beuthat, L. R. (1995). Indigenous fermented foods. In: Reed, G., Nagodawithana, T. W. (Eds), *Enzymes, Biomass, Food and Feed. Biotechnology,* vol. 9. VCH press, Weinheim, pp. 523-525.

Chou, C. C. (1998). 'Innovation and current issues in sufu production'. *IFT Annual Meeting.* Technical Program Abstracts 43–2.

Chou, C.C., Ho, F. M., and Tsai, C. S. (1988). Effects of temperature and relative humidity on the growth of and enzyme production by *Actinomucor taiwanensis* during sufu pehtze preparation. *Appl. Environ. Microbiol.* 54: 688–692.

Chou, C. C. and Hwan, C. H. (1994). Effect of ethanol on the hydrolysis of protein and lipid during the ageing of a Chinese fermentation soy bean curd-sufu. *J. Sci. Food Agric.* 66: 393–398.

Chou, C. C., Yu, R. C., and Tsai, C. T. (1993). Production of glutaminase by *Actinomucor elegans*, *Actinomucor taiwanensis* and *Aspergillus oryzae*. *J. Chinese Agric Chem. Soc.* 31: 78–86.

Chou, C. C. (1998). 'Innovation and current issues in sufu production. IFT Annual Meeting'. Technical Program Abstracts 43–2.

Chou, C. C. and Hwan, C. H. (1994). 'Effect of ethanol on the hydrolysis of protein and lipid during the ageing of a Chinese fermentation soy bean curd-sufu'. *J. Sci. Food Agric.* 66: 393–398Han, B. Z., Rombouts, F. M., & Nout M. J. R. (2001). 'A Chinese fermented soybean food'. *International Journal of Food Microbiology*, 65:1-10.

Fu, J. Q. (1997). Red sufu-the new technique. Agric New Tech. 1997(1) 39–40.

Fukushima, D. (1979). Fermented vegetable (soybean) protein and related foods of Japan and China. *J. Am. Oil Chem. Soc.* 56: 357–362.

Fukushima, D. (1981). Soy proteins for foods centering around soy sauce and tofu. *JAOCS* 3: 346–354.

Fukushima, D. (1985). Fermented vegetable protein and related foods of Japan and China. *Food Rev. Int.* 1: 149–209.

Ferng, S. and Chiou, R. Y-Y. (1993). Characterization of white particulates in brine of indigenous fermented foods. *J. Food Protect.* 56: 243–245.

Fung, D. Y. C., Phebus, R. K., Kang, D. H., and Kastner, C. L. (1995). Effect of alcohol-flaming on meat cutting knives. *J. Rapid Meth. Automat. Microbiol.* 3: 237–243.

Hesseltine, C. W. (1965). A millennium of fungi, food, and fermentation. *Mycologia* 57: 163–168.

Hesseltine, C. W. and Wang, H. L. (1967). Traditional fermented foods. *Biotechnol. Bioengin.* 9: 275–288.

Hesseltine, C. W. and Wang, H. L. (1978). Sufu. In: A. K. Smith, and S. J. Circle, eds. *Soybeans: Chemistry and Technology.* AVI Press. Westport, CT. pp.410–413

Hesseltine, C. W. and Wang, H. L. (1979). Fermented foods. *Chem. and Ind.* 16: 393–399.

Hesseltine, C. W. and Wang, H. L. (1980). The importance of traditional fermented foods. *BioScience* 30: 402–404.

Hwan, C. H. and Chou, C. C. (1994). Content of nucleotides, organic acids and sugars as well as some physical properties of sufu prepared with different starter. *Food Sci.* 21: 124–133.

Jackson, G. J. (1990). Public health and research perspectives on the microbial contamination of foods. *J. Anim. Sci.* 68: 884–891.

Kovats, S. K., Doyle, M. P., and Tanaka, N. (1984). Evaluation of the microbiological safety of tofu. *J. Food Protect.* 47: 618–622.

Kuba, M., Tanaka, K., Tawata, S., Takeda, Y., and Yasuda, M., (2003) 'Angiotensin I-converting enzyme inhibitory peptides isolated from tofuyo fermented soybean food'. *Biosci. Biotechnol. Biochem.*, 67:1278-1283.

Lin, L. P.; Chen, S. H. and Chan, H. C. (1982). 'Relationships of mold growth, chemical components and microstructure during manufacture of sufu'. *Research Report, College of Agric.,* Natl. Taiwan Univ. 22: 76–91.

Li, Y. M. (1991). Inoculated different species of microorganisms for sufu making. *Chinese Fermentation* 1991(1) 28–32, 42.

Liu, P. W. and Chen, S. H. (1965). Study on the isolation, purification and properties of the proteolytic enzyme (protease) produced by Mucor sufu. I. The isolation and purification of protease from *Mucor sufu* by use of repeating ammonium sufuate salting out and dialysis. *J. Chinese Agric. Chem. Soc.* 3: 110–116.

Liu, Y. H. and Chou, C. C. (1992). Change of enzymeactivity in pehtze and some chemical components in aging infusion during the fermentation of sufu with *Actinomucor* species. *Food Sci.* 19: 57–67.

Liu, Y. H. and Chou, C. C. (1994). Contents of various types of protein and water soluble peptides in sufu during aging and the amino acid composition of tasty oligopeptides. *J. Chinese Agric. Chem. Soc.* 32: 276–283.

Lu, J. M., Yu, R. C., and Chou, C. C. (1995). Growth and production of glutaminase by the submerged liquid culture of *Actinomucor taiwanensis. J. Chinese Agric. Chem. Soc.* 33: 404–411.

Lu, Y. F. (1997). The questions about sufu making. *Chinese Fermentation* 1997(3) 28–32, 41.

Niki, E. (1998). Molecular mechanism of free radical-induced injuries. In *Recent Development of Food Factors for Disease Prvention;* Niki, E., Yoshikawa, T., Osawa, T., Eds.; CMC Books: Tokyo, pp 1-6.

Osawa, T.; Yoshida, A.; Kawwakishi, S.; Yamashita, K.; Ochi, H. (1995). Protective role of dietary antioxidants in oxidative stress. In *Oxidative Stress and Aging;* Cutler, R.G., Packer, L., Bertram, J., Mori, A. Eds.; Birkhauser Verlag: Basl, Switzerland, pp 367-377.

Pao, S. H. (1995). Halophillic organisms in sufu, Chinese cheese. P*h*D. dissectation, The Ohio State University. *Diss. Abstr. Int.* 55: 4190– B.

Rao, P. F., Li, J. C., Wei, Z. (1996). Study on substances with SOD-like activity in four traditional Chinese foods. Journal of the Chinese Cereals and Oils Association. 1996, 11(2), 18-22.

Quan MH, Li LT, Cheng YQ, Deng CP. (2006) 'Changes in antioxidant activity of sufu during fermentation'. *Chinese Condiment,* 10:22-26.

Rehberger, T. G., Wilson, L. A., and Glatz, B. A. (1984). Microbiological quality of commercial tofu. *J. Food Protect.* 47: 177–181.Schmidt, R. H. and Morris, H. A. (1984). Gelation properties of milk proteins, soy proteins, and blended protein systems. *Food Tech.* 5: 85–96.

Shaw, J. L. and Chou, C. C. (1990). Effect of ethanol on the hydrolases activity of *Actinomucor taiwanensis* and enzymatic hydrolysis of tofu substrate. *Food Sci.* 17: 160–166.

Su, Y. C. (1986). Sufu. In: N. R. Reddy, M. D. Pierson and D. K. Salunkhe, eds. *Legume Based Fermented Food.* CRC Press, pp, 69–83

Szabo, R. A. and Jarvis G. A. et al. (1989). Microbiological quality of tofu and related products in Canada. *J. Food Protect.* 52: 727–730.

Tang, Y. C. (1977). Dietary aflatoxins and liver cancer. *Proc. of Workshop on Food Safety, Food Industry Research and Development Institute,* Hsinchu, Taiwan, Oct. 17 to 19, p. 193

Troller, J. A. (1986). Water relation of foodborne bacterial pathogens-an updated review. *J. Food Protect.* 49: 656–670.

Tsai, S. J., Lan, C. Y., Kao, C. S., and Chen, S. C. (1981). Studies on the yield and quality characteristics of tofu. *J. Food Sci.* 46: 1734–1777.

Tuitemwong, K. and Fung, D. Y. C. (1991). Microbiological study of tofu. *J. Food Protect.* 54: 212–216.

Wanasundara, P.K.J.P.D.; Shukla, V.K.S.; Shahidi, F. (1997). 'Endogenous Antioxidants from Oilseeds and Edible Oils'. *Food Rev. Int.* 13 (2):225-292.

Wang, H. L. and Hesseltine, C. W. (1970). 'Sufu and lao-chao'. *J. Agric. Food Chem.* 18: 572–575.

Wang, H. L. and Hesseltine, C. W. (1979). Moldmodified foods. In: H. J. Peppler and D. Perlman, eds. Microbial Technology, Fermentation Technology. Academic Press, New York. pp. 95–129.

Wang, H. L., Fang, S. F. (1986). History of Chinese fermented foods. In: Hesseltine, C. W., Wang, H. L. (Eds). Indigenous Fermented Food of Non-western Origin. J. Cramer Press, Berlin, pp. 23-25.

Wang Lijun, Masayoshi Saito, Eizo Tatsumi and Li Lite. (2003) 'Antioxidative and Angiotensin I-coverting Enzyme Inhibitory Active of Sufu (Fermented Tofu) Extracts'. *JARQ,* 37 (2):129-132.

Zhang XF, Li LT, Cheng YQ, Fan X, Eizo T, Saito M. (2006), 'ntioxidant activity of sufu'. *Machinery for Cereals Oil and Food Processing*. 6: 17-19.

Wai, N. S. (1968). Investigation of the various processes used in preparing Chinese cheese by the fermentation of soybean curd with *Mucor* and other fungi. Final Tech. Rep., UR-A6-(40)–1, U.S. Department of Agriculture, Washington, DC.

Wang, H. L. (1967a). Productions from soybeans. *Food Tech.* 21: 115–116.

Wang, H. L. (1967b). Release of proteinase from mycelium of Mucor hiemalis. *J. Bacteriol.* 93: 1794–1799.

Wang, H. L., Ellis, J. J., and Hesseltine, C. W. (1972). Antibacterial activity produced by molds commonly used in oriental food fermentations. *Mycologia* 64: 218–221.

Wang, H. L. and Hesseltine, C. W. (1970). Sufu and lao-chao. *J. Agric. Food Chem.* 18: 572–575.

Wang, H. L. and Hesseltine, C. W. (1979). Moldmodified foods. In: H. J. Peppler and D. Perlman, eds. *Microbial Technology, Fermentation Technology*. Academic Press, New York. pp. 95–129

Wang, H. L., Ruttle, D. I., and Hesseltine, C. W. (1969). Antibacterial compound from a soybean product fermented by *Rhizopus oligosporus. Proc. Soc. Exp. Biol.* 131: 579–583.

Wang, H. L., Swain, E. W., and Hesseltine, C. W. (1980). Phytase of molds in oriental food fermentation. *J. Food Sci.* 45: 1262–1266.

Wang, H. L., Vespa, J. B., and Hesseltine, C. W. (1974). Acid protease production by fungi used in soybean food fermentation. *Appl. Microbiol.* 27: 906–911.

Whitaker, J. R. (1978). Biochemical changes occurring during the fermentation of high-protein foods. *Food Tech.* May: 175–179.

Zhang, X. Q. (1997). The technique of Jizhou sufu. *Agric. Tech. Bull.* 1997(10): 39–40.

Zhao, D. A. (1990). Chinese cabbage sufu. *Chinese Fermentation* 1990(2): 26–28.

Zhou, G. F., Gou, Y. Y., and Kang, Y. (1990). The study of amino acid contents changing and influence affect during sufu processing. *Res. Dev. Nat. Food Prod.* 2: 49–53

Chapter 7

DOUCHI

1. INTRODUCTION

1.1. History

Douchi is one of the earliest microbe-fermented foods in China. It was named *youshu* by Chinese people in the old times due to *youshu* meaning boiled soybean fermented in an airtight environment. This kind of traditional Chinese food was called *douchi* in the Qin Dynasty. Douchi is a kind of seasoning food with a unique flavor. It can be used as not only seasoning but also a directly-eaten food. It is made from various legumes such as soybeans, black beans and so on. Douchi production had been greatly developed in the Han Dynasty due to ancient recordation. Douchi ginger was found as a res buried in *Mawangdui* West Han Grave in Changsha City of Hunan Province, in the middle of China. This discovery demonstrated that douchi had been a favorite food in China since 200 BC. Nowadays douchi is widely eaten by people in many southern provinces in China such as Zhejiang, Fujian, Sichuan, Hunan, Hubei, Jiangsu, Jiangxi and the north of China. There are lots of douchi varieties with special flavors such as Babao Douchi in Linyin City, Shandong Province, Yongchuan Douchi and Tongchuan Douchi in Sichuan Province, *Yi Pin Xiang* Douchi and Liuyang Douchi in Hunan Province, Watermelon Douchi in Henan Province, Yangjiang Douchi in Guangdong Province and Huangyao Douchi in Guangxi Province and so on. In addition, douchi is also very popular in Japan and the countries of southeast of Asia. Douchi in China is mostly made from soybeans and black beans, especially soybeans. Douchi had been popular since 2000 years ago due to its smooth luster, exquisite character, unique flavor and high nutrient value. Douchi is not only a kind of fermented food with high nutrient value, but also has been a kind of specific medication since the early ages. In the Han Dynasty, there was a literature named *Talking About Typhoid* written by Zhang Zhongjing who was one of the most famous physicians in Chinese history. The literature had introduced a prescription *Soup Made of Douchi and Gardenia* which had been described like this: a kind of soup was gained through decocting douchi and gardenia; the soup can be drunk by patients to cure cold, anorexia and so on. In the Ming Dynasty, *Ben Cao Gang Mu*, which was written by Li Shizhen who was another distinguished physician in Chinese history, is another very famous part of medical literature in Chinese history. It had descriptions as follows: douchi had a lot of curative effects such as improving digesting, stimulating appetite, removing fidget, asthma and advancing diaphoresis. After being introduced into Japan, through certain studies, Japanese researchers reported that douchi was conducive to digestion, improved the disease-fighting ability of humans, retarded insenescence, eliminated tiredness and even prevented cancers.

1.2. Definition

Douchi, a kind of traditional fermented food based on soybeans or black beans, has been consumed for more than 2,000 years in China. In the Tang Dynasty, *Jian Zhen*, who was a famed Chinese monk, brought douchi to Japan; since then, douchi had been increasingly developed to become a staple food which is now called Natto in Japan. At the same time, the producing technology of douchi was also introduced into Indonesia, North Korea, Philippines and other Southeast Asian countries. Tempeh, which is mostly similar to Chinese douchi, is very popular in Indonesia now. Tempeh has recently been called a "*Substitute of Meat*" by Americans, and "*Meat from Crops*" by the Japanese. Though different areas have different kinds of Douchi, their producing technologies are similar to each other. First, the soybeans are soaked in water, and then boiled till soft but not fragmentized, after that, the boiled soybean is firstly fermented with a natural species of microbes. The fermented soybeans are laid out for a certain number of days and then processed in the secondary fermentation. The final product should be a kind of fermented food with black-brown or yellow-brown color and full-granules.

In the past twenty years, many scholars in the world have been dedicated to the research about soybean fermented foods such as douchi and sauce. They got the conclusion that douchi contains soybean protein, linolic acid, phosphatides, dietary fibre, various mineral elements and vitamins. Besides, douchi also contains physiologically-active ingredients such as soybean peptides, oligosaccharides, isoflavones, melanoidin and so on. All these nutritional substances and physiologically-active ingredients have unique health-keeping effects. It has the functions of banishing tiredness, decreasing cholesterol, lowering blood pressure, accommodating the level of insulin, enhancing intestinal function to improve immunity and inhibiting the production of venomous substances in the intestines. Melanoidin of douchi has a strong antioxidative function and can be used to inhibit the synthesization of nitrosamine which is a kind of intense carcinogen in the stomach. Ssoybean isoflavone in douchi is an anti-tumour, anti-aging composition, and can be used to prevent the capillary vessels from becoming brittle.

1.3. Classification

Douchi could be categorized on different roles such as microorganisms, main material or state of the products.

Douchi can be sorted to mucor douchi, aspergillus douchi and bacteria douchi based on the variety of microorganisms used in fermentation.

Mucor douchi is produced in cold winter when the temperature is 5–10 °C. It takes mucors from the air and the environment to make starters. Most of douchi produced in Sichuan Province are this kind of food.

Aspergillus douchi can be produced year round. The cultivating temperature of aspergillus is higher than that of mucor. The temperature of making starters for aspergillus douchi is 26–35 °C, and the condition is feasible to take advantage of environmental temperature and to keep the temperature for making starters at a certain level. In Shanghai, Hubei and Jiangsu Provinces, aspergillus Hu Liang 3.042 is often used to make starters in a ventilated environment for douchi production.

Bacteria douchi is processed by covering the boiled soybeans or black beans with haulms or pumpkin leaves to make bacteria reproduce on the beans. The process of making starters ends as soon as the appearance of glutinous substances. The temperature of making starters for bacteria douchi is relatively low. Bacteria douchi is mainly produced in Shandong, Yunnan, Guizhou, and Sichuan Provinces.

Douchi can be categorized according to the main materials used for production. Most of Douchi is made of soybeans but some of douchi in Jiangxi, Hunan, Shandong, and Sichuan Provinces are also made from black beans.

Douchi may be categorized for their different seasoning. Most of douchi varieties produced in China are salty douchi which have a strong taste. Salt is added to fermented soybeans to get them pickled during the producing of salty douchi. Traditionally, the salt of douchi was very high (over 14%), but the realization of the negative effects of high salt content on health, especially on hypertension has led those producing douchi to lower the salt-content. Now salty douchi is mainly used as flavoring in view of its excellent smack. The kind of douchi in which salt is not added entirely or the addition is very limited after secondary fermentation is categorized as saltless douchi.

In the market, four kinds of douchi are sold according to the moisture content. One kind is dried douchi. Fermented douchi is dried and the moisture content of the finished product is 25%–30%. This kind of dried douchi is incompact and in the shape of granules. Compared with the low moisture content of dried douchi, the moisture content of wet douchi is higher. The finished product is soft and pressed together. There is another water douchi, which is soaked in a supersaturated and liquid mixture, and fermented for a long time to make the finished products to be granule-shaped while in the dipping state. To produce lumpish douchi, soybeans are cooked firstly into a lumpish state. Lumpish douchi for making starters and fermenting are processed at the same time, and then fumed for a certain period after fermentation. Lumpish douchi is extremely tasty due to its unique flavor characteristics and fumed flavor. The taste of chipped lumpish douchi which is braized or stir-fried is extremely excellent.

Douchi could be categorized based on different tastes resulting from seasoning materials on the market, too. Alcohol douchi, ginger douchi, capsicum douchi, aubergine douchi, sauce douchi, shallot douchi, sesame oil douche, and so on, are normally sold.

1.4. Consumption

There is not much information on production and consumption even though the quantity is quite a large amount and the number of producers may be over 3000, if small homemakers are also counted. There are more than 100 canneries to produce can of "Diced Fish in Black Bean Sauce" (it is called "Diced Fish in Black Bean Sauce" too) (Figure 7-1). The sales just for 20 canneries in Guangdong are over 10 billion Yuan (about 1.5 billion US dollars) each year. The contents are mainly fried fish and douchi. It is usually used to prepare a lettuce dish (Figure 7-2), and it can be found in Chinese restaurants almost all around the world. The production is so important that the national standard on canned "Diced Fish in douchi" was issued in 1999 (QB/T 3605-1999).

The salt content was higher usually than 12%, sometimes even over 18% for preserving, so the consumption had been affected by the worry about hypertension. As a result of research, the salt content can be as low as 6~8% now and the consumption is expected to enlarge further.

The consumption of douchi is affected heavily by habit and region. Some people will not eat douchi at all, but others may use it every day for dinner. The most popular regions for douchi dishes include Sichun, Hunan, Hubei, Jianxi, Fujian, Guandong and Shandong Provinces.

Figure 7-1. Picture of cans of "Diced Fish in douchi".

Figure 7-2. Vegetable dish with douche.

2. Materials for the Production

Materials for douchi processing are relatively simple and mainly include soybeans, microorganisms and salt. Of course, seasonings could also be added according to the needs.

2.1. Soybeans

The only main raw material for douchi is the soybean. It is said about 1 kg of soybeans can produce 1.5 kg douchi. The soybean decides the color, the size and sometimes the compositions of flavor and function. Douchi can be divided into black bean douchi, soybean douchi from the soybean type, and big particle douchi or small particle douchi from the size of soybean particles.

Protein is one of the most significant components in soybeans and the hydrolysate results in the production of flavor compositions such as amino acids. The protein of Chinese soybeans is mainly about 40% while some special variety can reach 50%. Most of the protein in soybeans (about 80%) is globulin and is water-soluble protein (86–88%) In addition, the digesting rate of soybean protein is as high as 92–100%, which is significantly higher than that of animal protein. Besides soybean protein, which can decrease serum cholesterol content, soybean also contains many other physiologically active components such as soy isoflavone, lecithin and so on. Moreover, it is found that the amino acid contents and isoflavone in douchi are correlated closely with the soybean variety.

Song et al. (2003) researched the variation of amino acid contents in black soybeans and douchi (Table 7-1). It is clear that most of the amino acid contents increased in the fermentation for douchi production, except for the decreasing of tryptophane.

Table 7-1. Variations of amino acid contents in soybeans and douchi (mg/100g)

	Lysine	Threonine	Leucin	Isoleucin	Tryptophane	Methionine	Valine	Phenylalanine
Soybean	5.94	3.65	7.36	4.46	1.01	1.20	5.03	5.32
Douchi	20.4	9.8	20.4	14.1	-	19.2	21.9	34.8

Table 7-2. The weight and volume variations of black soybeans soaked in 25 °C water

Soaking time *(h)*		0	1	2	3	4	5	6
Yangjiang black soybean	Weight (g)	100	161	178	199	212	222	226
	Volume (ml)	135	262	274	310	335	335	335
Guangxi black soybean	Weight (g)	100	157	171	189	214	220	229
	Volume (ml)	138	250	280	310	345	350	350
Henan black soybean	Weight (g)	100	156	165	188	211	219	222
	Volume (ml)	135	240	270	298	325	335	335
Jiangxi black soybean	Weight (g)	100	144	164	190	209	219	237
	Volume (ml)	134	218	253	310	330	355	355
Vietnam black soybean	Weight (g)	100	124	135	149	173	181	197
	Volume (ml)	124	150	174	220	250	269	295

It is said that trypsin which is a kind of inhibitor of protein digestion is destroyed and the molecular weight of protein decreased during fermentation.

There is not much research on the soybean for douchi processing. But the varieties of soybean have significant effects on the quality of douchi. Li et al. (2005) reported that black soybeans planted in different places have a different absorption speed of water. Black soybean planted in Yangjiang, Guangdong Province showed the highest speed of water absorption while that of Vietnam has the lowest (Table 7-2). It is said speedy absorption of water is beneficial to suitable hardness of douchi. It was also shown that the total amino acids of douchi made from Yangjiang black soybeans was better than others made from Guangxi, Henan, Jianxi provinces and from Vietnam (Table 7-3).

The compositions of amino acid of douchi are also nalyzed in the report, and the results are affected by the production area of raw soybeans for douchi making. Douchi made from black soybeans of Yangjiang and Guangxi had higher contents of glycine, serine and alanine which can increase the sweet taste and decrease the bitterness of douchi.

Table 7-3. The effects of black soybeans from different areas on compositions of douchi (100g, d.b)

Production areas	Yangjiang	Guangxi	Henan	Jiangxi	Vietnam
Total acids (g)	1.69	1.72	1.91	1.82	1.90
Total amino acids (mg)	28.02	27.30	25.64	26.03	25.23
Amino acid nitrogen	0.74	0.70	0.65	0.62	0.63

Generally, soybeans used to make douchi need to be fully mature, plump, fresh and with high protein, without impurities. Tannin and indican-species substances in the coating of soybeans, after long-time storage, are easy to hydrolyze and oxidate due to enzymes and increase bitterness and acerbity in the product's flavor.

2.2. Microorganisms in Douchi Production

The microorganism is another important factor in douchi production, during which boiled soybeans take advantage of enzymes produced by microorganisms to decompose protein, starch and so on to get complicated compounds which react with each other to form the final unique flavor of douchi. Microorganisms used to produce douchi are mainly mucor, aspergillus oryzae and bacteria.

Mucor in douchi is mainly mucor racemosus. Traditionally, making starters with mucor under natural environment can be processed only in winter because of the need for low temperature (room temperature 2–6°C, production temperature 5–12 °C). Moreover, the time for making starters with mucor under the natural environment is generally 15–21 days. To solve the problems, pure mucor M.R.C-1 has been separated in Sichuan Province, from natural douchi starters. After cultivation through heating, M.R.C-1 showed good properties of blooming mycelia, strong adaptability and high enzyme activity of main enzymes such as proteinase and β-amylase. The time for making starters has been significantly shortened to 3–

4 days and douchi can be produced all year. Furthermore, qualities of sense, physical and chemical properties of douchi made from with pure mucor are improved significantly.

Aspergillus is a middle-temperature microorganism and aspergillus douchi is often produced with natural inoculum in many rural areas and with pure inoculum in relatively large factories. Making starters with natural inoculum usually is restricted by environmental conditions such as season and temperature. Inoculum concentration in producing pure aspergillus douchi is high while aspergillus grows quickly and the fermentation period is short. But the alcohol flavor or ester flavor of aspergillus douchi, such as Liuyang Douchi in Hunan Province and Yangjiang Douchi in Guangdong Province, is stronger than that of mucor douchi.

Natto, a kind of fermentation soybean food in Japan is also one of bacteria douchi. Bacterium used in douchi production is part of bacillus subtilis. It is found that Natto bacillus was oxygen-needed G+ bacterium. The ability of excreting various out-of-cell enzymes such as proteinase, amylase, γ-GTP, saccharase and phytase is a distinct characteristic of bacillus subtilis. Enzyme activities of proteinase and γ-GTP excreted by Natto bacillus are 15–20 times and 80 times those of other bacillus subtilis respectively. Same characteristics can be expected for douchi because the fermentation process is almost the same.

There have been many researches on microorganisms in douchi production recently and the activity of microorganisms has been much improved.

2.3. Supplementary Materials

The main supplementary materials for douchi production include capsicum, watermelon pulp juice, wheat flour, salt, fennel, fresh ginger flake (thread), almond, perilla leaves, zanthoxylum, liquor, sesame oil and so on. In most of cases, these seasoning materials are fitted together for producing a certain variety of douchi. For instance, watermelon pulp juice and wheat flour are used to produce watermelon douchi in Kaifeng City in Henan Province in China. The major supplementary materials for Hunan pungent douchi are ginger and capsicum which give douchi of unique pungent flavor.

3. Processing Technology of Douchi

Most douchi is still processed with traditional processing technology, and it is different from modern processing in some aspects. Different states and kinds of douchi may also be used in some processing.

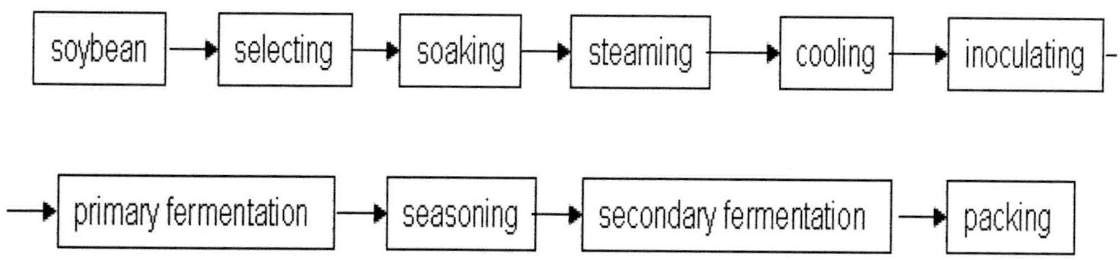

Figure 7-3. Scheme of traditional processing of douche.

Figure 7-4. Soaking of soybeans.

3.1. Traditional Processing Technology

Traditional technologies of douchi production usually use mucor, aspergillus or bacterium to depolymerize soybean protein to a certain degree during fermentation, and then the fermentation process is stopped or slowed down by inhibiting enzyme activities through adding of salt and alcohol, drying and so on. Traditional technologies of douchi production are not good for stable quality control of final product, while enzyme activity is low and fermentation period is long.

Processing of douchi is explained in Figure 7-3.

3.1.1. Soaking

Soaking increases the water absorption of soybeans before steam boiling. The requirements of soaking are a little different from that for tofu production. Soybean particles must be completely without wrinkle of soybean coat and the moisture content was better controlled in about 45%~50%. So the soaking time is a little shorter than that for tofu production. The former may be adjusted in the range of 2~5 h while the latter must be 12~24 h according to soaking temperature. It is necessary to replace the soaking water one time to decrease bubbles on the surface of the soaking water. Soaked soybeans need to be washed several times to clean all mud and soil conglutinated to them again. It is said that over-soaking may result in the decomposing of koji and let the douchi lose the bright and slippery skin. Soybean is soaked in a pool with water, in which the surface of the water should be approx 30cm higher than that of the beans just as shown in Figure 7-4.

3.1.2. Steam or Boiling

Soaked soybeans used to be boiled in the countryside traditionally for douchi production. Soybeans are boiled for 2~3 h usually until they are entirely cooked. If the cooking is not enough, the denaturalization and digestion of protein would be affected, hardness of douchi increases and taste is degraded. But over-cooking would result in the over denaturalization and the decreasing of texture of douchi. Moreover, water absorption of soybeans is difficult to control by boiling so it is replaced by steaming in most factories, and boiling is just used in family work. Soaked, clean soybeans are steamed under normal atmosphere for 4h and 0.1 MP for 45min until soybeans become cooked-soft, easily-broken and have a flavor characteristic of soybeans. The judging standard is that there is no hard heart when soybean is nipped to two pieces. Final water content of steamed soybeans is of great importance in douchi production. If water content is too low, microorganism growth would be affected and production of enzymes may not be enough. While if water content is too high, temperature controlling during fermentation would become difficult and would result in the increase of useless bacteria and soybean decomposition. It is better to control the moisture content of steamed soybean in 55~57%.

3.1.3. Primary Fermentation (Making Douchi Qu)

The traditional method of producing douchi qu is usually by natural inoculation. Primary fermentation is known as making douchi qu (koji). Suitable conditions including temperature, humidity of environment and moisture content of soybeans facilitate microbes propagating for douchi fermentation, production of complicated enzymes and different kinds of metabolized outcomes which endow douchi flavor and taste. Methods of making douchi qu are a little different due to the microbes used.

1) Primary fermentation with aspergillus

Primary fermentation is known as making douchi qu (koji) with aspergillus. Steamed soybeans are cooled to 35 °C at room temperature, encased into bamboo dustpans (the thickness of beans is about 2–3cm, the entourage is thicker than the center) and then moved to the fermentation room (Figure 7-6), the temperature of the environment and the product are kept at 26–30 °C and 25–35 °C, respectively. After 24 h fermentation, the temperature of the soybeans begins to increase and a few small agglomerations appear. After about 48 h fermentation, the temperature of the soybeans may increase to about 37 °C; the soybeans are covered with mycelia, and there are many big agglomerations.

Figure 7-5. Steaming of soybeans.

Figure 7-6. Looking into a fermentation room.

It is necessary to mix the douchi qu most of which is agglomeration-shaped soybeans, into grains. The mixing makes the temperature of the douchi even and helps in propagating spores.

The secondary turnovering should be processed when the above soybeans get agglomerations and appear to have yellow-green spores. The douchi qu must be kept at 35–37 °C for two days and then cooled to about 28–30 °C by ventilation. After fermentation of 6~7 days, the processing of douchi qu finished. The moisture content of matured qu is about 21%, soybeans have furrows and yellow-green spores which can fly when being rubbed with hands, and mycelia can be seen in soybeans while it is divided.

2) Primary fermentation with mucors

Steamed soybeans are cooled to 30–35°C and set on mats with a thickness of 3–5 cm. The temperature of the fermentation room and douchi qu should be kept at 2–6°C and 5–12°C, respectively. The period of primary fermentation with mucors is generally 15–21 days. White mildews can be found in soybeans after 3–4 days. There will be orderly mycelia and a few brown spores in soybeans in 8–12 days. After fermentation of 16–21 days, mucors in douchi qu become mature and the color of mycelia, with tight texture, change from white to grey, and they are erect with a height of 0.3—0.5 cm. Meanwhile, there are green thalli tightly attached to soybean surfaces.

3) Primary fermentation with bacteria

In China, bacterium douchi (douchi produced with bacteria) is also called water douchi due to its high moisture content during fermentation. Both filtrate and steamed soybeans can be used to make qu. The detailed process is to make use of microbes from the air to inoculate naturally. Microbes in the system are complex but bacillus subtilis and lactic acid bacteria are major bacteria.

The process of making qu with steamed soybean filtrate is a little different from that with soybean particles. Pouring the filtrate into a big open container, keeping it at room temperature for 2–3 days, and agitating once when there is a little douchi flavor. And then it is cultivated for another 2–3days. Douchi juice is ready when we can smell dense douchi flavor and ammonia and we can see long mycelia being suspended when nipping douchi juice up with chopsticks.

Making qu with steamed soybeans, it is processed in bamboo baskets (Figure 7-7). First, fresh flat pu grass (its popular name is douchi leaves) is under-laid on the bottom of the bamboo basket, 10–15cm-thickness of steamed soybeans are spreaded on the douchi leaves, another 10cm-thickness douchi leaves are spread on the steamed soybeans and then steamed soybeans are cultivated in the cultivating room. Soybeans should be turned over once after 2–3days fermentation and then make them cultivate continually for another 3–4 days till the mature qu is ready. There is a layer of thick lumps enwrapping steamed soybeans and dense douchi flavor. The period of making qu with bacteria is about 6–7days.

Figure 7-7. Bamboo baskets for primary fermentation with bacteria.

3.1.4. Secondary Fermentation

Secondary fermentation is necessary for douchi processing and the methods are different with various primary fermentation. Sometimes, primarily fermented soybeans were mixed with seasonings such as capsicum, salt, fennel, fresh ginger, perilla leaves, liquor, and sesame oil and so on, but sometimes it is washed to clean off all attached material including aspergillus. Secondary fermentation may vary significantly with different factories or products.

1) Dry Douchi Produced with Aspergillus

Washing is a special process for producing aspergillus douchi. The purpose of washing with water is to wash away spores, mycelia and some enzymes attached to the soybeans and to limit the hydrolyzation. Proteins and starchs in soybeans can be decomposed to amino acids, sugars, alcohols, acids and esters, all of which compose flavor substances of douchi under certain conditions. But over-hydrolyzation may lead to the increase of soluble substances flowing out of the soybeans and result in soybean surfaces, that are coarse, deformed and dim. So washing is necessary to produce douchi with full grain, lucent figuration as well as unique flavor. In addition, washing can remove mycelia and spores to avoid a bitter flavor.

Though a washing machine for douchi production has been developed, the traditional method is still used widely. Primary fermented douchi is poured into a pool containing warm water to wash exterior conidiophores and mycelia. And then the douchi is moved to bamboo baskets to get rinsed by water till there is no mycelia and spores attached. The period of washing should be controlled at 10 min. or so. Fermented soybeans will get rotted easily once they are washed too long, which results in too high water absorption of soybeans.

Water-washed douchi is piled and fitfully sprinkled with water to increase the moisture content to 45%. It is important to control the moisture content. If the moisture content is too high, desquamate of soybean particles increase, and they easily rot and lose luster, also it is disadvantageous to secondary fermentation and may result in the increasing of hardness of final products.

The water adjusteded douchi qu is covered with plastic film and kept warm. The product temperature would reach about 55 °C after 6–7 h. Salt, capsicum and ginger can be added and mixed with douchi qu when there are mycelia in douchi qu and unique douchi flavor can be smelled.

After being mixed with seasoning, the douchi would be fermented a second time. It should be pressed tightly layer by layer in the jar (Figure 7-8) and, covered with salt and plastic film, sealed and fermented at normal temperature for 4–6 months.

Fermented and mature douchi is then cooled and dried till its moisture content is below 30% (Table 7-9). Processing of dried douchi is then finished.

Figure 7-8. Secondary fermentation of douchi qu.

Figure 7-9. Cooling and drying of douche.

2) Seasoned and Watered Douchi Produced with Aspergillus

Water douchi is usually produced with aspergillus. Matured qu is dried in the sun to reduce its moisture content, which is useful for removing spores attached to mature qu and avoiding a bitter taste. When drying in the sun, ultraviolet radiation can kill harmful microbes in mature qu to improve secondary fermentation.

Seasonings such as watermelon flesh juice, salt and flavor are mixed firstly and then mixed with the matured qu. One example of additions for watermelon douchi are 100kg soybeans, 125kg watermelon flesh juice, 25kg salt, suitable amount of fennel, ginger and other minor meterials.

The above douchi mixture is encased in a container, sealed and set in the sun for fermentation for 40–50days. With seasoning substitutes, i.e. apple or tomato juice for watermelon flesh juice, the final product is called apple-juice douchi or tomato-juice douchi.

3) Mucor Douchi

Mature qu is poured into mixing pool, dispersed, added with a certain amounts of salt and water, and then fully mixed and stewed for 24h. After that, distilled spirit, yellow wine, flavors and so on are added and mixed.

The above mixed materials are encased in a jar. While encasing, the materials should be tightly pressed layer by layer till the volume of materials is 80% that of the jar, and the surface of the materials is even, covered with plastic film and sealed.

The proportion of materials is as follows:100 kg soybeans,18 kg salt,3 kg distilled spirit (>50%,v/v), 4k g yellow wine and 6–10 kg water used to accommodate the moisture content of the mixture 45%.

3.1.5. Secondary Fermentation without Salt

Of course, secondary fermentation can be carried out without salt and other seasonings. The fermentation period can be reduced to 3–4 days because there is no inhibition of salt on enzyme activities.

1) Fermentation of Aspergillus Oryzae Douchi without Salt

Mature qu is washed by warm water to remove mycelia and spores attached to mature qu, drained, poured into a pool and mixed. Then hot water (65°C) is spilled into the mature qu till its moisture content is 45%. After that, the qu is encased in the fermentation jar kept heated, covered with plastic film and then sealed to keep product temperature at 55–60°C. Necessary time for fermentation is 56–57 h. Once there is no container with thermal retardation, hot water should be mixed into mature qu till its moisture content is about 45%. Then mature qu is added with 4% distilled spirit (>50%, v/v), covered with plastic film and other materials for thermal retardation.

2) Fermentation of Mucor Douchi without Salt

Hot water (about 65 °C) is also necessary for adjusting the moisture content of douchi qu to 45%.Then distilled spirit and yellow wine are immediately mixed in to make it piled and get its temperature higher. And then douchi qu is encased into the fermentation jar with thermal retardation. The temperature should be kept at 55–60 °C for 56–72 h.

3.2. Modern Processing Technology

Douchi products in the present market are based on traditional douchi which is often handmade in the countryside. Nowadays production of douchi takes advantage of modern technologies and equipment which assures qualities and quantities of douchi on the basis of traditional technologies. The two have similar characteristics such as color, flavor and taste. The differences of modern processing technologies are described.

1) Steaming

Revolving and high-pressured skillet is often used to steam soybeans in industrial production of douchi. It is reported 1 h is enough for soybeans to be steamed well under 0.1 MPa pressure.

2) Making Douchi Qu with Pure Inoculums

Industrial production of douchi uses pure inoculums other than natural inoculums, which are beneficial in controlling the quality of douchi in large-scale production.

When pure aspergillus is used, steamed soybeans are cooled to 35 °C, inoculated with 0.3% ($^{w}/_{w}$) aspergillus, and mixed and cased into bamboo dustpans with a thickness of 2cm.The room and product temperatures are kept at 25 °C and 25–35 °C, respectively. After being fermented for about 22 h, white mycelia can be found all over the soybeans, the soybean particles are agglomerated, and the product temperature will have risen to about 35 °C. After 72 h there are red aspergillus and yellow-green spores here and there in the soybeans and the douchi qu gets mature. Many kinds of aspergillus including Huliang 3.042, Aspergillus3.798 have been developed recently and are widely used.

When pure mucors are used for douchi production, steamed soybeans are cooled a little to 30 °C, inoculated with 0.5% pure mucors, then the inoculated soybeans are encased into sterilized dustpans with a thickness of soybeans 3–5cm and set in the cultivation room. Mucors are cultivated under the condition of the product temperature 23–27°C.There are small white mucor colonies after 24 h. Luxuriant mucors, erect mycelia having changed from white to French grey after 48h. During the process, there are more and more spores found in the steamed soybeans. The general period of making douchi qu with pure mucors is 3 days.

The best content of inoculums is 1–3% (w/w) of soybeans when pure bacteria was used to produce douchi. The stationary and decline phases of douchi *bacillaceae* are 14–20h and >20h, respectively under the conditions of pH7. 0.2%NaCl and 40 °C. So the best period for inoculating with douchi *bacillaceae* is 14–20h later.

3) Secondary Fermentation, Sterilization and Packing

Industrial secondary fermentation is almost the same as traditional methods except advanced equipment is used. The hot douchi is mixed with 0.08% sodium benzoate and then encased into jars and sealed. In this way, the douchi can be stored for a longer time.

Packaging varieties contain plastic bags, compound plastic bags, paper bags containing plastic bags, glass bottles gallipots and so on (Figure 7-10). Containers for packaging must be clean, hygienic and processed through sterilization.

Figure 7-10. Packaging of douche.

Table 7-4. The physicochemical indexes for douche

Item	*Douchi*	*Dry douchi*
Moisture content/ (g/100g)	≤45.00	≤20.00
Total acid (according to lactic acid)/ (g/100g)	≤2.00	≤3.00
Nitrogen in amino acid form/ (g/100g)	≥0.60	≥1.20
Protein/ (g/100g)	≥20.00	≥35.00
NaCl/ (g/100g)	≤12.00	—
Pb/ (g/100g)	≤1.0	≤1.0
As/ (g/100g)	≤0.5	≤0.5
Additive content	According to GB2760—1996	According to GB2760—1996
Aflatoxin B_1 / (μg/kg)	<5	<5
Coliform	≤30 coliforms/100g	≤30 coliforms/100g
Pathogen	can not be detected	can not be detected

3.3. Standards for Qualities of Douchi

Qualities of douchi should accord with the national standard GB 2712-81 (Hygienic Standards for Fermented Soybean–based Products). Sensory and microbiological indexes include the following items:

1. Water douchi—Canary yellow, fragrance peculiar to water douchi, delicious, piquancy, ginger flavored, without impurity, mildew and abnormal flavour such as sour, bitter and astringent, with suitable thickness.
2. Other douchi varieties—grainy, snuff color or black-brown, fragrance peculiar to douchi, delicious, with no impurity, mildew and abnormal flavours.
3. Microbiological level of coliform group needs to be below 30/100g, no pathogenic bacterium can be found in douchi.

All physicochemical indexes for douchi are shown in table 7-4.

3.4. Progress on the Technology

The traditional method of cooking soybeans in douchi production is mainly to boil them in water. While making douchi, after being boiled to a soft consistency, soybeans are fished out of the boiler and drained, and the filtrates with abundant nutrient ingredients are often squandered. Presently, steaming is used to cook soybeans since it preserves the most nutrients contained in the soybeans as well as simplifying the technology of cooking the soybeans. Traditional douchi production involves fermentation with various microbes in natural circumstances involving temperature, humidity and ventilation conditions.. Different kinds of microbes give birth to different kinds of proteases, amylase and other enzymes which hydrolyze soybean ingredients into different products which endow douchi with different flavors. Chanceful natural circumstances go against the usual growth of microbes and make the final product unstable. So some makers now take advantage of pure-species fermentation to ensure douchi with unified quality. The previous usually-used tools for making douchi are bamboo retort, bamboo dustpan and so on which are not easily cleaned and insanitary for douchi production. At present, stainless-steel instruments are widely applied to make douchi. In packaging of products, there have also been a lot of improvements taken in making douchi, such as sealed-and-plastic packages, tin packages, glass-bottle packages and box packages which protect the douchi from the harmful impacts of air, light and moisture. The past packaging containers are mainly crocks or jars which are difficult to seal well, and the upper layer of douchi often become rotted. The last, but also the most important improvement, is that nowadays automatic equipment is largely and widely used almost in every procedure of making douchi. The automatic equipment not only increases the production efficiency and decreases the cost, but also keeps the final products unified.

4. RESEARCH ON DOUCHE

4.1. Research on the Technology of Douchi Production

There is not much research on the technology of douchi production though similar researches are much higher in number for Natto or Tempeh. But it is reported that douchi made from family workshops or large-scale factories in taste is almost the same except that the former has a higher content of sulfur compound because the boiling time of the former is usually longer than that of the latter.

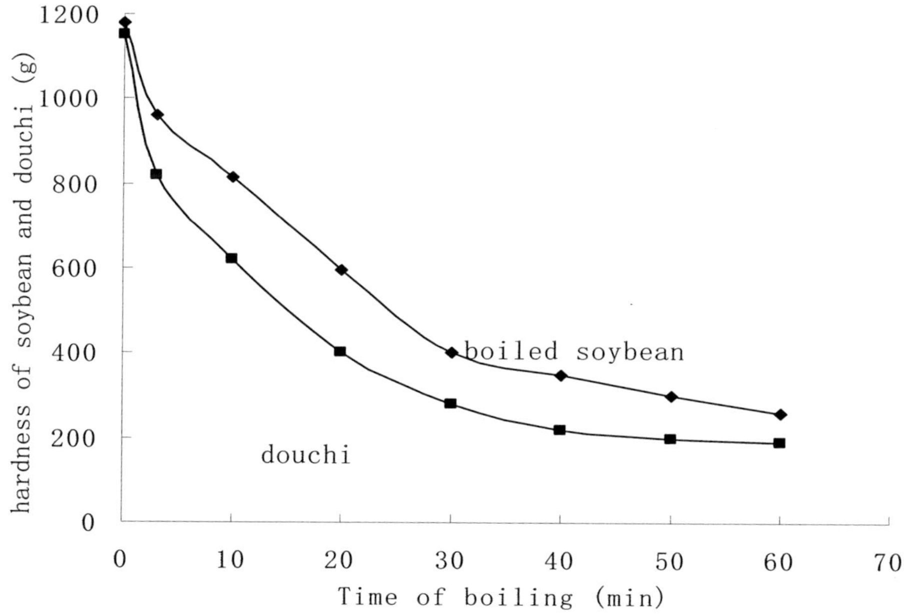

Figure 7-11. Relationship between hardness of soybean, douchi and boiling time (Cai and Zhao, 1997).

Cai and Zhao (1997) [Relationship between quality of douchi with boiling of soybeans. Chinese Condiment, 1997(3):12-14] reported boiling time had a significant effect on the hardness of douche, and there is a good relationship between the hardness of douchi and the soybeans just as shown in Figure 7-11. Preferred hardness of douchi is about 200~300 g, so boiling time may be controlled to 30~40 min.

Wang et al. (2005) reported the difference of temperature of douchi was affected significantly by quantity of aspergillus in primary fermentation of douchi (Figure 7-12).

She pointed out that the hardness of the douchi is higher at first but decreased at last when 10^6 /g aspergillus was used compared with 10^4 /g which was also used (Wang et al. 2005).

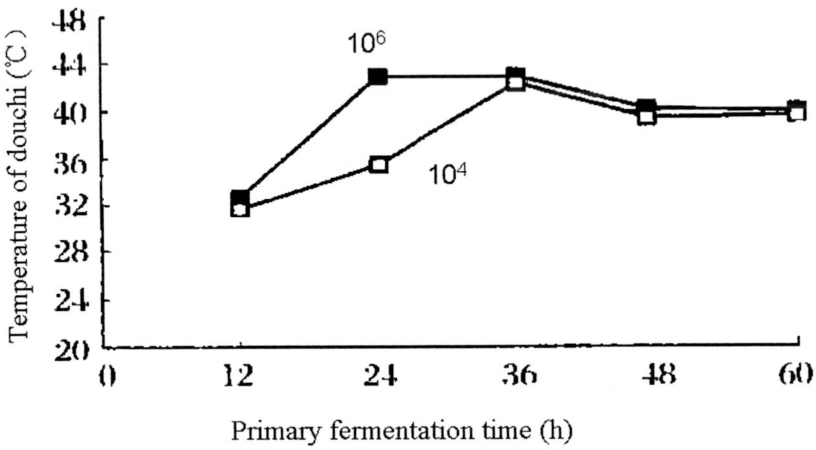

Figure 7-12. Quantity of aspergillus (/g) on the temperature of douchi during primary fermentation (Wang et al. 2005).

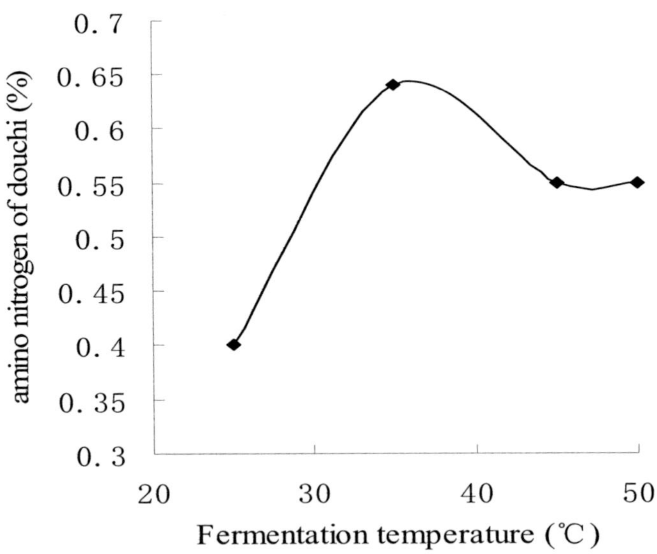

Figure 7-13. The effect of fermentation temperature on the amino nitrogen of douchi (Wang et al, 2006).

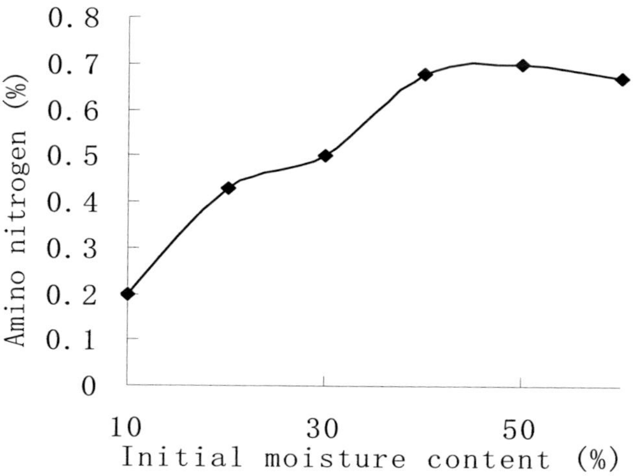

Figure 7-14. The effect of initial moisture content on the amino nitrogen of douchi (Wang et al., 2006).

Wang et al. (2006) studied the effects of fermentation conditions on the quality of douchi. It was reported that fermentation time, temperature and moisture content significantly affected the amino nitrogen contents. It is shown that fermentation temperature has a significant affect on the amino nitrogen content of douche, which decides the taste of the douchi (Figure 7-13). The most suitable temperature is about 35 °C and low temperature will result in lowering of amino nitrogen contents. Of course, amino nitrogen contents increased with fermentation time in suitable range. Increasing of moisture content could improve the taste of douchi but if moisture content is too high, the effect would be small (Figure 7-14).

Zou et al. (2006) researched the effect of addition of ethanol in ripening on the polypeptide content and antioxidant of douchi. As shown in Figure 7-15, ethanol may inhibit the production of polypeptide which is the main composition of douchi.

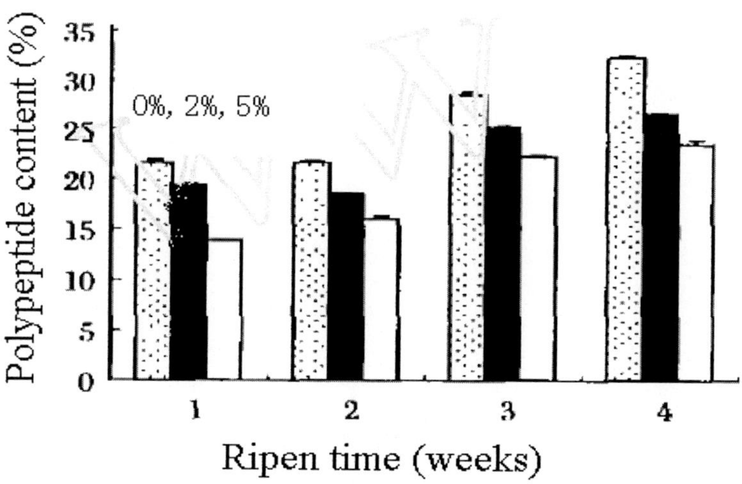

Figure 7-15. The effect of ethanol addition on the polypeptide content in different ripen times of douchi (Zou et al. 2006).

Table 7-5. Effects of different aspergilluses on the hydrolyzation ratio of soybean protein and casein (%) (Lin and Li, 1998)

Kinds of aspergillus	A	B	C	D	E
Soybean protein	100	101	96	93	87
Casein	100	102	97	96	90

Sometimes, the surface of the douchi soybeans showed some white dots that affected the quality of the douchi. Lin and Li (1998) reported the white dots mainly come from tyrosine and can be avoided by improving aspergillus. They used 5 kinds of aspergillus to process douchi and measured the hydrolyzing ability on soybean protein and casein. As shown in table 7-5, aspergillus E can decrease the hydrolyzation ratio 13% and 10% respectively.

4.2. Research on the Function of Douchi

Douchi without salt is one of 88 varieties of plants which can be used as food and medicine homologically. So the health function has been known from ancient times, though the mechanism is not clear, and there are many mentions on the functions in ancient books. Recently, the research increased and the functions of clearing the free radicals, fibrinolytic function, antioxidant activity and so on have been confirmed.

Zhang et al. (2006) reported douchi contains angiotensin I-converting enzyme (ACE) inhibitors which may lower blood pressure. The results showed that ACE inhibitory activities were improved following the fermentation. ACE inhibitory activities of 48 h-primary-fermented douchi qu did not change dramatically after preincubation with ACE, but increased greatly after preincubation with gastrointestinal proteases. The results suggest they were pro-drug-type or a mixture of pro-drug-type and inhibitor-type inhibitors. The ACE inhibitors in 48 h-fermented douchi qu were fractionated into four major peaks by gel filtration chromatography on Sephadex G-25. Peak 2, which had the highest activity, had only one peptide, composed of phenylalanine, isoleucine and glycine with a ratio of 1:2:5.

Table 7-6. The contents of isoflavones in soybean and douche (Cui et al., 2007)

Number of samples	Contents of soybean	Contents of douchi
1	4.801	9.943
2	4.934	9.662
3	4.639	9.987
4	4.728	10.120
5	4.890	9.987
6	4.713	9.603
Average (mg/g)	4.784	9.884
RSD (%)	2.4	1.9

Cui et al. (2007) measured isoflavones in soybeans and their fermented douchi and showed the content in douchi was much higher than that in the soybeans (Table 7-6).

Sun et al. (2000) in China determined isoflavone content in several typical soybean products, and the results showed that isoflavone in un-fermented products mainly exists in the form of β-glucoside whereas isoflavone in fermented soy products was completely decomposed to be daidzein and genistein which significantly improved physiological activity of isoflavone. In addition, the loss of isoflavone could be 16.3%, 28% and 36%, respectively, when it is soaked, steamed and deepfried. Meanwhile, some studies indicated that microbes which could excrete glucosidase in fermented soy products were mainly mucor, aspergillus, rhizopus, epiphytes belonged to saccharomyces. Mao et al. (2000) found that dissociative genistein and soybean flavin content in thin douchi was (230.64±9.14) μg/ g and (264.26 ± 4.22) μg/g, respectively, and that after being hydrolyzed with hydrochloric acid was (276.00± 7.81) μg/g and (287.65±5.70) μg/g, respectively, in the quantitative analysis of isoflavone in thin douchi by HPLC.

Wu et al. (2000) came to the conclusion after study on La Ba Soybean that fermentation had no effect on isoflavone content because the total content of soy isoflavone before and after fermentation was 538 μg/g and 561 μg/g, respectively, but the content of each component of isoflavone had changed a lot. Before fermentation, soy isoflavone mainly existed in the forms of daidzin, glycitin and genistin, all of which accounted for 86 % of total content; after mucor fermentation, dissociative glucoside sources obviously increased to 535 μg/g, accounting for 95% of total content, meanwhile, almost no glucoside-linked daidzin and glycitin could be detected.

It is said douchi contains a kind of melanin named melanoidin, which is produced by the Mailand reaction between soybean protein, its hydrolyzed peptides and reducing sugar. Melanoidin possesses a very powerful ability of antioxidation due to its inner stable free radical structures which can catch and collect free radicals in solutions. Meanwhile, melanoidin combines with ferrum, copper and other metal ions to form insoluble compounds separated out later. Melanoidin also has functions which are similar to dietary fiber, accommodating glucose level in blood, inhibition on ACE activity and so on. Kan et al. (1999) had studied douchi melanoidin and found that un-dialyzed melanoidin in mucor douchi bore relatively strong ability to remove free radicals, relatively obvious antioxidation on pig fat in the dry system and strong inhibition on the composition of N-dimethyl nitrosamine. They also determined melanoidin content in Yongchuan Douchi in Chongqing City was 3.61%. The effects of Vc and melanoidin were also compared and later showed higher scavenging ability of free radical in a suitable concentration (Figure 7-16).

Kang and Ding (2006) researched the composition of melanoidin and reported a part of the melanoidin skeleton was composed of peptide structure, in which the most reactive amino acids residues to form melanoidins mainly were asparticacid, glutamic acid, arginine, lysine and proline.

Researchers found that douchi also contained enzymes (douchikinase) which had the ability to dissolve thrombus similar to nattokinase. Douchikinase is a kind of neutral serine protease which can effectively decompose the main components of thrombus, i.e., fibrin and substrate for fibrinolysin HD-Val-Leu-Lys-pNA (S-225), and that douchikinase directly acts on cross-linked fibrinolysin, is not sensitive on fibrinolysin so that douchikinase will not result in bleeding. Douchikinase was found to possess a great function of dissolving thrombus either in vivo or in vitro through the experiment that douchikinase extract was dipped on fibrous slab and opaque circles in the slab became transparent, and douchikinase had the ability to dissolve thrombus through mainline or being taken orally on the animal thrombus model. In addition, the ability of douchikinase to dissolve thrombus was significantly stronger than fibrinolysin and elastin protease. Furthermore, the period of dissolving euglobulin was obviously decreased, of euglobulin was improved, the period of improvement on fibre-dissolving activity could last for 2~8h and douchikinase could accelerate liver and vein endothelium cells to produce t-PA and improve thrombus-dissolving activity 4 days later, for healthy people after taking douchi or douchikinase intestine-soluble capsules. Douchi contains certain protein which can dissolve fibre, i.e., enzyme for dissolving thrombus as the major component for anti-thrombus which is especially the same with patients with thrombus, dense blood, brain infarct and inferior health.

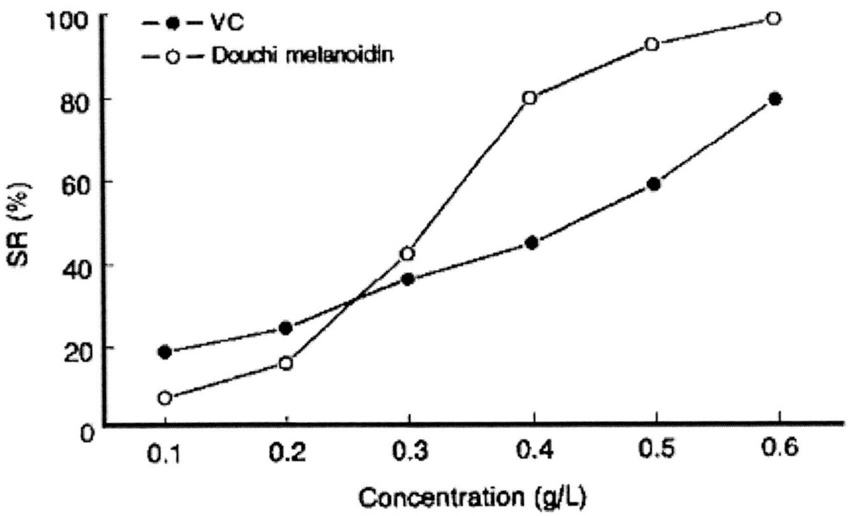

Figure 7-16. Effect of douchi nondialyzable melanoidin on active oxygens (Kan et al., 1999).

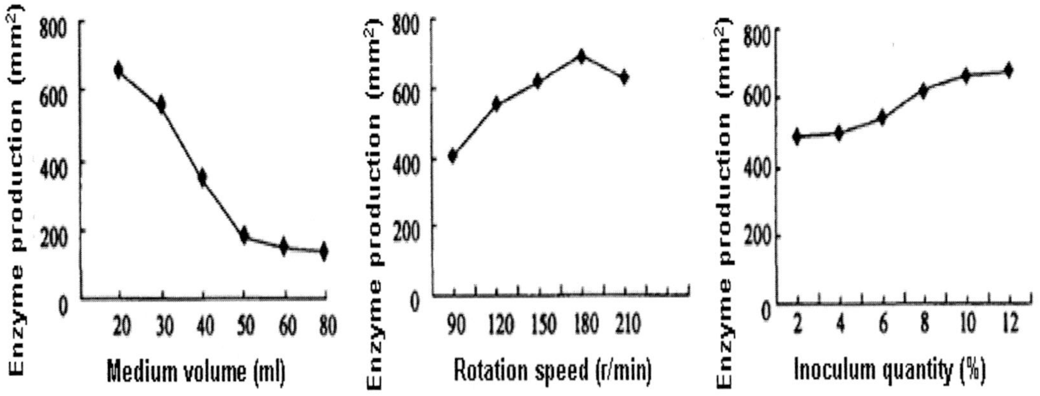

Figure 7-17. Effects of medium volume, rotation time and inoculum quantity on enzyme production (Yao et al., 2007).

The researches on the douchikinase are plentiful and some are listed in Table 7-7.

There are many researches on the methods of enhancing activity of fibrinolytic enzyme. Fan et al. (2006) selected a douchi sample which had higher fibrinolytic activity in corrected douchi samples and treated with ultraviolet rays and HNO_2. Three mutants were found to have high yield of fibrinolytic enzyme and high stability, and their enzyme production increased by 3.6, 3.7 and 4.75 times as compared with those of the original strain, respectively. Yao et al. (2007) researched the technology of liquid state fermentation of douchi fibrinolysin, and presented the effects of fermentation conditions on the enzyme production as shown in Figure 7-17.

Table 7-7. Researches on the douchikinase in China

No.	Authors	Title	Magazine	Date
1	Liu Xiaolin et al.	Study on the dissolve thrombus function of douchi	Chinese Journal of Gerontology	2007, 27(3):452-453
2	Xiao Lu et al.	Fermentation of Douchi Fibrinolytic Enzyme Gene Engineering Strain and the Purification of Recombinant Enzyme	Food and fermentation industries	2005, 31 (1):66-71
3	Yao Xiaoling et al.	Study on extraction and purification of douchi fibrinolytic enzyme	Food research & development	2007, 28 (8):79-82
4	Yao Xiaoling et al.	Study on liquid-state fermentation technology of Douchi fibrinolysin	China Brewing	2007, 169(4):38-42
5	Lan Xinyin et al.	Experimental study on the dissolve thrombus function of douchi extracter	Chinese Journal of Gerontology	2006, 26(8): 1081-1082
6	Wang Chengtao et al.	Studies on tibrinolytic function of subtilisin and its mechanism	Acta Nutrimenta Sinica	2007, 29(6): 600-604
7	Mu Guangqing et al.	Optimization on production conditions of fibrinolysin by Bacillus subtilis SY-3	China Brewing	2007, 167(2): 30-34
8	Sun Yuan, Mu Guangqing	Screen and characterization of fibrinolyticenzyme producing strain for fermented soybean	Food research and development	2007, 28(1): 36-39
9	Jia Nan, Mu Guangqing	Research on the protective additive used for the freeze-drying of douchi fibrinolysin	China Brewing	2007, 166(1): 17-19
10	Mu Guangqing et al.	Separating and character of douchi fibrinolysin	Science and Technology of Food Industry	2007,7: 90-93
11	Wang Weidong, Sun Yuee	Analysis of Nutrients and Bioactive Substance in the Fermented Liquid of Fibrinolytic Enzyme-producing Strain from Douchi	Modern Food Science and Technology	2006, 22(2): 56-58
12	Luo Wenhua et al.	High2level Expression of Douchi Fibrinolytic Enzyme (DFE) in Bacillus subtilisWB800	China J Appl. Environ. Biol.	2007, 13 (4): 565 ~ 569
13	Wang Xichun et al.	Study on the solid state fermentation of douchi with response surface analysis	Food Science and Technology	2007,1: 121-125
14	Liang Huiyi, Guo Yong	Whole Genome Shuffling to Enhance Activity of Fibrinolytic Enzyme producing Strains	China Biotechnology	2007, 27 (10): 39 ~ 43
15	Luo Wenhua, Guo Yong	Reviews on Foodborne Fibrinolytic Enzyme	China Biotechnology	2006, 26(8): 111-114

It is shown that suitable conditions for higher enzyme production are fermented 3 days in about 30 °C and neutral condition. The enzyme production was explained as the acting area of certain quantity of enzyme.

It is said fibrinolytic enzyme is aerobic and if one increased the quantity of enzyme in a container, the activity decreased as shown in Figure 7-17. But inoculum quantity had no significant effect on the enzyme activity. In an enlarged test (15 L), the enzyme of douchi reached to 2050 U/ml liquid (urokinase unit) (Yao et al., 2007).

Douchikinase showed good anticoagulant function in an in vitro and in vivo study on rats (Table 7-8). The samples of small dose (10 ml/kg weight each day) and large dose (20 ml/kg weight each day) extended significantly the clotting time and bleeding time after giving water extracted liquid for 10 days. The clotting time was measured by glass capillary method.

Table 7-8. Effects of douchi extracted liquid on the anticoagulant function in rats (x±s, n=10) (Yao et al., 2007)

Group	*Dose*	*Clotting Time*	*Bleeding time*
Control	-	1.53±0.51	10.60±2.66
Low dose	10	2.25±0.50	16.13±6.42
High dose	20	2.70±0.57	19.42 ±4.25

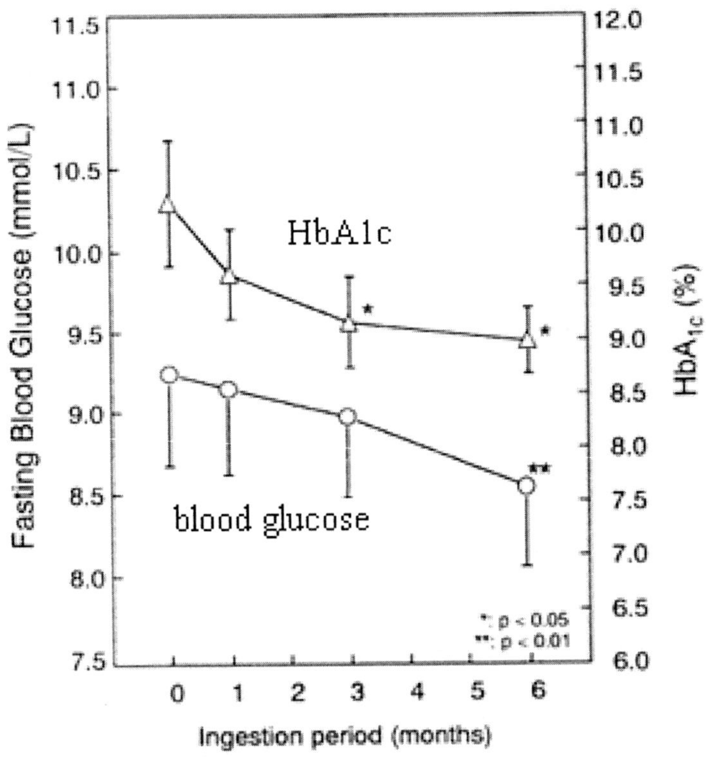

Figure 7-18. Effects of douchi-extract on the blood glucose and HbA_{1c} levels in diabetic subjects (Fujita et al., 2001).

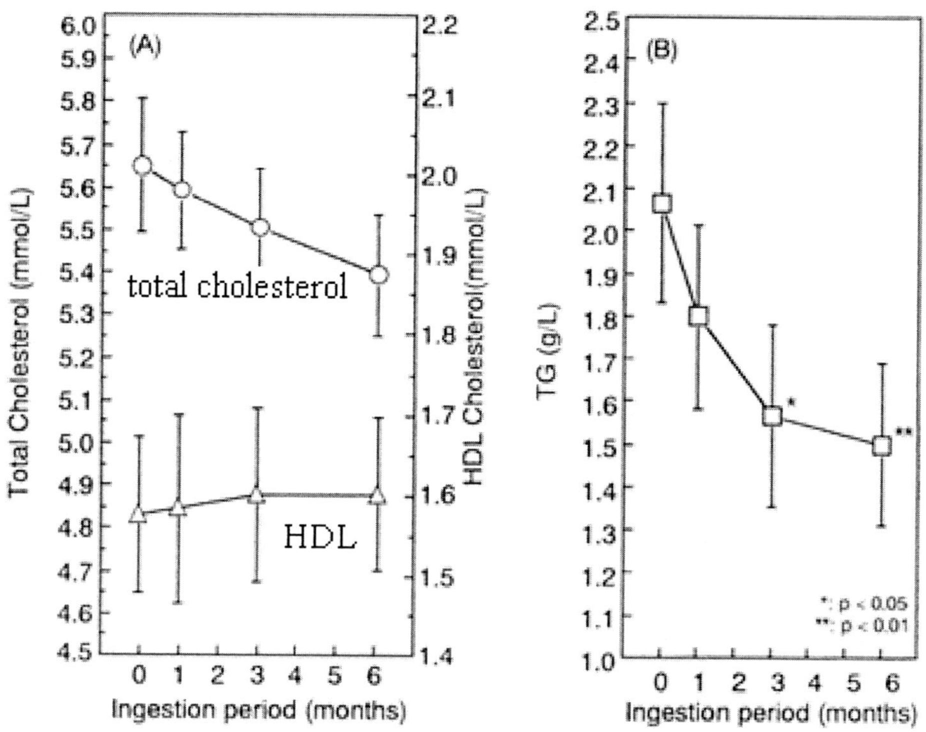

Figure 7-19. Effect of douchi-extract on serum lipid metabolism in diabetic subjects (Fujita et al., 2001).

During the fermentation of douchi production, microbe fermentation makes soy protein produce proteases which degrade soy protein into varieties of biologically-active peptides. It is also pointed out that the content of water-soluble protein, low or middle-molecular peptide and α-amino acid nitrogen in douchi increased by 300%~600%, 300%~800% and more than 10000% compared with soybeans, respectively, because denatured protein is propitious to microbes and increases the hydrolyzation rate of soy protein. It was said that peptides, among which there are special peptides with the function of inhibiting ACE activity, accounted for 68%~78% of total amino acids, and amino acids consisted of phenylalanine, isoleucine and glycin in Aspergillus Douchi. In addition, peptides by hydrolyzation of soy protein have effects of promoting the growth and metabolism of microbes such as lactobacillus, bifidobacterium, leavens and mycetes.

Douchi is sorted to three kinds of mucor douchi, aspergillus douchi and bacteria douchi based on the variety of microorganisms used in fermentation. Microbial starter is very important to decide the type, the functional compositions of douchi. There have been many researches on the microbial starters recently.

Though douchi is traditionally made from with single microbiao starter, Sun et al. (2007) tried to combine bacillus subtilis and aspergillus in douchi making. Results showed that the inoculating condition was similar with that of bacillus subtilis.

Fujita et al. (2001) studied the efficacy and safety of douchi. The report showed that water-extracted douchi exerted a strong inhibitory activity against rat intestinal a-glucosidase in foodstuffs. In borderline and developed diabetic subjects, 0.3 g of douchi-extract significantly inhibited postprandial blood glucose levels. And the safety was confirmed by using 9 healthy subjects and given 1 g of douchi extract before every meal (3 g/day) for 12 weeks. There is not any change in hematological and relevant biochemical parameters, body weight or BMI. In another study, 18 type-2 diabetic patients ingested 0.3 g of douchi extract before every meal (0.9 g/day) for 6 months (mo). The result showed that blood glucose (mean: 9.31± 0.71 mmol/L) and HbA1c (mean: 10.24± 0.58%) levels gradually decreased after 6 months and HbA1c after 3 and 6 months of post-ingestion of douchi extract (Figure 7-18). Indexes for serum lipids and total cholesterol level revealed moderate decreases with a slight increase in the high-density lipoprotein (HDL) level after douchi extract ingestion. However, triglyceride (TG) levels significantly decreased at 3 and 6 months of post-ingestion of douchi extract (Figure 7-19).

4.3. Research on Microbiological Safety of the Product

Most douchi uses natural fermentation and the period for production is long, so there are a large amount of mixed bacteria in final products. Zhang et al. (2006) researched on the diversity of microorganisms in aspergillus type douchi and reported the mean counts for mesophilic aerobic bacteria, mesophilic aerobic bacteria spores and moulds in two samples of douchi named as TMS and YPX douchi were 9.42lg CFU/g, 5.66 lg CFU/g, 7.15 log CFU/g and 8.93 lg CFU/g, 5.65 lg CFU/g, 7.11 log CFU/g, respectively. Yeast and salt- tolerant yeast in two douchi were very low in all samples. The ratios of aspergillus sp. to total moulds in two kinds of douchi qu were 90.4% and 95.1%, respectively. The main aspergillus strains in the two douchi qu were identified by their configuration of mycelial, hyphal and spores. The results showed that A. Egyptiacus Moub. And Moust., A.oryzae are the main aspergillus strains as shown in table 7-9.

The aspergillus and oryzae are the most dangerous epiphyte in douchi because aspergillus is the main infecting resource of immunodeficiency crowd while oryzae can produce fumitremorgins which has strong toxicity. So it is necessary to remember the danger especially in hot and dried area. In natural fermentation, it is difficult to control the microbiological infection in traditional processing and it is necessary to improve the traditional processing in the view of modern food processing and technology though douchi is a traditional Chinese food derived from fermented soybeans and has been eaten since very long ago.

It is reported that microorganisms were found in some Chinese medicines containing douchi even though there was not pathogenic bacteria and the contents were in permission. But Fujita et al. (2001) described in his study, other biochemical parameters excepting functional compositions were not affected in any of the patients, and no one complained of any side-effects or abdominal distension. This may be due to the lower inhibitory potency of douchi on a-glucosidase compared with currently employed therapeutic agents of similar mechanism of action, and this may account for the moderate effects in the small intestinal tract. As shown in table 7-10, abnormalities in hematological and relevant biochemical data were not observed, and the safety of douchi is thus clarified.

Table 7-9. The kinds and contents of aspergillus and moulds in douchi ($\times 10^5$) (Zhang et al. 2006)

		YPX	TMS	Total
Aspergillus	Egyptiacus Moub.	22	31	53
	Moust.	14	21	35
	Oryzae	11	0	11
	Parasiticus	3	2	5
	The others	8	12	20
Mucor	Mucor A	3	3	6
	Mucor B	0	4	4
Total		61	73	134
Ratio of aspergillus to total of moulds		95.1%	90.4%	

Table 7-10. Safety of douchi extract after long-term ingestion in diabetic patients (Fujita et al., 2001)

		Before	1 months	3 months	6 months
Hematology	White blood cell (cells/nL)	6.3±0.3	6.3±0.3	6.3±0.3	6.3±0.3
	Red blood cell (cells/pL)	4.71±0.12	4.67±0.11	4.62±0.10	4.65±0.10
	Hemoglobin (g/L)	1.44±2.7	1.44±2.7	1.43±2.4	1.42±2.6
	Hematocrit (%)	45.9±0.69	43.5±0.74	43.1±0.69	42.8±0.71
	MCV (fL)	93.4±1.45	92.8±1.19	92.8±1.18	92.0±1.15
	MCH (pg)	30.6±0.47	30.7±0.45	30.8±0.47	30.8±0.49
	MCHC (%)	32.8±0.22	33.0±0.23	33.1±0.21	33.5±0.17
	Platlet (cells/pL)	0.19±0.02	0.20±0.02	0.20±0.02	0.18±0.01
Biochemistry	GOT (U/L)	35.4±5.1	33.7±4.8	30.1±3.1	28.9±2.9
	GPT (U/L)	51.5±11.1	46.7±8.9	39.8±5.4	39.3±5.2
	ALP (U/L)	194.7±14.9	210.6±15.3	206.7±16.2	206.5±13.3
	g-GTP (U/L)	52.2±8.9	46.9±8.9	49.9±5.9	57.9±6.8
	Total protein (g/L)	73±1	74±1	73±1	74±1
	Amylase (U/L)	110.5±10.5	100.2±10.6	97.9±9.2	98.6±8.6
	Urea (g/L)	4.8±0.26	4.6±0.25	4.6±0.27	4.7±0.24
	Free fatty acid (mmol/L)	0.90±0.06	1.03±0.07	0.99±0.05	0.90±0.03
	Blood urea nitrogen (mmol/L)	5.67±0.31	5.89±0.02	5.82±0.27	5.67±0.27
	Creatinine (mmol/L)	79.6±5.3	79.6±2.7	79.6±5.3	79.6±5.3
	CRP (mg/L)	3.1±0.3	3.2±0.7	2.1±0.6	1.2±0.3

Eighteen diabetic patients ingested douchi extract (0.3 g) 3 times daily before meals for 6 months. The values were expressed as the mean 6 S.E.

Table 7-11. Sanitation and chemical indexes for douchi (GB 2712-2003)

Index	*Value*
Aflatoxin (μg /kg)	≤ 5
Coliform group (MPN/100g)	≤ 30
Pathogenic bacteria	Can not be detected

There are two sanitation indexes including coliform group and pathogenic bacteria and one chemical index of aflatoxin for douchi in China according to GB2712-2003 (table 7-11). The total number of bacteria and the level of coliform group index reflect the level of production techniques and control of sanitation and quality.

Douchi, as a traditional Chinese food, similar to natto in Japan and tempeh in Indonesia either in processing and function and nutritional contents, is known widely and widely. But compared with natto and tempeh, the research is still not enough and the position in diet is much inferior. In most cases, it is used just as a seasoning. It is necessary to improve the processing technology and lower salt content so as to widen the usage as a staple food as was done with natto.

REFERENCES

Cheng, L. J. S., Zhao, X. Cao, J. G. & Liang, H. Z.(2005). 'Analysis on the nitrogen components in Chinese traditional fermented soybean products'. *China Brewing*, 7: 47~49.

Cui Lijian, Huan Gun, Zhan Wenhong, Dou Yuhong, Wei Chune, (2007). 'Effect of fermentation process on total content of isoflavones in soybean'. *Soybean science*, 26(4):588-561.

De Lumen, O. B. (1992). 'Molecular strategies to improve protein quality and reduced flatulence in legumes:a review'. *Food Structure*, 11:33-46.

Fan, J. F., & Li, L. T. (2005). 'Biofunctional Activity of Traditional Fermented Soybean Foods'. *Food Science*, 26(1):250-254.

Fan Xiaodan, Xie Xiuzhen, Guo yong, (2006), 'Breeding of fibrinolytic enzyme producing strains isolated from douchi'. *Life Science Research*, 10(1):45-49.

Fujita Hiroyukia,*, Yamagami Tomohidea, Ohshima Kazunorib, (2001). 'Efficacy and safety of Touchi Extract, an a-glucosidase inhibitor derived from fermented soybeans, in non-insulin-dependent diabetic mellitus'. *Journal of Nutritional Biochemistry*, 12:351–356

He, G.Q. (2001). Technologies for Fermentation and Brewing of Food. *China Agricultural Press.* 324-326.

Jiang, L. W., Zhou, C. Y., & Li, Z. J. (2006). 'Discussion on quality and safety control of traditional Chinese fermented soybean products'. *China Brewing*, 1:1-3.

Kan Jianquan, Chen Zongdao, Shi Yisong, Wang Guangci. (1999). 'Study on Antioxidation and Inhibition of Nitrosamine Synthesis by DOUCHI Nondialyzable Melanoidin'. *Acta Nutrimenta Sinica*, 21(3):349-352.

Lin Ying, Li Guoji (1998). Avoiding white dots in spergillus douchi processing, *Chinese Condiment,* 2:19-20)

Li, P.L., and Wang, C.T. (2005). Safe production and quality control of fermented foodstuffs. *Chemical Industry Press*. 254-260.

Mao Junqin, Mi Heming, Lou Ziyang, Chai Yifeng, Yang Genjin, Liu Lili, Xing Wangxing. (2000), 'Detection of isoflavones contents in Seme sojae preparatum by HPLC'. *Academic Journal of Second Military Medical University*, 10:956-957

Megumi, K., Kumi, T., Shinkichi, T., Yasuhito, T. & Masaaki, Y. (2003). 'Angiotensin I-converting enzyme inhibitory peptides isolated from tofuyo fermented soybean food'. Bioscience, Biotechnology and Biochemistry, 67:1278-1283.

Pan, J.Q., & Li, L. (2005). 'Study on protein changes during the submerged fermentation of soy peptides'. *Science and Technology of Food Industry*, 26(5): 67-69.

Qin Likang, Ding Xiaolin (2006), 'Extractions of Melanoidins in the Long- Ripenned Douchiba (DCB) and Analyses of Amino Acid Compositions in Its Peptide Skeltons'. *Food Science*, 27(01):13-16

Shao, W., Xiong, Z., & He, X. W. (2005). 'Study on soybean peptide fermentation and its functions'. *Food Science and Technology*, 4:26-28.

Song, J.M. and Ju, H.R. (2002). The New Edition for Processing Technology of Soybean Food. *Shandong University Press*. 186-196.

Steggerda, F. R. (1968). 'Gastrointestinal gas following food consumption'. *Annual N.Y.Academic Science*, 50:57-66.

Sumi, H., Hamada, H., & Tsushima, H.(1987). 'A novel fibrinolytic enzyme (nattokinase) in the vegetable cheese Natto, a typical and popular soybean food in the Japanese diet'. *Experientia*, 43:1110-1111.

Sun Chengxin, Mu Guangqing, Sun Yuan, (2007), 'Research on the multistrain fermentation technology of douchi'. *China Condiment*, 3:43-46

Sun Meijun, Luo Lian, Shi Changying, Qian Yingyian (2000). 'Quantification Determination and Investigation of Isoflavones in Chinese soy Products'. *Food and Fermentation Industries*, 5:15-17

Tang, C. H. & Peng, Z. Y. (2000). 'Recent Development of Soybean Functional Components'. *China Oils and Fats*, 25(4):44-47.

The China State Bureau of Quality and Technical Supervision (2003). GB 2712-2003. 'Hygienic standard of fermented soybean product', National standard, P.R.C

Wang, L. J., Saito Tatsumi, M. E., & Li, L. T. (2003). 'Antioxidative and angiotensin I-coverting enzyme inhibitory activities of sufu (fermented tofu) extracts'. *JARQ*, 37: 129-132.

Zhang, J. H. (2003). 'Studies on Fermentation Mechanisms and Functions of Aspergillus Douchi'. A Doctoral Dissertation of China Agricultural University.

Yin, Y.G. and Liu, J.B. (2006). Technologies for Soybean Food. *Teaching Material Press Center of China Chemical Press*. 64-75.

Wang Lijun, Zou Lei, Li Lite, Eizo Tatsumi. (2005). 'Changes on Douchi Appearance, Temperature, Hardness and Color Tone during Fermentation'. *Food and fermentation industry*, 31(12):42-45.

Wang Yanjie, Su Qinghui, Li Zhijiang, Li Jing, Wu Tong, Dai Lingyan. (2006). 'Study on optimization fermentation technology of douchi and its quality analysis'. *Food and machinery*, 22(6):130-132.

Wu Yuxiang; Xia Yanbin, Liu Ailing (2002). 'Stability of Isoflavones During Labadou *Processing'.Journal of Hunan Agricultural University (Natural Sciences)*, 6: 237-238

Yao Xiaoling, Song Weijiang, Wu Sifang. (2007), 'Study on liquid-state fermentation technology of Douchi fibrinolysin'. *China Brewing*, 4:38-42.

Zhang Jianhua, Eizo Tatsumi, Ding Changhe, Li Lite (2006). 'Angiotensin I-converting enzyme inhibitory peptides in douchi, a Chinese traditional fermented soybean product'. *Food Chemistry*. 98: 551–557).

Zou Lei, Wang Lijun, Hu Qing, Li Lite, Yamaki Koji, (2006). 'Effect of Ethanol Supplement on Antioxidant Capacity of Douchi during the Ripen Time'. *Food and fermentation*

Chapter 8

VINEGAR

1. INRODUCTION

1.1. Definition of Vinegar

The word vinegar comes from the French word "vinaigre" which means "sour wine." Vinegar is a liquid processed from the fermentation of ethanol in a process that yields its key ingredient, acetic acid.

According to Codex Alimentarius Commission, vinegar is defined as "condiment made from alcohol fermentation and acetic acid fermentation".

The acetic acid concentration is the most important element of vinegar which typically ranges from 4~8% by volume and may be higher for pickling (up to 18%). The taste or sourness of vinegar is not decided only by the acetic acid concentration. Processing method and raw materials have a significant effect on the taste, too. Natural vinegars also contain small amounts of tartaric acid, citric acid, and so on. Vinegar has been used in China since ancient times, and is an important element in Western, European, Asian, and other traditional cuisines of the world. Every year over 26 million hectoliters of vinegar is produced in China. That is to say, more than 3.2 million liters of vinegar is consumed every day. Vinegar is considered as one of the favorite condiments, health products and sometimes even medicines.

1.2. The History of Vinegar

Vinegar was probably discovered by accident thousands of years ago—after a cask of wine had gone bad. When the wine was first made, natural sugars were fermented into alcohol. Over time, bacteria in the air transformed the alcohol into acetic acid, which gave the "sour wine" its bite. Similar folklore is well known in China that vinegar was first made by Du kang (2070 BC ~1600 BC, *Xia Dynasty*) when he was making alcohol in Shanxi province and then spread it to the southern area such as Zhenjiang. Although Du Kang is still used as a famous alcohol brand, there is no such vinegar brand now.

Recorded vinegar history starts around 5000 BC, when the Babylonians were using the fruit of the date palm to make wine and vinegar. They used vinegar as a food and as a preserving or pickling agent. Vinegar residues have been found in ancient Egyptian urns traced to 3000 BC. As well, China has a more than 5000 year history of producing vinegar in folklore while recorded vinegar history could date back to 1200 BC. But someone also said China probably has the oldest historical records about cereal vinegar in the world. The book "*Ceremony Notes*" (about 800 BC) (Xiao, 2000) recorded vinegar during the *Zhou dynasty* (1000 BC~256 BC). Also, in "*The Analects of Confucius*" (450 BC), there were many records about vinegar (Xiao, 2000). In "*Techniques of Qinese*" (533~544 AD), 23 methods for vinegar preparation are described in detail (Bao, 1985). Until the late *Qing Dynasty* (1644–1911 AD), when industrial-scale vinegar production appeared (Zhang, 2000), Chinese vinegar was only domestically produced on a small scale.

In 1394, a group of French vintners developed a continuous method for making vinegar called the Orleans method. In this method, oak barrels were used as fermentation vessels and the vinegar was siphoned off through a spigot at the bottom of the barrel. About 15% of the vinegar was left behind which contained the "mother of vinegar" and its concentrated bacteria floating on top. A new batch of cider or wine was carefully added to the barrel and the fermentation was quickly started by the remaining vinegar. The French vintners formed a guild of master vinegar makers, and using this Orleans method, they were better able to supply the lucrative vinegar market. The vinegar industry in Europe flourished during the Renaissance and many flavored vinegars were made from various spices, herbs, fruits and even flowers. By the eighteenth century, there were over one hundred varieties of infused vinegars available. In the middle of the 19th Century, Louis Pasteur published the first modern scientific research on vinegar, which is still used as a reference today.

Figure 8-1 is the copy of a certificate for Zhenjiang vinegar in 1840 by the *Qing Dynasty* and the company which kept this certificate is still the largest vinegar maker now in China.

The quality of vinegar is strongly determined by sensory properties as it may modify the overall appreciation of a given food or meal. High-quality vinegar already ranks among the cures of nature for over 3,000 years. The ancient health cultures of China, Egypt, Greece and Rome recorded its versatile function. During the health boom in the latter part of the last millenium, it was revived with scientific support from 800 BC—Written record of "Vinegar" in China. A lot of research has been done around the world in the past 20 years.

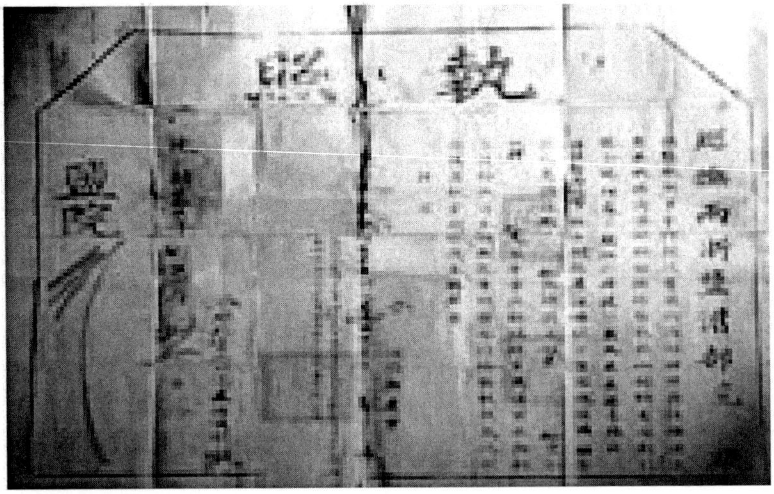

Figure 8-1. Certificate for Zhenjiang vinegar in 1840 by Qing Dynasty.

The Chinese made vinegar mainly from grains such as barley, millet, rice, and rarely from fruits including dates, cherries, and peaches or from honey, too. Among vinegar of this "versatile liquid", texts of many ancient books in *Chou Dynasty* (221 B.C.E.) mentioned a kind of food named "Liu" as a seasoning. Liu was usually interpreted as "vinegar," but its manufacture and ingredients remain unknown. In the 4th and 5th centuries C.E., the Japanese borrowed the techniques for making rice vinegar and rice wine from the Chinese. During the *Sung Dynasty* (960 ~1279C.E.), Sushi made from rice, vinegar, oil, and meat or raw fish was popular in China, but eventually became better known as a Japanese specialty. In the *Tang Dynasty* (618~907 C.E.), vinegar was used as a preservative. Sometimes its flavor was enhanced with the addition of leaves from kumquat trees and peach blossoms. During the *Qing Dynasty* (1644~1911 C.E.), Yuan Mei, a scholar who was interested in food and its principles, stressed the understanding of the natural properties of a given food in his recipe book, "*Shih tan*". He said the choice of condiments was important: soy, oils, and vinegars all have their own attributes and defects.

The manufacturing process of Chinese vinegar is a cross between the techniques of making rice wine and soy sauce. It involves six to seven months of soaking, steaming, and cooling of the chosen grain, primary fermentation with yeast, secondary fermentation with vinegar acid, maturing and concentrating of sunning during the daytime, and cooling at night, followed by straining and pasteurization before bottling. Chinese vinegars are lighter and slightly sweeter than Western white or cider vinegars even though the former has higher acetic acid concentrations.

1.3. Categories of Vinegar in China

Vinegar may be categorized according to its processing method, main raw material or even the color. But in most cases, categorizing is according to the main raw material. The raw material for vinegar is very rich and colorful so that almost all agricultural and horticultural products can be used for vinegar production. Various kinds of cereals including rice, wheat, buckwheat, oats, millet, barley, corn, sorghum, various fruits and berries, nuts, red dates and so on are all used in vinegar production.

According to the color, vinegar can be sorted to white vinegar, black vinegar and red vinegar. Figure 8-2 is the categories considering the processing and raw material. Synthetic vinegar is requested to mark its processing and mostly used for food processing or operation.

Shanxi province and Jiangshu province (Zhenjiang city) are the most famous producing areas of vinegar. Mature vinegar and aromatic vinegar are the main products of Shanxi and Zhengjian, respectively. The former, mainly made from sorghum, has a very long mature process (over 1 year), while the latter is mainly prepared from sticky rice. At the same time, there are many other brands of vinegar produced in different areas including Changde aromatic vinegar (Hunan province), Jilin rice vinegar (Jilin province), Baonin aromatic vinegar (Sichuan Province) in which some plants can be used as Chinese traditional medicine are added usually, Yongchun mature vinegar made from sticky rice and so on.

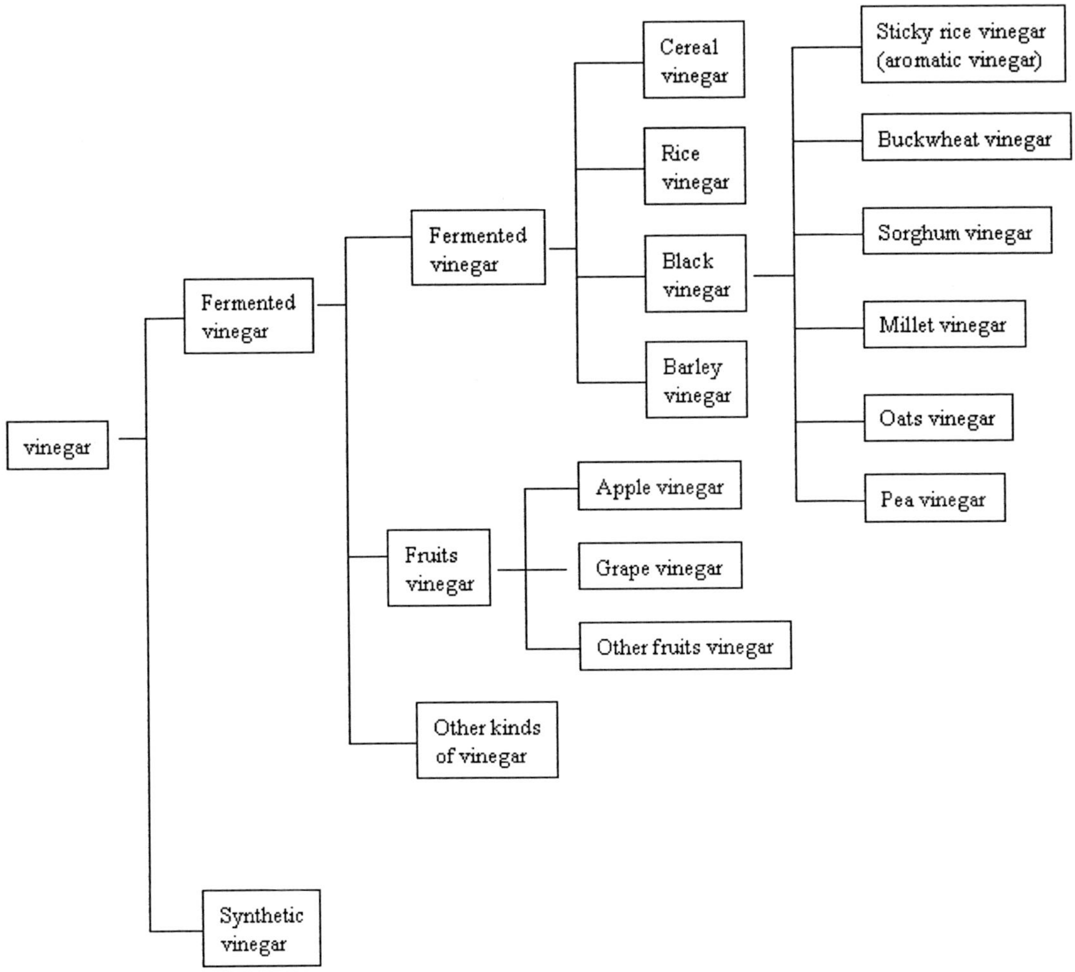

Figure 8-2. The categories of vinegars in China.

But the category of vinegar is still not clear because the naming system is different and complicated. As Shanxi mature vinegar, the names in English include mature vinegar, over mature vinegar, old mature vinegar and so on. There still is not a generally accepted appellation now.

1.4. Production and Consumption of Vinegar

There are over 6000 vinegar factories in China and the production of vinegar is about 3 million tons each year. The production of vinegar increased rapidly (Table 8-1) and the international trade of vinegar has a similar trend (Table 8-2) (Xin 2004). The output is mainly Zhengjian vinegar and Shanxi vinegar while the main input is white vinegar. In fact, the production and producer may be more because many small ones did not contribute to the data.

Table 8-1. Vinegar production and consumption in China (1000 ton)) (Xin 2004)

Year	*1980*	*1985*	*1990*	*2000*
Production of vinegar (1000 ton)	470	730	1000	1800
Vinegar consumption for each person every year (kg)	0.47	0.7	0.9	1.5

Table 8-2. The international trade on vinegar in China (Xin 2004)

Year	*2000*	*2001*	*2002*	*2003*
Output of vinegar (m3)	10000	10015	11650	12750
Input of vinegar (m3)	675	1250	2000	2010
The sum of output (1000 $)	7250	6750	10000	10250
The sum of input (1000 $)	750	1250	2000	1850

Vinegar can be used alone as a seasoning or with salt or sugar together. The concentration of detection threshold is higher than bitterness but is much lower than other seasonings (Table 8-3). It is said 0.1% acetic acid and 5~10% sugar can get the most welcome taste, so vinegar and sugar are used at the same time in a lot of Chinese cooking,

Vinegar can not only adjust the taste of the cuisine, but also can improve the storage properties. Vitamin C can be stored in vinegar longer than in water or acetic acid solution (Table 8-4). Many foods used vinegar for preservation in China.

As the functions of vinegar are becoming well-known recently, vinegar has been spreading from seasoning to drinks and health foods. There are some teahouses or bars as shown in Figure 8-3 where vinegar is the main drink in many big cities of China recently and it can be booked in almost all restaurants in China. The extract or concentration is encapsuled as health foods and the export to Japan increased significantly in these years.

Table 8-3. The detection thresholds of basic seasonings (%, W/V) (Higashi, 2004)

Solvent	Sugar	NaCl	Tartaric acid	Quinine sulfate	MSG
water	0.086	0.0037	0.00094	0.000049	0.012
0.094%(5mM)MSG solution	0.086	0.0037	0.0019	0.000049	-
0.25%(5mM)IMP solution	0.068	0.0037	0.03	0.0002	0.00019

Table 8-4. Storage properties (residue ratio) of vitamin C in vinegar, acetic acid and water (%)

	20°C			37°C			
Time (h)	1	5	24	0.3	1	3	5
Vinegar	9.7	9.2	9.5	8.6	8.1	6.1	4.7
Acetic acid solution	9.9	9.7	9.2	9.6	9.3	8.7	8.2
Distilled water	9.6	7.6	1.5	8.4	6.8	2.7	0.9

Figure 8-3. Vinegar drink house in Chengdu city, Sichun province.

2. Raw Materials for Vinegar Processing

Chinese vinegars have specific local features. Every region has its own manufacturers, who produce vinegar in specific processes, using particular raw materials (Table 8-5) (Liu 2004).

Table 8-5. The main types of Chinese vinegar (Liu 2004)

Main materials	Sticky rice	Sticky rice	Sticky rice	Sorghum	Wheat bran	rice
Saccharifying agent	Enzyme	Red koji	Wheat koji	Great koji	Herb koji	No
Process type	SSF	LSF	SSF	SSF	SSF	LSF
Production period (days)	100	>1000	60	>500	>400	160
Annual yield (kg)		3.0×10^5	6.0×10^7	2.1×10^8		
References	Hang and Cai, 1999	Huang and Yin, 2000a; Liu, 1982	Hang and Cai, 1999; Ren and Jian,2000	Huang and Yin,2000a; Wang, 1997	Hang and Cai,1999; Li,1999	Hang and Cai,1999

2.1. Main Raw Materials

The main raw materials for production of vinegars are starch, sugar-containing and ethanol-containing materials. The traditional Chinese vinegar is usually made from starch materials such as sorghum, rice or other starch materials and fruits. In the south of China, the main raw materials for producing vinegar are rice and sticky rice while the sorghum and corn are the main raw materials in the north. Wheat, oats, barley and peas are usually used for koji preparation (Chen, 2000). Table 8-6 lists the main starch materials used for Chinese vinegar (Chen, 2002).

Potato materials have high yield and with high starch content. Their starch granules are so big that they can easily cause starch gelatinization during cooking. There are some bad flavors produced during the fermentation. Therefore, the vinegars made from potato materials are not as good as those made from starch materials. Table 3 lists the main potato materials used for Chinese vinegar (Chen, 2002). Some by-products of the processing of agricultural products with high content of starch, sugar or ethanol are other starch materials for vinegar production. In this way, the amount of raw materials can be reduced so the costs can be lower. Major by-product starch materials are wheat bran, rice hull and foxtail millet (Huang and Cai, 1999).

Sugar-containing materials including fruits and vegetables containing fermentable sugar, minerals and vitamins are suitable materials for vinegar production especially for fruit vinegars. Major fruits used in vinegar production are pear, persimmon, apple, pineapple and their by-products during processing, such as the peel. Major vegetables used in vinegar production are the tomato, helianthus tuberosus and yam (Chen, 2000).

Table 8-6. Composition of major starch materials used in vinegar production (Chen, 2002)

Name	Moisture (%)	Carbohydrate (%)	Protein (%)	Lipid (%)	Cellulose (%)	Ash (%)
Sticky rice	13-15	69-73	5-8	2.4-3.2	0.6-1.0	0.8-1
Rice	12-14	72-75	7-10	0.1-1.3	1.5-1.8	0.4-1.2
Sorghum	10-14	62-68	8-15	3-5	1-3	1.5-3
Corn	11-19	62-70	8-16	3-5.9	1.5-3.5	1.2-2.6
Barley	10-12	58-65	12-18	1.8-3.7	1.8-9	1.5-5

Table 8-7. Composition of major potato used in vinegar production (Chen, 2002)

Name	Moisture (%)	Carbohydrate (%)	Protein (%)	Lipid (%)	Cellulose (%)	Ash (%)
Potato	12.0	65-75	6	0.5	1.4	2.4
Sweet potato	12.8	68.5	3.8	1	2.3	3.0
Tapioca	12.0	67-72	3-9.5	0.9-1.2	1	2.0

Vinegars made from different raw materials have different flavors. For example, vinegars made from sticky rice have a better sweet taste for there is more dextrin and oligosaccharides left in the vinegar product; vinegar made from rice is much clearer for there is less content of protein and impurities in rice; vinegars made from sorghum have a better flavor for there is some content of tannin in sorghum; vinegars made from corn also have a better sweet taste for there is more phytic acid which can promote the formation of alcohol materials during the fermentation. Vinegar made from different fruits has different fruit flavors (Chen, 2000).

Recently, special type cereals such as oats, buckwheat, tatarbuckwheat and so on have also been used in vinegar making for their special nutrition and health function.

2.2. Filling Materials and Accessory Materials

Chinese traditional vinegars are mainly processed by a solid-state fermentation method (SSF) and the process needs filling materials to transport oxygen from the atmosphere to the cell surface so that the oxidation of ethanol can be accelerated. A large number of accessory materials are also needed to provide the nutrition of the microbial growth or to increase the content of carbohydrate and amino acids (Chen, 2002). Major filling materials used in vinegar are hulls of cereals such as sorghum and corn while major accessory materials used in vinegar are wheat bran, rice hull and okara of soymilk (Chen, 2000). Wheat bran is the main carrier of microorganisms in the SSF system; besides being a source of starch, it also provides other nutrients, such as proteins, to microorganisms. Rice hull has less nutritional value. It merely acts as a carrier of microorganisms and spacer to increase porosity of the substrate mixture (Liu and Li, 1992). But there is still not enough research on the effects of filling material and accessory material.

3. Nutrition and Taste of Vinegar

Vinegar is abundant in nutrients including organic acid, amino acid and minerals. The smell and taste is also special for its high content of amino acid. But the nutrition and taste are different significantly with the kinds and processing methods of vinegars. Table 8-8 gives the general composition and properties such as pH, total acids, sugar and salt of several famous vinegars in China (Bao, 1995). Though the pH of most Chinese vinegar is much higher than that of Japan, the taste is softer and blander relating to the amino acid content.

As a contrast, compositions of some Japanese vinegars are also given in table 8-9. It is clear that the compositions of vinegars in China or in Japan varied significantly with the types of vinegar but the total acid, total sugar, ash and density of Chinese vinegar were higher than that of Japanese vinegar. But high ash content which may result from the processing is still a problem.

Table 8-8. General compositions and properties of famous vinegars (Bao, 1995)

Name of vinegar	Density	pH	Total acid (%)	Reducing sugar (%)	Total sugar (%)	saltfree solids (%)	Salt (%)	Nitrogen (%)	Ash (%)
Changde vinegar	1.094	3.68	5.88	2.79	3.45	12.15	3.86	0.71	5.03
Jilin rice vinegar	1.072	3.65	5.13	2.12	8.91	12.79	0.02	0.32	1.14
Zhangdemature vinegar	1.128	3.69	8.85	3.30	3.81	19.74	4.65	0.97	6.62
Beijing smoked vinegar	1.056	3.87	6.15	0.63	0.83	9.73	0.84	0.64	1.87
Beijing sticky rice vinegar	1.086	3.74	6.82	2.59	2.79	13.77	1.59	0.79	2.91
Zhengjian vinegar	1.086	3.73	6.82	1.50	1.84	11.91	3.18	0.69	4.39
Shanxi smoked vinegar	1.141	3.82	7.99	8.51	8.73	21.39	3.97	0.56	6.65
Shanxi mature vinegar	1.194	3.87	10.38	11.25	12.82	30.47	5.33	1.22	9.42
Sichuan Sanhui vinegar	1.114	3.83	7.18	4.50	7.32	21.37	1.47	1.25	3.39
Fujian rice vinegar	1.012	2.87	6.33	0.17	0.18	0.25	0.006	0.007	0.003

Table 8-9. The compositions of Japanese vinegar (Bao, 1995)

	Rice vinegar	Seed residues	Alcohol vinegar	Speedy fermented alcohol vinegar	Mature alcohol vinegar	Fujian rice vinegar made in Japan
Total acid (%)	4.60	4.59	4.73	5.33	10.1	4.45
non-volatile acid (%)	0.37	0.22	0.11	0.21	0.29	0.31
Ethanol (%)	0.15	0.18	0.033	0.36	0.14	0.31
Total sugar (%)	4.97	1.30	1.14	1.84	0.15	
Reduction sugar (%)	3.0	0.018	0.53	0.69	0.06	0.09
Total nitrogen (%)	0.035	0.018	0.032	0.010	0.009	0.113
Amino nitrogen (%)	0.017	0.008		0.0008	0.001	0.003
Solid (%)	5.86	1.71	1.30	0.64	0.31	2.06
Ash (%)	0.72	0.58	0.02	0.40		0.47
Density	1.049	1.018	1.001	1.011	1.018	1.013
pH	2.70	2.65		2.61	2.25	3.12

Table 8-10. Amino acid of famous Chinese vinegars (mg/100ml) (Bao, 1995)

	Changde vinegar	Jilin rice vinegar	Zhangde mature vinegar	Beijing smoked vinegar	Beijing sticky rice vinegar	Zhengjian vinegar	Shanxi smoked vinegar	Shanxi mature vinegar	Sichuan Sanhui vinegar
Aspartic acid	134.2	23.0	134.8	62.	150.7	94.6	141.2	171.4	160.0
Threonine	57.9	30.9	84.0	64.8	74.7	41.8	60.8	76.3	78.6
Serine	92.2	38.2	124.8	96.2	100.3	69.0	92.1	107.1	122.0
Glutamic acid	249.4	69.4	408.7	118.6	274.9	85.3	273.0	283.1	335.0
Proline	28.3	10.6	31.0	24.7	31.0	12.4	45.0	51.1	304.0
Glycine	74.7	36.7	105.9	104.4	96.0	45.7	76.7	87.0	128.0
Alanine	204.3	152.	371.5	364.9	323.5	160.1	291.4	376.0	577.0
Valine	133.5	56.3	187.0	153.9	174.6	92.2	137.8	185.2	246.0
Methionine	56.3	19.5	56.3	47.8	59.3	27.2	46.2	53.8	75.2
Isoleucine	80.4	42.0	136.4	99.7	110.3	62.0	84.0	119.1	350.0
Leucine	192.5	74.5	239.5	199.5	215.6	150.2	200.3	236.0	178.0
Tyrosine	51.0	22.3	60.2	16.2	37.7	32.1	53.8	57.8	92.0
Phenyl-alanine	81.9	16.9	54.5	64.1	77.0	41.0	67.3	76.3	164.0
Histidine	40.5	4.1	46.9	8.2	22.5	8.8	23.2	25.5	49.5
Lysine	193.5	62.7	210.7	115.9	210.0	121.8	144.5	180.4	165.0
Arginine	149.9	16.6	65.0	5.7	25.8	71.1	148.2	158.8	101.0
Total amino acid	1820.5	674.2	2331.7	1546.9	1990.1	1116.8	1894.3	2270	3125.3

Bao (1995) also reported that one important character of Chinese vinegars is the high content of amino acid (Table 8-10). Most vinegars in China contained 1000 mg amino acid in 100ml vinegar while there is almost fewer than 100 mg in Japanese vinegars. The mature processing is the main reason for higher amino acid content.

From table 8-11 and 8-12, we can find the aromatic compositions of vinegar are very complicated and may be affected by raw materials of vinegar and processing. It is found that vinegar of rice or okara and malt has high acetoin content which results in the vinegar having a bad smell.

4. Manufacture of Chinese Vinegar

Vinegar is consumed worldwide as a food condiment and preservative, especially in the Chinese diet. Generally, fermentation of Chinese vinegar can be divided into solid-state fermentation (SSF) and liquid fermentation. There are many types of Chinese vinegar on the market which are best sold at home and abroad. And most Chinese vinegar is produced by SSF, such as Shanxi old mature vinegar, Zhenjiang aromatic vinegar, Sichuan bran Vinegar (Chen, 2002). This chapter will firstly introduce the SSF process for Chinese vinegar and the liquid fermentation will be discussed.

Table 8-11. Alcohol and aromatic compositions of Japanese vinegars (ppm) (Bao, 1995)

	Rice vinegar	Okara vinegar	Apple vinegar	Malt vinegar	Grape vinegar	Alcohol vinegar	High acid content vinegar
Methanol	84	2~250	4~22	12~27	27	0~22	
Ether alcohol	270~3800	660~1600	1400~ 12000	260~3000	1500~ 3000	830~1300	
Propanol	0.5	0.6	trace	0.3~0.32	0.2~0.5	0.4	
2- Propanol	4.5	0.3~4.8	2.7~4.1	1.7~4.1	1.9~4.1	0~3.5	
Butylalcohol	trace	Trace~2.2	Trace~1.8	1.2~110	Trace~2.1	Trace~4.4	
Isobutylalcohol	43	12~84	15~240	23~840	28~37	3~34	
Pentanol	1.4	0.9~5.6	Trace~34	1.6~89	0.7~1.0	Trace~4.1	
Isobutan	25	0.7~5.1	0.4~8.5	2.8~18.2	4.9	0.4~1.8	
Octanol	75	14~40	6~1000	19~28	27~40	5~24	
2,3-butandaiol		2	2~25	trace	0	Trace~32	
Acetaldehyde	15~48	5~130	13~130	5~1000	21~50	45	5~55
Propionicaldehyde	Trace~30	Trace~20	7~46	4~25	9~52	2~26	8
Phenylaldehyde	0~trace	0~5	0.2~4	Trace~3	0	0.4	0~trace
Acetate ether	22~280	47~665	65~1000	0~242	50~200	53	5~205
Acetoin	15~355	8~706	17~73	17~9600	21~51	45	Trace~24
Diacetyl	4~15	0.8~84	4~9	3~117	2~3	2~3	0~2
Furfural	Trace~4	Trace~8	Trace~7	6~7	6~	0.4~0.6	4~16

Table 8-12. Alcohol and aromatic compositions of Chinese vinegars (ppm) (Bao, 1995)

	Changde vinegar	Jilin rice vinegar	Zhangde mature vinegar	Beijing smoked vinegar	Beijing sticky rice vinegar	Zhengjian vinegar	Shanxi smoked vinegar	Shanxi mature vinegar
Ethanol	369.0	2070.1	1333.3	2857.1	1152.8	465.5	324.6	187.1
Propyl alcohol	544.6	113.2	918.7	271.1	610.9	513.5	461.8	222.6
Isopropyl alcohol	758.3	-	1808.8	14.3	1674.8	1055.9	140.8	-
Isobutyl alcohol	-	343.5	379.0	214.3	340.8	42.11	-	-
Butane-diol	61.6	45.0	111.5	135.7	451.6	219.3	275.9	466.2
Acetoin	368.4	400.5	351.6	444.7	746.7	203.4	214.8	572.8
Diacetyl	45.2	38.2	41.7	30.6	34.8	34.1	51.7	3.7
Acet-aldehyde	4.5	-	23.4	8.5	13.9	19.9	-	-
Ethyl acetate	265.5	235.7	431.8	191.2	245.0	216.3	584.3	1119.3
Propyl acetate	337.3	456.3	196.2	77.1	30.8	122.3	-	24.1
Butyl acetate	9634.4	7080.8	6765.2	3127.1	4636.2	4240.7	5850..0	6506.9
Isoamyl-acetate	111.8	94.3	288.0	680.3	1103.0	271.7	98.2	1132.0
Furfural	82.0	203.7	85.6	243.7	118.1	21.9	47.9	106.8
Ethyl benzoate	12.7	13.3	6.1	5.3	9.2	-	2.5	13.6
Total	12634	11145	12838	8412	11281	7840	8073	103429

4.1. Solid-state Fermentation of Chinese Vinegar

Although vinegar is produced and consumed worldwide, there are few publications on vinegar production by SSF processes these days. Hubert (1976) reviewed vinegar history including fermentation technology, but did not include Chinese and Japanese SSF processes. The Japanese SSF processes are rooted in and similar to relevant processes in China (Bao, 1988).

The SSF of Chinese vinegar consists of four distinct stages, namely koji preparation, saccharification of starch and alcohol fermentation, acetic-acid fermentation, and maturation (Huang and Yin, 2000a). Figure 8-4 is a general flow chart of Chinese vinegar SSF.

Traditional production processes of Chinese vinegar are typically empirical and craftlike. In the past, processes were developed by trial and error, and handed down by mouth and hands of the veterans in the plants. A great body of experience and practical knowledge accumulated over time, but the underlying microbiological and biochemical mechanisms were not investigated systematically. Consequently, many different production processes exist and significant improvements have been scarce. There are two basic SSF processes in Chinese vinegar production, based on the treatment method of the raw material, namely the ''cook method'' and the ''steam method'' (Huang and Cai, 1999). In the ''cook method'', saccharification and alcohol fermentation occur in the liquid state (Figure 8-5A), whereas the acetic acid fermentation is SSF. In the ''steam method,'' the whole process takes place in the solid state (Figure 8-5B), where rice hull and part of the wheat bran are mixed with the main raw material (grain) immediately after soaking and are steamed together.

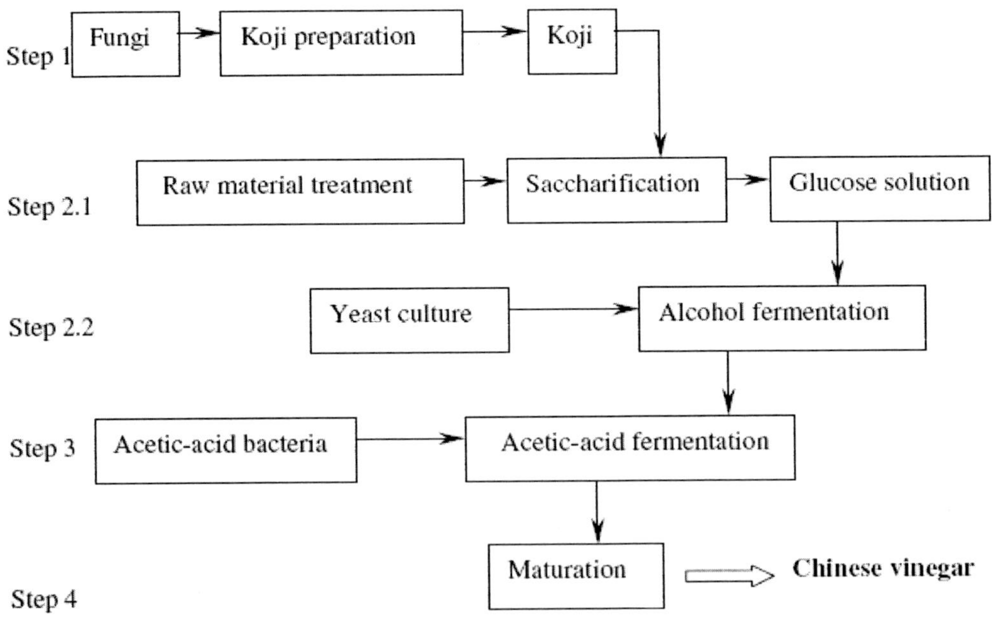

Figure 8-4. general flow chart of SSF Chinese vinegar depicting the four steps (Liu et al., 2004).

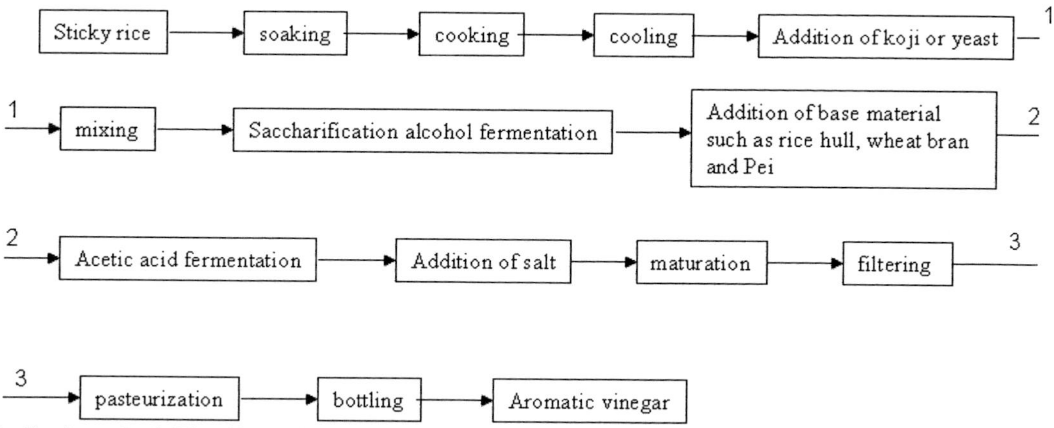

A. Cook method (Zhenjiang vinegar).

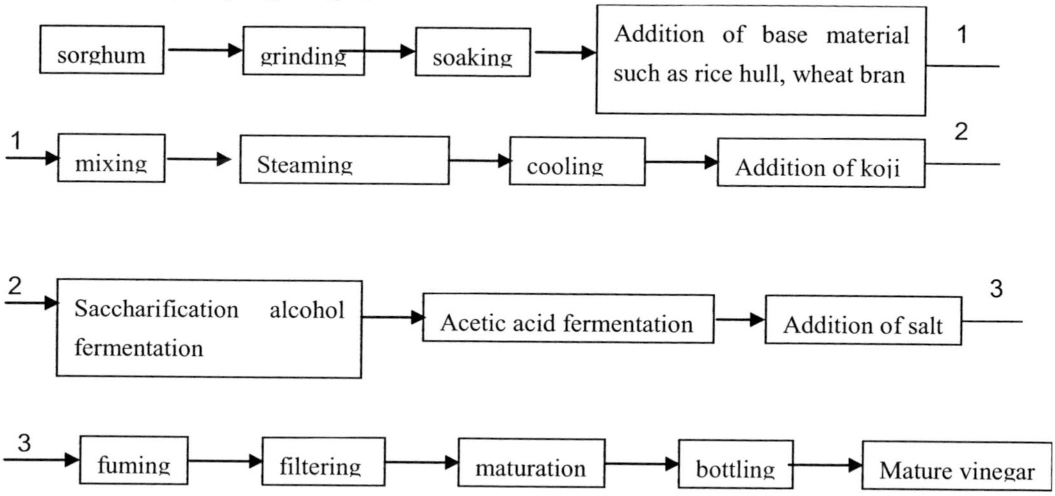

B. Steam method (Shanxi mature vinegar).

Figure 8-5. Flowchart of solid-state fermentation of Chinese vinegar.

4.1.1. Koji Preparation

Koji is a solid enzyme-rich mixture. It serves as a source of enzymes for the saccharification of starch in the vinegar process much as malt serves a similar purpose in Western beer manufacture. The koji process was developed centuries ago in the Orient. Its preparation method originated in China and then was widely used in Japan and other Southeast Asian countries (Fukashima, 1989).

In China, the production of distilled spirit (alcohol), soy sauce and vinegar all use koji for saccharification. Koji can be classified as great koji, mini koji, bran koji, wheat koji, herb koji, and red-rice koji according to its preparation method (Liu et al., 2004). Koji contains many kinds of enzymes and microbes. However, various moulds, such as *aspergillus*, *rhizopus*, *mucor* and *penicillium* are the dominant microorganisms in koji (Huang and Yin, 2000). Different types of koji have different microbial floras. For example, most moulds in great koji are *rhizopus* and *mucor*, while few of them is *aspergillus* (Zhang, 2000). The preparation of koji is laborious and time consuming. Some manufacturers tried to replace the saccharifying action of koji with enzyme mixtures. However, they found that the quality of the vinegar was not good enough, sometimes even very poor (Hou et al., 1996).

The preparation of great koji and mini koji, introduced as follows, are widely used in making traditional Chinese vinegar. Other kinds of koji such as wheat and bran koji are a pure-culture product and much simpler to prepare (Kitamoto, 2002).

4.1.2. Great Koji Preparation

Great koji is mainly made from barley and pea. The process of great koji preparation needs more than 10 steps and more than 21 days (Chen, 2002). A rich multi-enzyme system and metabolites are obvious advantages of great koji. Liquefaction and saccharification ability of great koji are so powerful that great koji can translate starch into glucose and other sugar by starch enzyme hydrolysis and protein into amino acid by protein enzyme hydrolysis, ferment glucose into grain alcohol, and also produce many kinds of organic acid during the brewing process (Shan et al., 2005). In processing of Shanxi mature vinegar, great koji is not only an important source of enzyme and microorganisms, it is also an important resource of raw material for the high addition ratio of great koji. In normal Shanxi mature vinegar making, the addition ratio of great koji may be as high as 50~60% of sorghum (Shan et al., 2005).

The steps of the process are schematized in Figure 8-6 (Liu et al., 2004). But oats and buckwheat is used for koji processing now to increase the health function. Raw materials for making great koji are usually composed of 70% barley and 30% pea. Barley and pea are ground into a powder and producers make sure the coarse powder (diameter is over 1mm) is 30–45% of the whole powder. Then, the powder is mixed with some water to configurate koji cakes which is 3.5 kg. The ratio of powder to water is 1-0.5~0.55. Koji cakes are piled in a line in the "koji house" where the temperature is about 25 °C. The temperature goes up when the windows are closed, and the koji cakes get moldy. The aeration involves opening the windows and expanding the spacing among the koji cakes to cool the cakes (Zhang, 2000). There are three stages in the fermentation and all of them are at a different temperature. The temperature varies from 30°C to 48°C and is controlled by adjusting the space among the cakes and opening or closing windows (Wang, 1995). The maturation lasts for 2 days and the temperature of great koji is controlled at 28°C~30°C (Zhang, 2000). The most important processes are metabolic heat production, heat convection, and water evaporation (Liu et al., 2004). The high-quality great koji should not contain too much water (15%) in its center and the content of total acid should be lower than 0.5% (Zhang, 2000).

4.1.3. Mini Koji Preparation

Mini koji preparation is similar to that of great koji. Raw materials for making mini koji are mainly barley and rice. The steps of the process are schematized in Figure 8-7 (Zhang, 2000).

The rice is ground and then dipped in some cold water. Then it is put in boiling water. The pure rice hull and koji seeds are also ground. The proportion of the pure bran is 87–92%, the rice powder is 5–10%, Quzhong (resource of koji, Qu species) is 3%. Water is 64–70% weight of the raw materials (Zhang, 2000). The raw materials are mixed 4–5 times to make koji cakes. The fermentation of mini koji preparation is almost the same as great koji, but the temperatures of the three stages vary from 22°C to 39°C. The koji cakes are dried until the proportion of water is lower than 11% (Zhang, 2000).

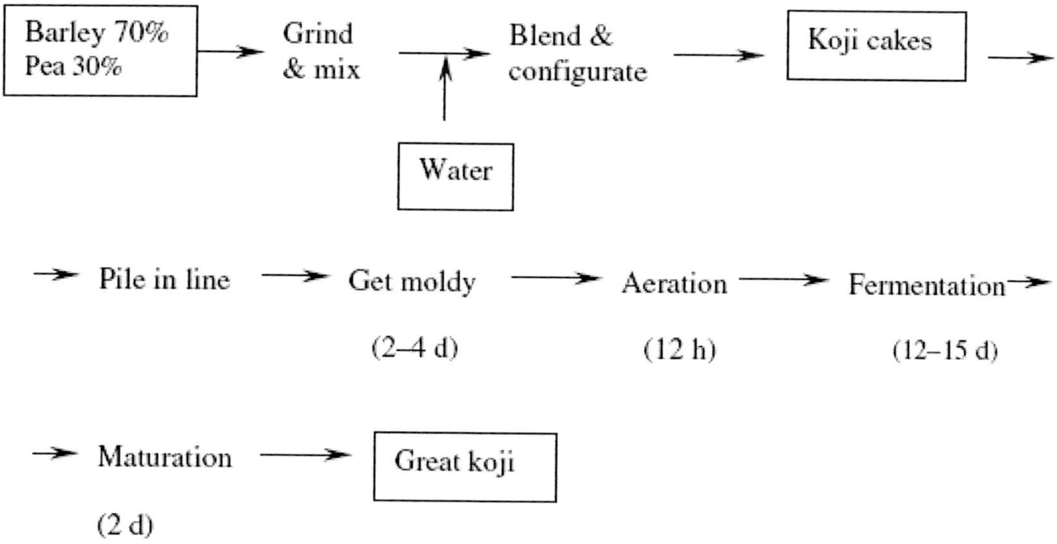

Figure 8-6. Typical flowchart of preparation of Chinese great koji (Liu et al. 2004).

4.1.4. Saccharification and Alcohol Fermentation

In traditional processes, there are two different heat treating methods of the raw materials, namely the "cook method" and the "steam method" (Huang and Cai, 1999). No matter which method is chosen to treat with raw materials, there are some things similar between two methods.

(1) All the microorganisms in saccharification come spontaneously from the environment or are indigenous to the koji and create so many kinds of products that provide a unique flavor for the traditional Chinese vinegar (Lei, 2000).
(2) Alcohol fermentation starts as soon as the liquefaction and saccharification of starch. In some traditional Chinese vinegar processes, acetic-acid fermentation, alcohol fermentation, liquefaction and saccharification of starch starts at the same time.
(3) The utilization of raw materials is usually insufficient.
(4) The average duration of saccharification is 5~7days (Zhang, 2000).

However, in recent years, some traditional Chinese vinegar manufacturers replaced koji with enzyme mixtures to improve this process. By adding yeast, the starch of the raw material is liquefied and then saccharified completely (Chen and Chen, 1996).

Figure 8-7 is a typical alcohol fermentation of vinegar processing. The temperature during alcohol fermentation must be lower than 28°C so that the evaporation of ethanol can be minimized. And the adjusting of temperature is by opening or closing the slipcover and the window of the workshop. The whole process of saccharification and alcohol fermentation needs 10 days (Huang and Cai, 1999). The product of saccharification and alcohol fermentation of SSF has a better flavor than others (Zhang, 2000).

Figure 8-7. Alcohol fermentation of vinegar processing.

4.1.5. Acetic-Acid Fermentation

In traditional processes, the fermented materials from the alcohol fermentation are mixed with wheat bran and vinegar Pei (which is a kind of material from the last batch of acetic-acid fermentation). Some manufacturers, however, take an inoculum of acetic-acid producing bacteria for vinegar seeds (Liu, 2004). The ratio of the main substrate to the bran and rice hull is generally 1:0.5~0.7:0.8.The mixed substrate contains 60%~65% water and 4~5mL/100g ethanol (Zhang, 2000). The average duration of acetic-acid fermentation is 8~9 days.

In SSF, keeping the proper content of starch and water in the mixed substrate is very important for acetic-acid fermentation. High content of starch results in low utilization of raw materials and the temperature of the fermentation becomes hard to control. Generally, the content of starch in the mixed substrate is 14%~16% (lower in summer, higher in winter). If the content of water is too low, the duration of the maturation will be longer. By contrast, if the content of water is too high, utilization of raw materials would be affected for shortage of substrate.

In the acetic-acid fermentation (Figure 8-8), the rate of ethanol oxidized to acetic acid depends on the temperature and availability of oxygen. Temperature and the supply of the oxygen are controlled by stirring and turning the substrate (Liu, 1982). If there is sufficient oxygen, bacteria will propagate and metabolize rapidly and produce a large amount of metabolic heat. When the temperature becomes too high, the operator will "turn up" or mix the substrate to disperse heat, and then press the surface of the container to reduce the oxygen supply, thereby decreasing the rate of ethanol consumption and heat production (Liu et al., 2004). Generally, stirring and turning up the substrate is carried out once per day and the temperature is usually kept at 38 °C to 40 °C (Guo, 2000).

Figure 8-8. The sightseeing of acetic acid fermentation of vinegar processing.

4.1.6. Fuming of Vinegar

After acetic acid fermentation, the substrate is fumed and filtered to extract vinegar especially for Shanxi mature vinegar (Figure 8-9). The fuming of the substrate takes about 6~7 days and the extract of vinegar becomes deep brown and the flavors becomes excellent. The temperature of the substrate is about 70 °C during fuming and the variations of chemical compositions are very complicated but there is still not any research on the chemical reactions during fuming which result in the variation of vinegar in taste, color and so on.

Figure 8-9. Fuming of vinegar.

A. Maturation of Zhengjian vinegar.

B. Maturation of Shanxi mature vinegar.

8-10. The container of vinegar during maturation.

Figure 4.1.7. Maturation (Aging)

Most traditional production processes of Chinese vinegar feature a maturation period in which many flavor substances, such as esters, are formed by chemical reactions in the period. Salt is added to vinegar as soon as the acetic-acid fermentation ends to prevent microbial overoxidation of acetic acid to carbon dioxide. The content of added salt is usually 1.5%~2% of the vinegar Pei (higher in summer, lower in winter) (Chen, 2000). In the maturation period, there is always a long time of storage, which varies in the different traditional Chinese vinegar processes. For example, manufacturers of Zhenjiang vinegar usually mature their vinegar by adding salt and sealing the surface of the fermented solid substrate with clay (Figure 8-10 A); thus, the maturation process is anaerobic (Guo, 2000). The manufacturers of Shanxi mature vinegar, however, use solarization (in the summer) and ice removal (in the winter) to mature and to concentrate the vinegar in open clay (Figure 8-10 B); thus, the course of the latter is mainly aerobic (Liu, 2004). The final concentration during maturation of Shanxi mature vinegar will be increased over 3 times in comparison with the original concentration after about one year of maturation. Not only does the concentration of the vinegar vary, but also there are many chemical reactions during the aging process. The levels of concentration, dissolvable solid substance, non-volatile acid, deoxidization sugar, and amino acid are increased. The long aging period makes Shanxi mature vinegar keep an excellent flavor, very good sense and mild acidity (Shan et al. 2005).

4.2. Liquid Fermentation

Most of Chinese traditional vinegars are made by SSF and they have a better flavor and aroma properties than liquid fermentation (Table 8-13). But large-scale application of SSF is still limited for it takes a long time and needs a lot of space. In recent years, many kinds of vinegar prepared by liquid fermentation have appeared on the market.

Table 8-13. Flavors of different types of vinegar (Wang, 2006)

Category	Non-volatile acid g/100mL	Lactic Acid g/100mL	Ethyl Lactate g/100mL	Amino acid N g/100mL
Vinegar by SSF	0.88	0.745	0.00462	0.08~0.12
Vinegar by liquid fermentation	0.135	0.191	—	—

The liquid fermentation of Chinese vinegar consists of two distinct processes. The first step is an ethanol fermentation performed mostly by yeasts. And the second step, acetogenic fermentation is carried out by acetic acid bacteria (Robert, 2006)

4.2.1. Ethanolic Fermentation

The first step of liquid fermentation is relevant to saccharification and alcohol fermentation of SSF. But the koji is replaced by some pure yeasts, which are also widely used in the modern distilled spirit fermentations, such as Rasse II yeast, Rasse □ yeast, k-tuple yeast (introduced from Japan), number 1300 yeast, 1308 yeast (Zhang, 2000).

4.2.2. Acetogenic Fermentation

The second step of liquid fermentation, conversion of ethanol substrate to acetic acid, can be performed by one of several methods, all of which rely on oxidative fermentation of ethanol by acetic acid bacteria. These processes are generally referred to as the open vat method, the trickling generator process, and the submerged fermentation process (SMF). The latter two are widely used now, especially for large vinegar manufactures since they can be performed in continuous mode and can reduce the fermentation time from several weeks to a matter of days (Robert, 2006).

The ability of acetic acid bacteria to perform the acetic acid fermentation is limited primarily by the diffusion or transport of oxygen from the atmosphere to the cell surface. It is possible to significantly accelerate the oxidation of ethanol by increasing the surface area to which oxygen is exposed.

Most of the vinegar produced worldwide is now made using submerged fermentation systems. Some kinds of Chinese traditional vinegar are prepared by combined SSF and SMF, such as Jiangzhe rose vinegar and Fujian rice vinegar (Liu, 1982; Huang and Cai, 1999). The most important feature of SMF is their ability to provide rapid and efficient aeration. For example, turbine is equipped to mix the liquid with air or oxygen and deliver the aerated mixture at very high rates inside the fermentor. The aeration system's ability to break up air bubbles and facilitate transfer of oxygen molecules from the gas phase to the liquid phase is essential, since the success of SMF relies on the transfer of oxygen from the medium to the bacteria.

5. Research and Technological Advances in Vinegar

5.1. Technological Advances in SSF

At present, in SSF, specific enzymes are usually added to improve efficiency, and have shortened the vinegar fermentation cycle. Adding enzyme mixtures makes the starch material liquefied and saccharified more quickly and completely. Different α-amylase, glucoamylase and acid resistant proteinase are widely used in saccharification and alcohol fermentation (Li, 2006).

There are two kinds of α-amylase with high temperature and medium temperature resistant used in vinegar production. Medium temperature resistant α-amylase has a better effect on the starch material liquefied quality and a shorter duration. It also has a lower dosage than high temperature resistant α-amylase, and is easier to control in appropriate temperature and pH value (Table 8-14).

Table 8-14. The effects of high and medium temperature resistant α-amylase on liquefied quality of starch material (Li, 2006)

Types of α-amylase	Appropriate temperature (°C)	Appropriate pH value	Dosage of one g materials (IU)	Duration of liquefied (min)	Quality of liquefied
high temperature resistant α-amylase	90-110	5.0-5.8	6	60-100	saccharified completely
medium temperature resistant α-amylase	65-90	5.5-5.7	2-4	40-60	saccharified incompletely

Table 8-15. Effect of adding acid resistant proteinase on alcohol fermentation (Li, 2006)

Additive	Alcohol (V/V) after 36 hours fermentation	Alcohol (V/V) after 48 hours fermentation	Liquor Clarity
Control	7.7	9.4	general clear
Acid resistant proteinase	10.2	10.7	Very Clear

Table 8-16. Flavors of vinegar with different additive during maturation (Li, 2006)

Additive	Amino acid N (g/100mL)	Carbohydrate (g/100mL)	Total Acid (g/100mL)	Non-volatile acid (g/100mL)
Control	0.075	1.45	5.10	0.38
2% bran koji	0.09	1.26	5.05	0.83
Glucoamylase and acid resistant proteinase	0.099	1.48	5.05	0.89

Acid resistant proteinase converts protein into amino acid and peptide. Adding acid resistant proteinase can promote the protein hydrolysis of the raw materials in alcohol fermentation. It also can increase absorbed nitrogen for the yeasts to promote yeast growth and reproduction. Therefore, the duration of alcohol fermentation is shortened and yield increases (Lei, 2000). After 36 hours of alcohol fermentation, the content of alcohol in the vinegar product by adding acid resistant proteinase is higher than the content in the control after 48 hours of alcohol fermentation. The utilization of raw materials and equipment turnover rate are improved as shown in Table 8-15 (Li, 2006). Vinegar added glucoamylase and acid resistant proteinase in the maturation process is reported having a better flavor and aroma properties than others (Table 8-16) too.

5.2. Advances in Liquid Fermentation

Vinegar made by SSF has a better flavor and aroma properties than liquid fermentation because of different methods of fermentation or fermentation cycle. The former has a longer fermentation cycle so the products have a better flavor. However, the latter has high efficiency so it is suitable for large-scale production.

In the alcohol fermentation stage, adding 3–5% of the common lactic acid bacteria together with yeast, the content of acid in vinegar can be increased. At the same time, the content of non-volatile acid increases by about three-fold, as well as the creation of favorable conditions for ethyl lactate, which is the major aroma property in the vinegar (Table 8-17) (Fan, 2004).

Adding 2% of the flavor-producing yeast in the alcohol fermentation can increase the content of esters as well as non-volatile acid. Different genera and species of yeast have different production capacity of esters (Table 8-18) (Fan, 2004).

Table 8-17. Flavors of vinegar with added lactic acid bacteria (Fan, 2004)

Additive	Total Acid (g/100mL)	Volatile acid (g/100mL)	Non-volatile acid (g/100mL)	Total Esters
Control	5.53	5.44	0.135	24.1
Latic acid bateria together with yeasts	5.26	4.8	0.696	40.18

Table 8-18. Quality of vinegars with different yeasts added (Fan, 2004)

Name	Alcohol volume fraction (%)	Total Acid (%)	Total Esters (%)	Acidity
AS2300	4.8	0.0406	0.3588	0.357
1312	3.9	0.0444	0.4231	0.299
1342	3.2	0.0486	0.3290	0.318
1343	2.8	0.0463	0.4674	0.328
⍰Alcohol yeast	9.9	0.0104	0.0380	0.212

It was reported that using acetate-acetic acid bacteria to ferment could increase the total acid of vinegar 5% (Fan, 2004).

The other effect of technological advances in liquid fermentation is that they appropriately extend the fermentation process, such as an appropriate extension of alcohol fermentation and storage life cycle. When the alcohol fermentation cycle is extended to 7–9 days, the quality of the vinegar is improved (Wang, 2006). The container of vinegar during maturation has some effects on the quality of the vinegar. If liquid fermented vinegar is stored in stainless steel containers for 3 months, the ethyl acetate of vinegar will be increased by 42%. However, the increase of ethyl acetate in vinegar will be 30% if stored in plastic buckets and 20% in bottles (Fan, 2004).

5.3. The Function of the Vinegar

It is said that brewed vinegar, a commonly used condiment of food, also has medicinal uses by virtue of its physiological effects, such as antibacterial, antioxidant activity, regulating blood pressure, blood glucose, antitumor, promoting recovery from exhaustion and promoting calcium absorption.

5.3.1. Antibacterial Effect

The bacteriostatic and bactericidal activity of vinegar is obvious. Vinegar exerted strong antibacterial effect against streptococcus, diplococcus pneumoniae, staphylococcus, influenza virus and other respiratory virus pathogenic microorganisms. In vinegar obtained through a secondary aerobic fermentation of, e.g., alcoholic fermented rice (East Asia) or from beer without added hops (Europe), the stored grains of cereals are metabolically in a resting state, which is primarily controlled by low water activity (aw%0.6, 14% moisture). In this state the constituents are not available for microorganisms, and the endogenous enzymes are inactive.

There are a lot of reports about the antibacterial effect of vinegar and acetic acid. Fumigation with acetic acid was found to be effective for *botrytis* and *penicillium* molds in preventing post-harvest decay of apples (Sholberg and Gaunce, 1995).

The combined effects of acetic acid, temperature and sodium chloride on *Escherichia coli* O157:H7 inactivation was examined in processed vinegar. Shigetomo et al. (1998) proposed quadratic polynomial models to express the bactericidal effect of processed vinegar against *Escherichia coli* O157:H7. *Escherichia coli* O157:H7, which is recognized as an important cause of hemorrhagic colitis and hemolytic uremic syndrome, is disseminated by food and person-to-person contact. One of the physiological characteristics of *E.coli* O157:H7 is its acid tolerance. In addition, organic acids are often used in food-processing procedures designed to eliminate contaminating bacteria, whereas *E.coli* O157:H7 has an extreme resistance to organic acids. On the other hand, when vinegar (acetic acid) was used in combination with sodium chloride, a predominant synergism on the bactericidal activity against *E.coli* was found. The antibacterial effect of acetic acid has been explained by the ability of undissociated acetic acid molecules to pass through the cell membrane. Their study showed that the acid tolerance of *E.coli* O157:H7 varies from its growth conditions. On the other hand, cells grown in acidic medium could induce more acid-tolerant properties. In addition, the acid tolerance of *E.coli* O157:H7 varies among the strains.

Molina et al. (2005) compared the antibacterial effect of acetic acid and citric acid, and concluded that acetic acid exerted a greater antibacterial effect than citric acid at the same concentration and pH. In his study, acetic or citric acid was added to Luria Bertani (LB) broth to reach different final concentrations. The pH was then adjusted to different values with 1 M HCl or 1 M NaOH. LB broth pH 7.0 served as a growth control. Broth of each pH/organic acid concentration was inoculated with each strain in stationary phase to provide an initial cell density of approximately $10^7 ml^{-1}$. Three strains (serotypes O91:H21, O111: H- and the O157:H7 clinical isolate) were incubated in apple cider vinegar, capsicum pickle brine and apple juice. All strains survived for<10 min in apple cider vinegar (0.8 m acetic acid) and <1 h in brine (0.4 m acetic acid). When exposed to apple juices, these serotypes displayed different behavior indicating that the O157:H7 strain had the lowest acidic resistance.

Their experiments demonstrate that other than O157:H7 VTEC serotypes develop diverse stress resistance when challenged by acidic environment, although previous researchers mention the extreme acid resistance of *E.coli* O157:H7 (Diez and Russell, 1999).Under several acidic conditions, VTEC O91:H21 strain showed a stress resistance higher than the two O157:H7 strains tested, normally used as the target organism for HACCP regulation.

Lu et al. (2007) reported Chinese mature vinegar can kill almost all of *V ibrio parahaem olyticus* in 5 min (Table 8-19). The effect also depended on the temperature, and high temperature is preferred.

Table 8-19. Efficacy of vinegar in killing *V ibrio parahaem olyticus* (Lu et al., 2007)

Concentration of vinegar	2 min	5 min	10 min	15 min
100% vinegar	99.99	100.00	100.00	100.00
50%vinegar	99.38	99.52	99.82	100.00
25%vinegar	99.20	99.52	99.66	99.79

5.3.2. Antioxidant Activity

Xu et al. (2007) researched four Zhenjiang aromatic vinegars and demonstrated that Zhenjiang aromatic vinegar extracts (Table 8-20), and individual fractions of extract, vinegar melanoidins have DPPH radical scavenging activity, reducing power, and hydroxyl radical scavenging activity. They held that vinegar melanoidins are important for antiradical properties of vinegar. In their study, the antioxidant activity of high-molecular-weight fractions (MW>3500 Da) separated from ethanol supernatant extraction of concentrated Zhenjiang aromatic vinegar was evaluated by different in vitro tests. The DPPH radical scavenging activity, the reducing power, total phenol content, and the inhibitory effect on hydroxyl radical were analyzed. Each individual fraction was found to have antioxidant activity in all the model systems tested. But as shown in Table 8-21, the insolvable material of ethanol and water extract have the highest DPPH radical scavenging activity. At the same time, Decoction, storing and aging may affect vinegar's antioxidant activity.

Table 8-20. Yields and total phenolic content of different polar Zhenjiang aromatic vinegar fractions (Xu et al., 2007)

	Yield (g/L)	*Total phenolic content (μg/mg)*
Ethyl acetate extract	16.88	22.79
Acetone	19.54	22.49
Solution of ethanol	13.69	23.96
Insolvable material of ethanol	71.33	53.70
Water extract	16.05	42.89
Insolvable material of water	25.66	--

Table 8-21. DPPH radical scavenging activity (%) of different polar Zhenjiang aromatic vinegar fractions (Xu et al., 2007)

Concentration of solution	*150 μg/ml*	*250 μg/ml*
Ethyl acetate extract	11.98	29.16
Acetone	13.68	29.37
Solution of ethanol	13.26	30.43
Insolvable material of ethanol	56.31	71.44
Water extract	59.42	71.44
Insolvable material of water	-	-

Xu et al. (2006) researched the relationship between total phenol content and their DPPH radical scavenging activity of 26 kinds of Chinese vinegar. They found DPPH radical scavenging activity of vinegar is almost linear related with the total phenol content (Figure 8-10).

In Japan, some researchers have been studying the antioxidant activity of rice vinegar, such as Kurosu (unpolished rice vinegar). Shoko Nishidai et al. (2000) found that the antioxidative activity of an ethyl acetate extract of Kurosu (EK, most similar to Chinese vinegar) was much higher than that of the extracts of other types of vinegar in vitro, comparing the antioxidative activity of Kurosu, rice vinegar, grain vinegar, apple vinegar and wine vinegar. They also demonstrated that EK showed anti-inflammatory and antitumor promoting effects in mouse skin. Shimoji et al. (2002) isolated dihydroferulic acid (DFA) and dihydrosinapic acid (DSA) from Kurosu (unpolished rice vinegar) and suggested they are the major constituents responsible for Kurosu's radical scavenging activity. The levels of antioxidative activity of DFA and DSA in DPPH radical scavenging were higher than those of their respective structurally related compounds, ferulic acid and sinapic acid. The concentrations of DFA and DSA were low in common rice vinegar (polished rice vinegar), suggesting that Kurosu is more advantageous than rice vinegars as an antioxidative food item. As the concentrations of DFA and DSA were low in unpolished rice, too, these acids are thought to be produced in Kurosu through the process of the fermentation from ferulic acid and sinapic acid, respectively. The reason Kurosu exhibits stronger DPPH radical scavenging activity than other vinegars is that it contains higher concentrations of phenolic compounds than the other vinegars.

Our researches also showed different Shanxi mature vinegars made from different kinds of cereals such as oats, black wheat, tatary buckwheat and black sorghum and so on have much higher activities contents compared with sold rice vinegar (Table 8-21) and the main antioxidant compositions are also phenol and isoflavone (data is still waiting for publishing).

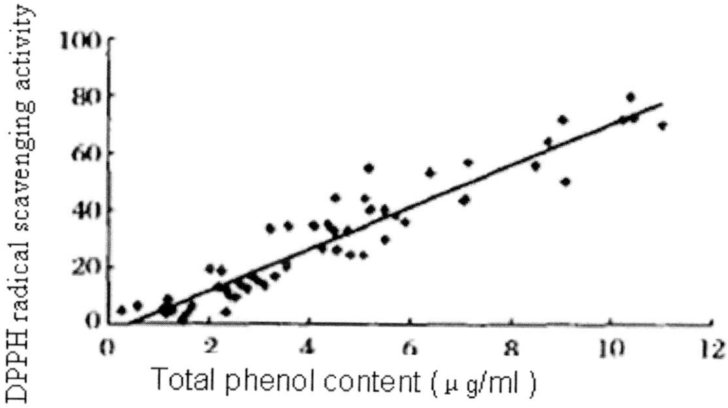

Figure 8-10. The relationship between total phenol content and its DPPH radical scavenging activity of vinegar (Xu et al., 2006).

Table 8-21. Polyphenolics and flavone contents of different kinds of Shanxi mature vinegar

Source	Polyphenolics (mg/ml)	Flavonoids (mg/ml)
Black sorghum vinegar	3.96±0.64	3.06±0.48
Tartary buckwheat vinegar	4.52±0.27	3.99±0.61
Oat vinegar	2.81±0.52	2.52±0.29
Black wheat vinegar	2.99±0.41	2.47±0.33
Black millet vinegar	1.69±0.16	0.93±0.07
Rice vinegar	0.38±0.02	0.18±0.01

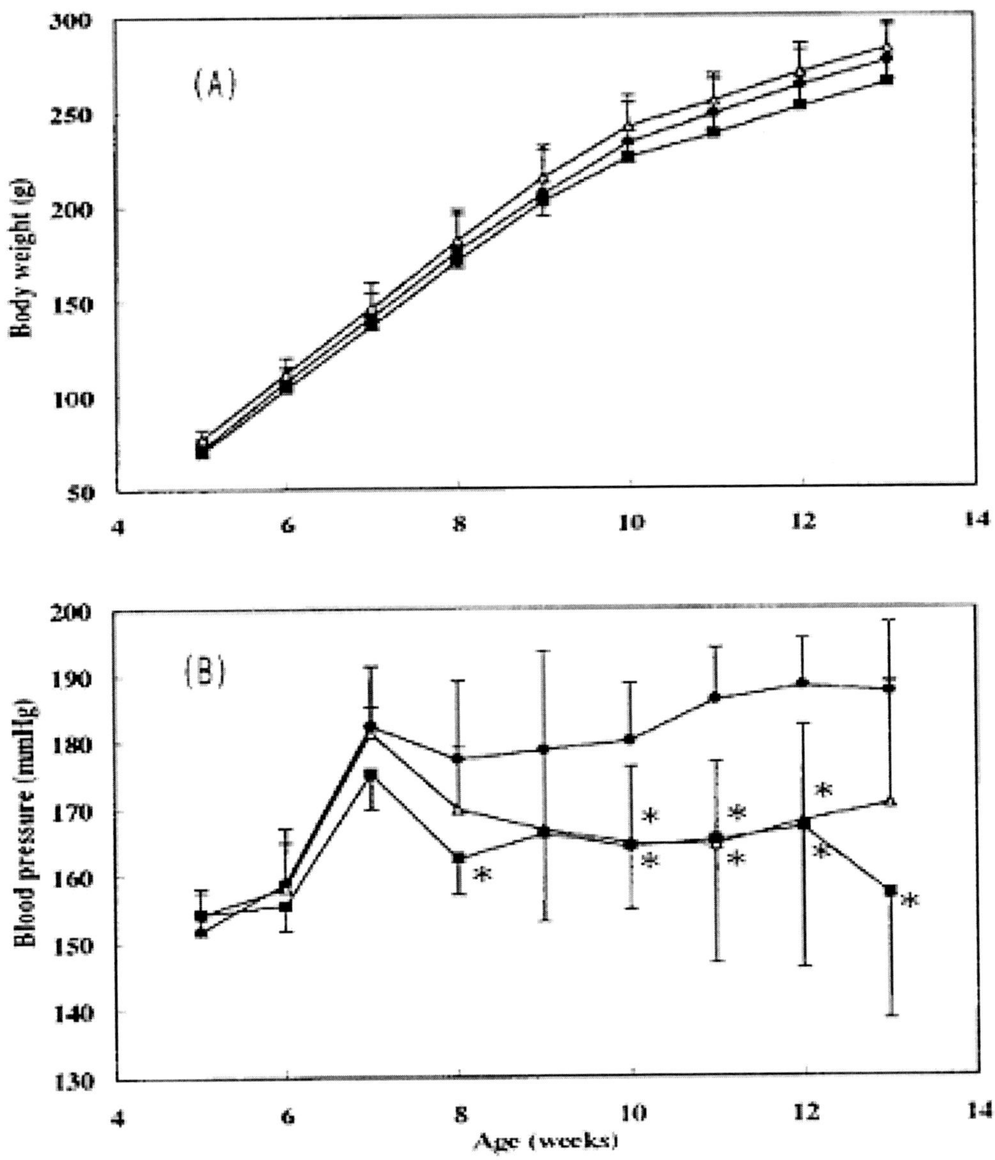

●, control group; □, acetic acid group; ■, rice vinegar group. Each point indicates the mean of six animals

Figure 8-11. The body weight (A) and the alteration in blood pressure (B) during feeding of a diet containing acetic acid or rice vinegar in SHR (Kondo et al., 2001).

5.3.3. Antihypertensive Effect

So far, it has been traditionally said that vinegar may play a role in lowering blood pressure, and many scientific attempts to test this have been done. Most were aimed at the rennin-angiotensin system, which is a typical blood pressure regulatory system, and these investigations were mainly done in vitro. In experiments in vivo, such as a single oral administration of ginseng vinegar to spontaneously hypertensive rats-stroke prone (SHRSP), which may show spontaneous hypertension, has been given. All these reports suggested that vinegar extracts, "minor" components of vinegar, play a role in blood pressure reduction. On the other hand, the main components of the commercial vinegar are acetic acid and saccharides, and these contents are 4–5% and 1–5%, respectively. Particularly, diets containing such a high concentration of acetic acid are only the various vinegars.

Ohnami et al. (1985) demonstrated that an ethanol extracted fraction of rice vinegar inhibited ACE activity in vitro. Moreover, Nishikawa et al.(2001) reported that blood pressure was reduced by the long-time administration of rice vinegar residues with acetic acid removed to SHR, suggesting that rice vinegar residues inhibit ACE activity in the rennin-angiotensin system.

To clarity the possibility of a preventive effect of dietary vinegar on blood pressure, long-term administration of vinegar or acetic acid to spontaneously hypertensive rats (SHR) was examined. In the study of Kondo et al. (2001), it was observed that acetic acid itself, the main component of vinegar, significantly reduced both blood pressure ($p<0.05$) and rennin activity ($p<0.01$) compared to controls given no acetic acid or vinegar as well as vinegar (Figure 8-11). There were no significant differences in angiotensin Ⅰ-converting enzyme activity in various organs. As for the mechanism of this function, it was suggested that reduction in blood pressure may be caused by the significant reduction in rennin activity and the subsequent decrease in angiotensin Ⅱ. From this study, it was also suggested that the antihypertensive effect of vinegar is mainly due to the acetic acid in it.

The results confirmed that the plasma rennin activities and plasma aldosterone contents of mice fed 10 weeks of acetic acid solution or rice vinegar are significantly different from that of mice fed a normal diet.

Our research also finds vinegar made from black sorghum has excellent effect on rabbit platelet aggregation in vitro and ability to dissolve plasma thrombus (data is still waiting for publishing) just as shown in Figure 8-12.

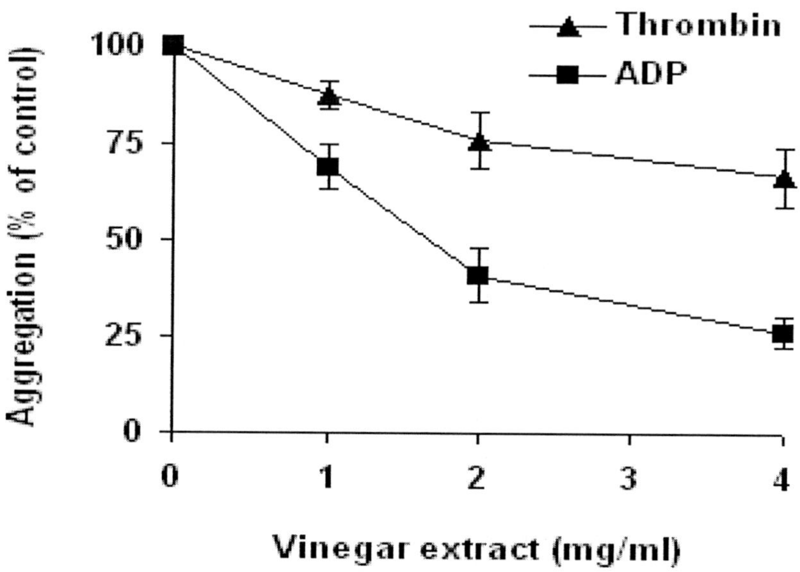

Figure 8-12. Effect of vinegar extract on rabbit platelet aggregation in vitro.

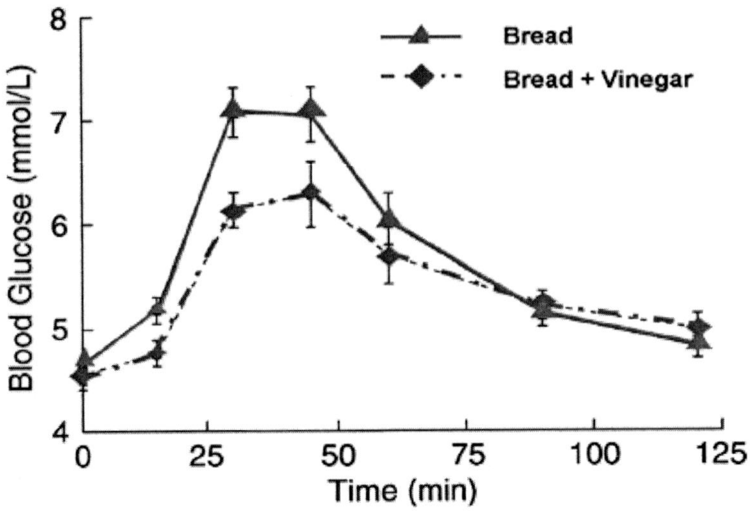

Figure 8-13. Vinegar Reduces Post-Prandial Glucose (James et al. 2008).

5.3.4. The Effect of Vinegar on Blood Glucose

Average mealtime glycemic load has been associated with risk of developing diabetes and obesity. Dietary glycemic load can be lowered either by reducing carbohydrate intake or by choosing carbohydrate foods that have a low glycemic index. A third possibility exists, however, to reduce postprandial glycemia that has not been well considered: complementary foods, such as vinegar. This approach is simpler than approaches that would require dietary change. The consumption of vinegar with meals was used as a home remedy for diabetes before the advent of pharmacologic glucose-lowering therapy. Indeed modern studies indicate that vinegar significantly reduces post-meal glycemia, probably because acetic acid slows gastric emptying and thus delays carbohydrate absorption and improves satiety. So, many studies focus on the effect of vinegar on blood glucose.

Based on preliminary data indicating that vinegar ingestion might favorably affect mealtime glycemia, Johnston and Buller (2005) examined whether vinegar products could lower postprandial glycemia without altering mealtime glycemic load. They concluded that the addition of vinegar products to a high–glycemic load meal significantly reduced postprandial glycemia. In their study, Eleven healthy subjects consumed two test meals (bagel and juice, glycemic load 81; or chicken and rice, glycemic load 48) under three conditions (control, vinegar, or peanut) using a randomized, crossover design. Vinegar ingestion reduced the 60-minute glucose response to both test meals by 55%, but these reductions were significant only for the high–glycemic load meal. After consumption of the high–glycemic load meal, energy consumption for the remainder of the day was weakly affected by the vinegar treatments, a reduction of 200 to 275 kcal (P.111). Regression analyses indicated that 60-minute glucose response to the test meals explained 11% to 16% of the variation in later energy consumption.

James et al. (2008) showed that 1 to 2 tablespoons of vinegar, when added to a meal containing high-glycemic-index foods such as white bread or white rice will lower post-prandial glucose by 25% to 35% (Figure 8-13). The addition of vinegar to a standard meal can improve the meal-induced oxidant stress by blunting the post-prandial glucose excursion, and increase and prolong satiety, which should help to reduce food cravings and lower caloric intake over the subsequent 2 to 4 h.

5.3.5. Antitumor Effect of Vinegar

It is said there is not any worker who had caught cancer in the largest vinegar factory near Taiyuan city during about hundred years even though the city is famous for heavy pollution. But the research on the antitumor activity of Chinese vinegar is much less than Japanese vinegar.

Oxidative stress is widely accepted as an important factor in the mechanisms for tumor promotion. Since extract of Kurosu (EK) inhibited the oxidative stress in mouse skin, Nishidai et al. (2000) evaluated in vivo the antitumor promoting activity of EK by a two-stage carcinogenesie experiment in mouse skin. The mice were treated with EK at the dose 100-or 1000-fold higher than that of TPA (1μl acetone) 40 min before each TPA treatment. The antirumor promoting activity was determined by both the tumor incidence and the number of tumors per mouse. EK at both a 100- and 1000- fold dose relative to TPA did not reduce the tumor incidence. In the group treated with EK at 1000-fold dose of TPA, the average number of tumors per mouse was markedly reduced by 36% ($P < 0.01$) at 15 weeks, although this effect was not significant from 16 weeks. The average number of tumors per mouse in the control group reached a maximum of 23 per mouse at 15 weeks and the increase stopped at this time. However, the tumor numbers in the EK-treated group increased slightly from 16 weeks and caught up with that in the control group. Thus, EK is suggested to delay skin tumor development. Their findings indicate that EK is effective not only for inhibiting TPA-induced acute inflammation but also for reducing the tumor development induced by TPA.

Fukuyama et al. (2006) examined the activity of Kurozu on colon cancer and investigated the mechanisms involved, focusing on active oxygen generation, apoptosis and metalloproteinases (MMPs). The results indicated that the administration of Kurozu inhibits the development of colon cancer in human colon cancer cell transplantation model in mice. Further, the administration of Kurozu inhibited nitrotyrosine production, decreased total MMP-2 and total MMP-9 levels, and inhibited activation of MMP-2 and MMP-9 in the lesion in this model.

Abe et al. (2007) also researched on the generation antitumor active neutral medium-sized α -glycan in apple vinegar fermentation. The result confirmed that vinegar has the ability of antitumor and the activity is affected by the fermentation.

5.3.6. Promoting Recovery from Exhaustion

Modern medicinal research shows that vinegar take effect in improving metabolism and promoting recovery from exhaustion. People feel tired due to acid environment in vivo. Normally, body fluid and urine are litmusless and alkalescence (pH 7.2-7.5). Lactic acid accumulates in vivo when working furiously or eating too many acidic food products. In this case, the generation of oxaloacetate can not decompose so much lactic acid and lactic acid remained in blood vessels and muscle changed to exhibit acidity. The intake of vinegar and organic acid, turned to citric acid by virtue of glucose, can promote the generation of CO_2 by tricaboxylic acid cycle (TCA cycle) to decrease lactic acid in vivo and eliminate fatigue.

Wu et al. (2007) reported that the supplement of Zhenjiang vinegar could accelerate the rate of dispelling blood lactic acid after exercise, improve the ability of blood buffering and mitigate the occurrence of body metabolic acidosis. So the supplement of Zhenjiang vinegar could promote the recovery of exercise fatigue. In the research, 18 excellent track and field athletes were divided into C group and V group and were fed with distillation water and Zhenjiang vinegar respectively after two different times of exhausted exercise with progressive increase in workload. The indexes of blood lactic acid, blood ammonia and blood gas transform were tested. It showed that there were significant differences between C group and V group in the indexes of the rate of dispelling blood lactic acid, the blood pH value and HCO3 concentration as well as the recovery of the concentrate of blood ammonia ($P < 0.05$) (Figure 8-14, Figure 8-15, Figure 8-16, Figure 8-17). In these Figures, R is in rest, Ex is just after exercise and other values in X axis is the time after exercise (min) while * in figure showed there were significant differences between C group and V group.

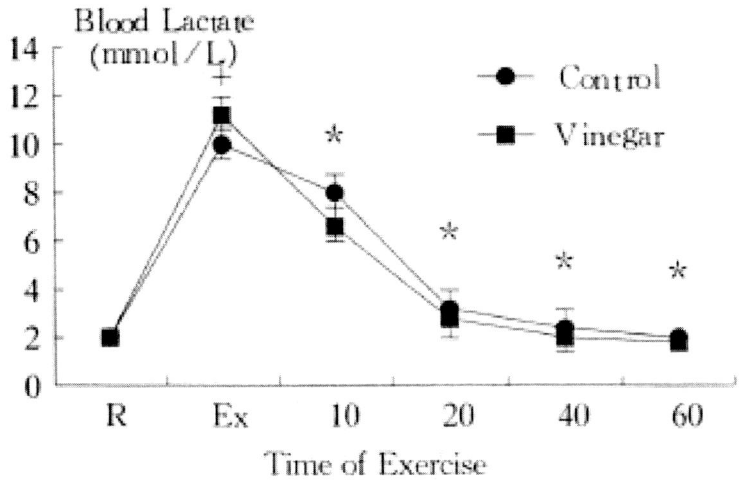

Figure 8-14. The effect of vinegar on the blood lactate contents after exercise (Wu et al., 2007).

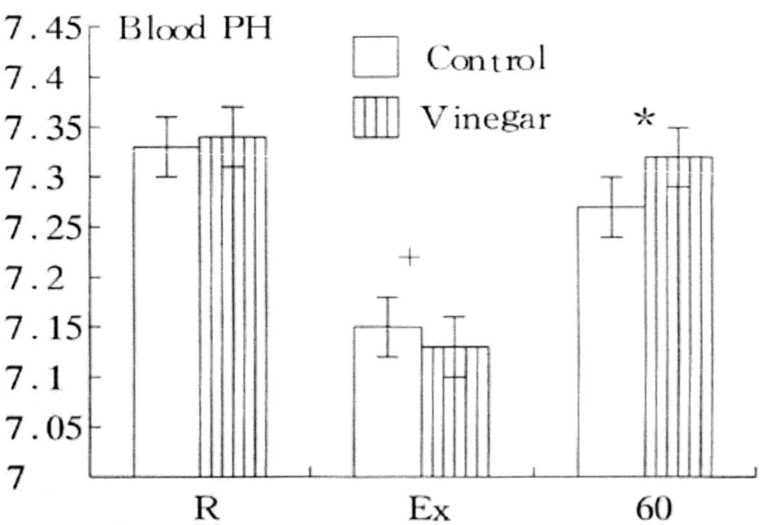

Figure 8-15. The effect of vinegar on the blood pH values after exercise (Wu et al., 2007).

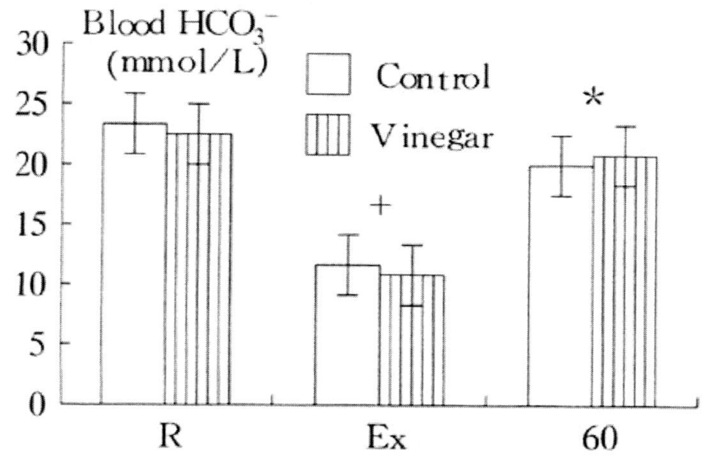

Figure 8-16. The effect of vinegar on the blood HCO_3 contents after exercise (Wu et al., 2007).

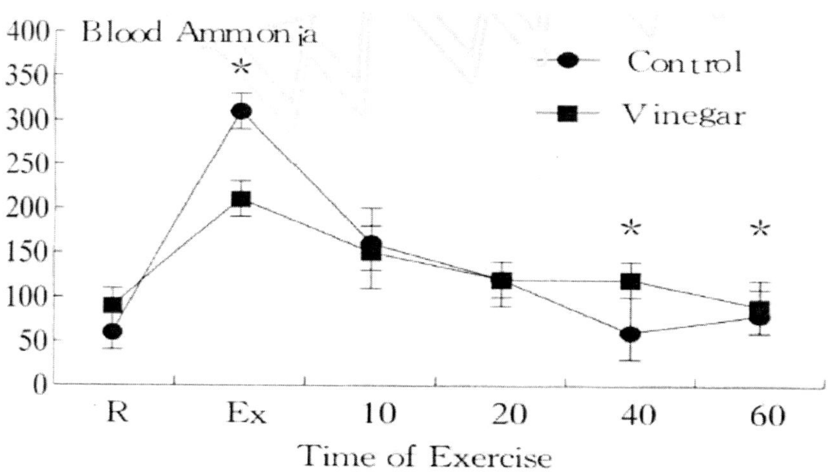

Figure 8-17. The effect of vinegar on the blood ammonia contents after exercise (Wu et al., 2007).

5.3.7. Preventing Osteoporosis Effect

Osteoporosis is the most common cause of age-related bone loss. In order to maintain bone mass, an increase in calcium intake is recommended for women who are at high risk for osteoporosis. Therefore, enhancing the calcium intake is an important issue.

Vinegar contains various compounds, but the main component is acetic acid. Acetic acid is also one of the major short-chain fatty acids produced by intestinal microbes, and it has been shown that these short-chain fatty acids can affect intestinal functions and metabolism. Recent studies have also suggested that short-chain fatty acids are involved in intestinal calcium absorption.

Mikiya Kishi et al. (1999) studied the effect of dietary vinegar on calcium absorption by using ovariectomized rats fed on a low-calcium diet. The apparent absorption of calcium was higher when the rats were fed on a diet containing 1.6% vinegar for 32 days than when fed on a diet without vinegar ($p<0.05$). The calcium content in the femur of the rats given diets containing 0.4% and 0.6% vinegar were also higher($p<0.05$). The serum parathyroid hormone level was lower and the crypt depth of the duodenum thicker in the rats fed on a diet containing 1.6% vinegar ($p<0.05$). These results suggest that dietary vinegar enhanced intestinal calcium absorption by improving calcium solubitity and by the trophic effect of the acetic acid contained in vinegar, which would reduce the bone turnover caused by ovariectomy and be helpful in preventing osteoporosis.

6. Quality Standards of Vinegar in China

6.1. Standards of Brewing Vinegar

Standards of brewing vinegar GB18187-2000 was issued by The China State Bureau of Quality and Technical Supervision on September 1st, 2000, and administered on September 1st, 2001. In the standard, sensory character, chemical compositions are requested as table 8-22 and 8-23, respectively.

6.2. Sanitation Standards of Vinegar

Sanitation standards of brewing vinegar are administered by GB2719-1996 which includes sensory criterion of vinegar as normal color, aroma and sapor, no acerbity, no peculiar smell, no peculiar solid content, clear and so on while sanitation standards of vinegar is shown in table 8-24.

It is predictable that the production and consumption of vinegar will still increase henceforward for the increased research on the health functions and varieties of vinegar. Mature vinegar, a kind of traditional food with a long history, would have a bright future for its special taste and functions.

Table 8-22. Sensory character of brewing vinegar (GB18187-2000)

Item	Request of sensory character	
	Solid fermentation	Liquid fermentation
Color	Amber and brown	Special color of liquid fermentation vinegar
Aroma	Special aroma of solid fermentation vinegar	Special aroma of liquid fermentation vinegar
Sapor	Soft sourness, no peculiar smell	Soft sourness, no peculiar smell
Posture	Clear	

Table 8-23. Chemical index of brewing vinegar(GB18187-2000)

Item	Request of index	
	Solid fermentation	Liquid fermentation
Total acidity (Acetic acid)/(g/100ml)	≥ 3.50	
Non-volatile acid (Lactic acid)/(g/100ml)	≥ 0.50	—
Soluble solids content of inorganic salt /(g/100ml)	≥ 1.00	≥ 0.50

Table 8-24. Sanitation standards of vinegar (GB2719-1996)

Item	Criterion
As /(mg/L) ≤	0.5
Pb /(mg/L) ≤	1.0
Free mine acid	Don't allowed to be checked
Aflatoxin B1/(ug/kg) ≤	5
Food additive	According to the rule of GB-2760
Total bacteria /(cfu/ml) ≤	10000
Escherichia coli /(cfu/100g) ≤	3
Pathogens (from intestines)	Don't allowed to be checked

REFERENCES

Abe, Kaoru; Kushibiki, Toshisada; Matsue, Hajime; Furukawa, Ken-Ichi; Motomura, Shigeru. (2007), 'Generation of antitumor active neutral medium-sized-glycan in apple vinegar fermentation'. *Bioscience, Biotechnology, and Biochemistry*. 71 (9):2124-2129.

Bao Qi'an (1988). 'Japanese vinegar fermentation: the fourth report on Japanese, food industry after visiting on invitation'. *Chinese Condiment* 10:1–9.

Bao Qi'an (1995) 'Researches on the flavours'. *Shanghai Condiment*, 11:1-14.

Chen X., Chen H. (1996). 'A novel production process of aromatic vinegar'. *Chinese Condiment* 6:13–16.

Chen Jie, (2002) *Processing technology and prescription of seasoning.* Science and Technology Literature Publishing House. Beijing.

Chen Jinpin, Zhang Yili (2000) Processing technology of seasoning. China Light Industry Press. Beijing.

Fan Guangpu (2004). 'Methods on improving the quality of vinegar by liquid fermentation'. *Chinese Condiment*, 6:29-33.

Fukashima, D. (1989). Industrialization of fermented soy sauce production centering around Japanese Shoyu. In: Streinkraus, K. H., ed. Industrialization of Indigenous Fermented Foods. New York: Marcel Dekker, pp. 2–7.

Guo, L. (2000). 'Analyses on the traditional production process of Zhenjiang aromatic vinegar'. *China Brewing* 4:1–4.

Higashi kazoo (2004) *Fermentation and Chewing* (1). Korin Co. Ltd, Tokyo, Japan.

Hou, H., Pang, Q., Hou, X. (1996). 'The effects of three different processes on vinegar production'. *China Brewing* 6:2–5.

Huang, Z., Yin, X. (2000a). *Questions and Answers About Vinegar Production.* 1st Ed. Beijing: Light Industry Publishing House of China, pp. 194–220.

Huang, Z., Yin, X. (2000b). **Questions and Answers About Vinegar Production**. 1st Ed. Beijing: Light Industry Publishing House of China, pp. 123–136.

Huang, Z., Cai, M. (1999). Vinegar. In: Shanghai Institute of Brewing, ed. *The Production Technology of Fermented Condiments.* Beijing: Light Industry Publishing House of China, pp. 289–419.

Hubert, A. C. (1976). Vinegar: its history and development. In: Periman, D, ed. *Advances in Applied Microbiology*, Vol. 20. New York: Academic Press, pp. 81–133.

James H. O'Keefe. (2008) 'Dietary Strategies for Improving Post-Prandial Glucose, Lipids, Inflammation, and Cardiovascular Health'. *Journal of the American College of Cardiology*, 51(3):249-255.

Kishi M; Fukaya M; Tsukamoto Y; Nagasawa T; Takehana K; Nishizawa N (1999), 'Enhancing effect of dietary vinegar on the intestinal absorption of calcium in ovariectomized rats'. *Bioscience, biotechnology, and biochemistry*, 63(5):905-910.

Kitamoto, K. (2002). Molecular biology of the koji molds. In: Laskin, A. I., ed. Advances in Applied Microbiology, Vol. 51. New York: Academic Press, pp. 129–153.

Kondo, Shino; Tayama, Kenji; Tsukamoto, Yoshinori; Ikeda, Katsumi; Yamori, Yukio. (2001), 'Antihypertensive effects of acetic acid and vinegar on spontaneously hypertensive rats'. *Bioscience, Biotechnology, and Biochemistry*. 65(12):2690-2694.

Lei, M. (2000). 'Function of microorganisms and enzymes in vinegar production by SSF'. *Chinese Condiment* 9:20–22.

Li Dajin, Wang Ruzhen (2006). 'The application of enzyme preparation in vinegar production'. *Chinese Condiment*, 2:38-43.

Liu Dengru, Zhu Yang, Rik Beeftink, Lydia Ooijkaas, Arjen Rinzema, Chen Jian, and Johannes Tramper (2004) 'Chinese Vinegar and its Solid-State Fermentation Process' *Food Reviews International*, 20(4): 407–424.

Liu, C., Li, Y. (1992). 'Study on amount reduction of rice hull in vinegar production with SSF'. *Jiangsu Condiment* 2:10–14.

Lu, Xiao-qing; Zhang, Chao-ying; Zhou, Xiao-bin; Wang, Qiu-zhen. (2007), 'Experimental study on efficacy of garlic solution and vinegar in killing Vibrio parahaemolyticus'. Chinese *Journal of Disinfection* 24(1):48-50.

P.M. Molina, M.E. Sanz, P.M. A. Lucchesi, N.L. Padola, A.E. Parma. (2005) 'Effects of acidic broth and juices on the growth and survival of verotoxin-producing Escherichia coli (VTEC)'. *Food Microbiology* 22 (2005):469–473.

Robert W.Hutkins, (2006) *Microbiology and Technology of Fermented Food,* Blackwell Publishing,USA.

Shan Fang, Wang Lixia, and Li Wende (2005), 'Spotlight on Shanxi very mature vinegar'. *Agricultural Science and Technology*, 2:23-28.

Shigetomo Tsujihata, Etsuzo Entani, Mito Asai, Yoshinori Tsukamoto, Michio Ohta. (1998) 'Mathematical modeling to predict the bactericidal effect of processed vinegar on Escherichia coli O157:H7'. *International Journal of Food Microbiology*, 43 (1998): 135–138.

Sholberg and Gaunce, (1995) 'Fumigation of Fruit with Acetic Acid to Prevent Postharvest Decay'. *HortScience*, 30(6):1271-1275.

Nishidai S., Nakamura Y.; Torikai K., Yamamoto M., Ishihara N., Mori H., Ohigashi H. Kurosu (2000), 'Traditional Vinegar Produced from Unpolished Rice, Suppresses Lipid Peroxidation in Vitro and in Mouse Skin'. *Biosci, Biotechnol, Biochem*, 64:1909-1914.

Wang Lijian, Wang Limei, Cheng jue. (2006), 'Technology to improve the quality of vinegar made with liquid fermentation method'. *Chinese Condiment*, 7:38-39

Wu Baoai, Wang Renwei, Nei Tengyou (2007), 'Effects of vinegar supplement on indexes of blood lactic acid, blood ammonia and blood gas during recovery of exercise fatigue'. *Journal of Wuhan Institute of Physical Education*. 41(11):47-50.

Xiao Fenzi (2000), 'The history and cultural of vinegar'. *Chinese brewing*. 4:31-32.

Xin Xiu (2004), ' The situation of production of sauce and cinegar in China'. *Global Food Industry Information*, 2004(8):22-23.

Xu Qingping, Tao Wenyi, Ao Zonghua (2006). 'The Comparison of the Antioxidant Activity in 26 Kinds of Vinegar'. *Food and Fermentation Industry*, 32 (1):95 - 98.

Xu Qingping; An Guangjie; Tao Wenyi. (2007), 'Antioxidant activity of different polar extracts in Zhenjiang aromatic vinegar'. *China Brewing*, 6:18-20.

Yumi Shimoji, Yoshitaka Tamura, Kumiko Nanda, Shoko Nishiidai, Yasushi Nishikawa, Nobuhiro Ishihara, Kazuo Uenakai, Hajime Ohigashi. (2002) 'Isolation and Identification of DPPH Radical Scavenging Compounds in Kurosu (Japanese Unpolished Rice Vinegar)'. *J. Agric. Food Chem.* 50:6501-6503.

Zhang, P. (2000). 'Vinegar and soy sauce: necessities in life'. *China Brewing* 4:33–34.

Zhang Shuinian, Liu Geng (2000). *Processing Technology of condiment.* South China University Technology Press.

The China State Bureau of Quality and Technical Supervision, (2000) *'Standards of brewing vinegar'* (GB18187-2000).

The China State Bureau of Quality and Technical Supervision, (1996) *'Sanitation standards of brewing vinegar'* (GB2719-1996).

INDEX

B

C

D

E

F

G

I

S

T

U